Principles and Methods of Harvesting Timber

The largest hardwood tree found in the American forests is the yellow poplar. This is a very distinctive and characteristic tree of the southern Appalachian hardwood region. For its size and height, it scales out as well as any tree in the American forests. Note the straightness, uniformity, and small amount of taper. Yellow poplar lumber is highly prized for its fine seasoning and other technical properties. *Courtesy U. S. Forest Service.*

Principles and Methods of Harvesting Timber

by

NELSON COURTLANDT BROWN

BIOTECH

2018

BIOTECH BOOKS®

ISBN 978-81-7622-078-1

Published by: **BIOTECH BOOKS®**
4762-63/23, Ansari Road, Darya Ganj,
New Delhi - 110 002
Phone: +91-011-23262132
E-mail: biotechbooks@yahoo.co.in

PRINTED IN INDIA

PREFACE

This book is a revision of two books previously published on the principles and methods of logging in the United States and Canada.

The history of lumbering is a great drama in the development of the United States and of Canada. It is an epic story of courage, great hazards and difficulties overcome, success and failure in a struggle to provide some of the necessities of mankind. The great prairies could scarcely have been settled and developed without the white pine from the Lake States and the southern pine from the South. The last frontier of virgin timber is being removed on the West Coast.

The same courage, risks, pioneering spirit, and enterprise are required in building new forests where nature originally provided an abundant and low-priced supply. The North American lumberman is meeting the challenge of this new problem with the pioneer spirit of the early days. Logging is now largely conducted in regrowth forests. Our future forests will be grown from natural stands rather than from planted forests by artificial reforestation.

The importance of logging is at once apparent when one considers the very large annual log output from the 60,000 or more woods operations throughout the United States and Canada. Saw logs vary in contents from 1 or less to 20 per thousand board feet in different regions. The log unit varies from the 4′ pulpwood bolt weighing a few pounds in the Northeast and Canada to logs ranging in length from 20′ to 100′ or more, weighing up to several tons each and containing individually up to several thousand feet, board measure, in the Pacific Northwest. Many hundred thousand men are annually employed in the woods to harvest the log product. One company employs about 15,000 men annually in its several woods operations.

The stability of employment and the permanence of woods operations are of vital importance to the social and economic welfare of the country. Logging constitutes a very large part of the practice of forestry in our North American forests. It may be good, poor, or indifferent forestry, but logging is the important part of forestry in which our woods are treated and a future supply of timber is assured or the areas are left in unproductive condition. Harvesting the forest requires more labor and capital than any other phase of forest work and is, therefore, the principal business of the woods. It constitutes a very large share of actual forest management.

Recently a more scientific approach has been made to the problems of logging, and the profitability of removing the logs of different species and sizes from our woods has been more accurately determined. Mechanized logging has made important advances since World War II, especially felling, bucking, and log loading. The tractor and truck have replaced many methods formerly used, especially power logging on the West Coast, logging railroads, flumes, chutes, stream driving, sled logging, animals, and wagons. As a result of the high cost and relative inefficiency of woods labor, mechanical devices have been developed to replace manual labor in all forest regions.

Logging comprises approximately one-half the total cost of production from the stump to the finished product. Methods of log transportation are in a state of continual change and improvement. Logging engineering has developed to a remarkable state of efficiency in the Pacific Coast region, where the physical difficulties of moving logs of long lengths and heavy weights from the forest to the sawmill present great problems.

In every section of the country, methods must be employed to suit the local conditions of topography, climate, density of timber, volume flow of logs, and size of individual trees. Progress in the development of logging methods during the last decade has been considerable, and it is likely that still further changes may be made in the harvesting of our forest.

Selective logging has created conditions which have vitally affected methods of logging. The accurate determination of the most profitable log sizes to be removed from the woods, together with the necessity of leaving the residual stand in favorable condition for continued growth and later logging, has brought the problem of logging to the forefront of the forest industries in the United States.

Labor shortages, strikes, and inefficiency, coupled with high wages, after the last World War have forced many woods superintendents and logging managers to mechanize their operations even more than before the war. The mechanization of logging has already reduced the man-power requirements for producing 1000 b.f. of logs from 15 to 20 man-hours down to about 5 to 12 man-hours over a period of about 25 years. This ratio of man-power requirements may be decreased still further.

Improved mechanical devices have been developed in connection with the following changing methods of logging:

1. Expanded use of tractors for skidding and loading, and as an auxiliary in several phases of transportation, such as assisting in truck operations on adverse grades; moving logs about landings; constructing, improving, and repairing woods roads.

2. Felling and bucking with power saws. This has been the most dramatic as well as one of the significant laborsaving developments in mechanization.

3. Use of motor trucks as the major method of log transportation in nearly all regions.

4. Skidding logs in tree lengths to landings for bucking and loading rather than the formerly prevailing practice of bucking logs in felling areas. The bringing of long-length logs to landings has definitely resulted in lower skidding costs, as well as reduced bucking and loading costs.

5. A great variety of improved loading devices to reduce the time required for loading and unloading. Among these are the Drott, Bouffard, Logger's Dream, Montague, and several pulpwood loaders with automatic conveyors.

Some of the best examples of efficiency in reducing logging costs are shown in southern pulpwood operations known as "pulpwood or slasher mills" where felling, skidding of tree lengths, and bucking and loading are done mechanically and are very closely associated together. All machinery used in these operations is quickly, easily, and inexpensively portable. Some of the newer machines, such as the Logger's Dream, combine skidding and loading which also results in lower costs.

Bringing in the logs required for the 50,000 sawmills in this country poses very great problems. There are probably 60,000 or more logging operations required to supply these sawmills. The production of more than 18 million cords of pulpwood also involves important logging problems. Sometimes there are two to ten or more separate logging operations to supply the requirements for one sawmill or pulp mill, and these operations may be widely scattered up to 100 miles apart.

The small mill has increased in importance in all regions, especially in the logging of the regrowth of small and scattered areas, such as farm wood lots; small tracts formerly not considered worth cutting; prelogging for pulpwood, veneer logs, poles, piling, and cross ties prior to saw-log cutting; and relogging areas formerly cut. Most of the methods emphasized in the text apply to small sawmill operations. The only exceptions are power logging as practiced on the West Coast and, to a less extent, flumes and some phases of water transportation, such as stream driving, barging, and ocean rafting. Even power yarding on the West Coast is used by relatively small sawmills.

Because of the great increase in stumpage prices in recent years, utilization standards and practices have changed and improved to a remarkable degree. Species and sizes formerly considered unmerchantable are now profitably harvested and marketed.

Progress in prelogging, relogging, integrated logging, and salvage cuttings has presented new phases of the industry that are important in handling our forest resources.

It is hoped that this text will help the student and possibly the operator to gain a better understanding of some of the logging methods employed in the United States and Canada.

NELSON COURTLANDT BROWN

Syracuse, N. Y.
January 1, 1949

ACKNOWLEDGMENTS

In preparing material for this text, the author has studied and inspected logging operations in all the timber regions of the United States and Canada. During these studies, he has received the cordial co-operation and assistance of logging superintendents and others, for which he is very grateful.

Chapter IX on Power Logging was prepared by Professor J. Kenneth Pearce, who has taught logging engineering for 16 years at the College of Forestry, University of Washington, Seattle, and has served as consultant to many companies. Material and photographs dealing with the felling, bucking, skidding, loading, and transport of southern pulpwood were furnished by G. E. Knapp and H. J. Malsberger of the Southern Pulpwood Conservation Association, Atlanta, Ga., and for several pulpwood regions by H. E. Brinckerhoff of the American Pulpwood Association of New York City and E. E. Esgate formerly with that association. J. E. McCaffrey of the Southern Kraft Division of the International Paper Company at Georgetown, S. C., has been especially helpful with information and data. A. M. Koroleff of the Canadian Pulp and Paper Association of Montreal and L. A. Nix, consulting forester of Toronto, have contributed information about logging conditions in Canada. William Hammerle and E. R. Schindler of the Southern Pine Association at New Orleans; R. E. Broderick of the Northeastern Lumber Manufacturers' Association, New York; Howard Hanlon of Cotton and Hanlon Lumber Company, Odessa, N. Y., C. Verne Holbrook of the Long Bell Lumber Company of Kansas City; and officials of the Pacific Lumber Company, Scotia, Calif.; Weyerhaeuser Timber Company of Tacoma, Wash.; Crossett-Watzek-Gates group of companies on the West Coast and in the South; and W. T. Smith Lumber Company of Chapman, Ala.; and many others have been especially helpful.

Several members of the U. S. Forest Service, especially Ira J. Mason, P. H. Bryan, E. E. Matson, Basil Wales, George Stoltz, L. K. Mays, F. C. Simmons, and members of the U. S. Forest Products Laboratory at Madison, Wisc., have been very helpful in supplying information. I am also indebted to the U. S. Forest Service and to many manufacturers of motor trucks and logging equipment for their willingness to supply data and photographs for the book, especially the Peterbilt Motor Truck Company of Oakland, Calif.; White Motor Company of Cleveland, Ohio; General Motors Company of Detroit; International Harvester Company of Chicago; Mack Truck Company of Allentown, Pa.; Chrysler Corporation of Detroit; and the Ford Motor Company of Detroit.

The following manufacturers of power felling and bucking saws have been very co-operative and helpful: the Mall Tool Company of Chicago; the Disston Saw Company of Philadelphia; the E. C. Atkins Company of Indianapolis; the Hansen Machine & Supply Company of Eureka, Calif.; the Lombard Governor Corporation of Ashland, Mass.; the Reed-Prentice Corporation, Worcester, Mass.; the Cummings Machine Works, Boston; Precision Parts, Montreal; Power Machinery Company, Seattle.

The following manufacturers of portable mechanized loading equipment have assisted with data and photographs: Taylor Machine Works, Louisville, Miss.; Drott Manufacturing Corporation, Milwaukee; B. L. Montague Company, Sumter, S. C.; Chain Belt Company, Milwaukee; American Hoist & Derrick Company, St. Paul, Minn.; Link-Belt Company, Chicago; the Blair Equipment, Ltd., of Ottawa, Canada (Bouffard loader); and the Camp Manufacturing Company of Franklin, Va. The illustrations for Chapter VIII were supplied by Skagit Steel & Iron Works, Sedro Wooley, Wash., and Young Iron Works, Seattle, Wash.

The following manufacturers of tractors have been especially helpful with data and photographs: The Caterpillar Tractor Company of Peoria, Ill.; Allis-Chalmers Company of Milwaukee; the Cleveland Tractor Company of Cleveland; the International Harvester Company of Chicago; and the Ford Motor Company of Detroit.

The following manufacturers of woods tools and equipment have been of much assistance: Warren Axe and Tool Company, the Kelly Axe and Tool Company, the Disston Saw Company, the Plumb Axe Company, the American Fork and Hoe Company, the Atkins Saw Company, the Simonds Saw Company and several others.

The *American Lumberman* of Chicago; the *Southern Lumberman* of Nashville; the *Timberman* of Portland, Ore.; the *Canadian Lumberman* of Toronto; *Timber of Canada* of Ottawa; the *West Coast Lumberman* of Seattle; the *Southern Lumber Journal* of Jacksonville,

Fla.; and other lumber trade journals have been the source of helpful literature on the subject.

The secretaries and managers of the lumber manufacturers associations, especially R. A. Colgan, R. E. Broderick, Burdett Green, H. C. Berckes, O. T. Swan and P. A. Hayward have been most helpful.

NELSON COURTLANDT BROWN

CONTENTS

PART I. GENERAL INTRODUCTION

PART I
GENERAL INTRODUCTION

CHAPTER I
Forest Utilization—General Considerations

DEFINITION AND SCOPE

Forest utilization is the conversion of standing timber into forms useful in human economy and arts. It includes the harvesting of the forest crop, transportation of the raw materials to market, their manufacture into various usable commodities, and their merchandising and distribution to their ultimate use.

Forests have always served mankind in multitudinous ways. The business of harvesting our timber has played a most important part in the economy of the Nation.

The main branches of forest utilization are:

1. Properties and uses of wood.
2. Logging.
3. Manufacturing, conditioning, and treatment.
4. Merchandising and distribution.

Logging may be divided into the following main categories:

1. The felling and preparation of logs for transport.
2. Minor transportation, or log assemblage.
3. Major transportation, or main log haul.

Logging has principally to do with the production of saw logs, which are the chief primary product of the forest. It includes also, however, the production of large quantities of pulpwood, cross ties, poles, piling, mine timbers, veneer logs, bolts, and miscellaneous other forms. Practically all are produced as logs in the round form of varying lengths. They are later converted into refined forms of the final product, generally in square or rectangular shapes.

Forest utilization constitutes the major part of the forestry operations practiced in the United States. Until our virgin forests are sufficiently depleted and prices have risen enough to compel the practice of forest management by private owners of timber on a sustained-yield basis, the great problem in our commercial forests will continue to be the complete and efficient utilization of our standing timber. With second-growth forests producing wood at a rapid rate, fire protection improved, per capita consumption of wood decreasing, and the use of substitute materials increasing, we may have sufficient supplies of wood for the needs of our people for at least several decades and perhaps forever.

The great distances between centers of production and of consumption with high costs of transportation involved have greatly intensified the problems of American forest utilization.

The perpetuation of our forests by cutting methods designed to assure the regeneration of our timberlands is of considerable national importance. Wood has played and always will play an exceedingly vital part in our national economy. It is important that sufficient forest products be grown not only for our domestic requirements but also for an extensive and profitable export lumber business. Until about 1930, those responsible for logging our forests have been little concerned with the continuation of one of our greatest natural and renewable resources.

Logging represents a considerable part of the cost of converting standing timber into usable forms. Stumpage prices are relatively low compared to the value of the finished product; they generally represent 10% to 22% of the value of the salable product, f.o.b. sawmill. Logging represents a charge of approximately 50% of the total cost of putting lumber on the market. With many species and in some regions this percentage may be considerably higher. Most lumber products are made from relatively few species and are used for dwellings and general construction purposes.

Since our more accessible and better-quality forests have been cut, logging is now being conducted in more inaccessible, scattered, and difficult locations. The timber now being cut, with the exceptions of virgin timber, is smaller in size and generally inferior in quality. Therefore, the tendency is toward more difficult and, consequently, more expensive logging on the larger operations, whereas for the 60,000 or more small operations the timber being cut consists mainly of small and scattered bodies of regrowth.

EARLY FORESTS—THEIR INFLUENCE IN THE DEVELOPMENT OF THE AMERICAN NATION

Nowhere in the world have forests more completely met the varied needs of mankind and proved more profitable than in North America.

As discovered by the early colonial settlers along the Atlantic Seaboard, our forests were rich in both hardwoods and softwoods to meet every utilitarian requirement. Qualities of strength, durability, workability, beauty, and fitness were found in these trees to meet the needs of either a primitive or an advanced type of civilization.

Although the forest was both a protection and a support of the early settler, it was also a great obstacle in the development of early colonial civilization. During a period of about 200 years, the more fertile and accessible valleys were cleared of timber by an exceedingly slow and extremely laborious process. The early settlers in Virginia, Massachusetts, New York, and Pennsylvania used the forest first for shelter and fuel. It harbored the hostile Indian and the savage beast and yet provided game for sustenance and furs and timbers for the first export business.

The forest occupied great areas of fertile land that could more profitably be devoted to agricultural purposes. Much land, however, was gradually cleared in the eastern states that has since proved to be unprofitable for food crops. Consequently, many thousand acres of farms have been reabandoned to forest growth since the Civil War in both the South and the East. Much land in the East and the South probably would never have been cleared originally, if the great fertile plains of the Middle West had been discovered first, since food crops can be so much more readily grown in that region.

The most important and abundant species of wood found in the New England and Middle Atlantic States was white pine. This proved to be the most useful tree in the American forest, and for nearly 300 years after the first settlement in Jamestown in 1607 it was continuously the premier lumber tree of the country. Because of the accessibility of large stands of white pine in Maine, this state quickly became the center of the American lumber industry. White pine also came into early prominence in Massachusetts, New York, and Pennsylvania.

It is reported that white pine was often found to be 130′ to 160′ in height and 40″ to 50″ or more in diameter. It is said that some trees even attained a height of 250′ and a diameter of 80″.

Maine continued the center of the American lumber industry until the early decades of the nineteenth century. In 1850 New York became the leading lumber state, producing about 20% of the total lumber supply with Albany as the great lumber mart of the Nation. From this center lumber was distributed to many points east, west, and south. With the facilities offered by the Erie Canal and the Hudson River, lumber was shipped and distributed along the Atlantic Seaboard and even to the export markets in Europe and the West Indies.

Equally important to the South was the southern pine, which played a prominent part in both domestic and export lumber business, as well as being used for naval stores.

The forest was a significant factor in the development of the American nation. From earliest days logging and the industries dependent on the forest furnished a livelihood for many settlers. Vast woodland stretched unbroken to the treeless prairies of the Middle West. A rugged and forceful civilization was developed as a result of this close association of our people with the forest.

Lumbering did not develop as a great industry, however, until the growing population forced the opening of the Northwest territory—now the Ohio Valley and the Lake States. Rapid increase in lumber production marked the entire nineteenth century. The United States became a great wood-using nation.

The forests of the Lake States, and later of the South, were considered inexhaustible. Little thought was given to efficient or complete utilization because timber was so plentiful and inexpensive. Enormous wastage was a natural concomitant of the colonial and western expansion phases of our national development.

EARLY LOGGING PRACTICES AND SAWMILLS

Naturally, the first logging operations and sawmills were located where good timber was available, where there was a demand for the product, and where streams were present for power purposes as well as for transportation to the markets.

This meant that the first sawmills were generally located in or near the centers of population. Streams were used to float logs to the mills, and lumber was commonly floated on rafts or carried on vessels to its destination. Often the local market consumed the output of these mills, but export trade developed very early. Large quantities of spars, ship timbers, and construction lumber were shipped to Great Britain, Holland, and the western European countries. Pitch and naval stores were early produced from the southern pineries.

The lumber industry developed primarily along readily available and inexpensive systems of water transportation. This pattern continued almost unbroken through colonial and pioneer stages until about 1860, when railroad facilities gradually began to replace water transportation for shipping large quantities of logs and lumber.

Lumber and log rafts were a dominant influence in the early development of commerce on the Maine and New Hampshire rivers as well as the Hudson, Delaware, and Susquehanna, and, still later, the Allegheny, Ohio, and Mississippi Rivers and their tributaries. Even after the coming of the railroad era, water continued for many years to be the chief medium of transportation. The opening of the Erie Canal furnished the impetus for the eastward movement of lumber after the Michigan forests were developed to a leading position in the industry.

The first definite records of the location of sawmills in the United States were at Jamestown, Va., in 1625 and Berwick, Me., where one was established in 1631. It has been alleged that the first mill was located at York, Me., in 1623. It is possible that sawmills may have been built soon after the settlement of the Jamestown colony in 1607. We do know, however, that the first large gang sawmill patterned after the European style was located in Berwick, Me., in 1650. Sawmills driven by windmills (as they are today in Holland) were reported established by the Dutch on Manhattan Island, now New York City, in 1625.

IMPORTANCE AND PLACE OF THE SUBJECT IN FORESTRY

Utilization phases of forestry are intimately related to the social and economic welfare of the country. The lumber and associated industries employ large numbers of men. Vast amounts of capital are invested in them. Large sums of money flow to and from the employees and corporations concerned with equipment for logging and sawmill operations and with transportation and distribution of the product. In many sections of the United States, notably in the Northwest, the entire economic structure is vitally interwoven with the timber business. That banking, export transportation, and trade are affected by the prosperity and condition of the lumber industry may be shown as follows:

1. Normally, more than one million men are directly employed in the lumber industry alone.

Relatively few men are required to plant, protect, and assist in growing trees, but large numbers are required to fell, convert, and distribute the products of the forest. For the production of 35 billion b.f. of lumber and a still larger quantity of other forest products over 5 million people are directly supported through our forest industries and probably an equal additional number through transportation, distribution, and the industries depending on wood as a raw material.

2. Ten billion dollars of capital are invested in the lumber and associated forest industries and many additional millions in industries depending on the forest as a natural resource.

3. More than $3600 million are represented in the value of the natural product of the industry and a considerable additional amount in fabricated products.

4. Normally, 50,000 or more sawmills are engaged in lumber production and a still larger number of logging operations are required to supply them with logs.

5. An important export trade in lumber contributes to employment and prosperity in the United States. It is believed that sufficient lumber and timber will be available to continue this export trade for many years. The relatively small percentage of production represented in the export of lumber may determine to some extent the general prosperity and activity of the domestic lumber business and, consequently, of the people.

6. Whereas foresters recognize more than 1000 species, at least 60 trees are commercially important, and 3, including southern pine, Douglas fir, and ponderosa pine, comprise about 60% of the lumber used in the United States. Due consideration, therefore, must be given to these three important species or groups which with hardwoods practically dominate our lumber industry. About 25 species comprise 93% of the lumber produced in the United States.

7. The forests of the United States cover more than 495 million acres, or about one fourth of the area of the country. Of this area, 461 million acres are commercial forests. They contain more than 1601 billion b.f. of merchantable timber.

8. The lumber industry employs more than 3 billion primary horsepower, the fourth largest of all the industries.

9. Rail shipments of forest products reached a peak of 3,744,883 carloads in one year. Of all such shipments recently, 40% was lumber, 24% saw logs, 11% pulpwood, 5% cross ties, and 20% other products.

LOGGING AS A SILVICULTURAL OPERATION

Continuity of production and use, or sustained yield, is the underlying principle of forestry practice. Forest productivity and permanence are fostered and aided by effective fire protection and insect control. To a limited extent they are being assisted by a small amount of planting. These policies have been helpful and, to a certain degree, effective. But the methods by which the mature and growing forests are logged are the important factor by which sustained yield may or may not be assured. Most of our forests are and will be reproduced naturally rather than by planting. The significant and essential difference between preforestry operations and an intentional plan for permanent productivity rests in the type of cutting operations practiced.

National Forest timber sales are designed to yield a revenue as well as to conform to the principles of continuity of yield. A definite and rapid advance has been apparent in the attitudes of organizations and groups toward continuity of yield as well as of individual and corporate owners of timberlands within the past few decades. During periods of economic prosperity positive progress is made in the development of the idea that our forests should be kept in a productive condition. Financial restrictions have been the outstanding barrier against the permanent and efficient management of forests by private companies.

Progress in improved logging methods as a silvicultural as well as a utilization and economic measure has recently been directed toward selective or partial logging. In the past, the object of selection has usually been to determine accurately the limits of profitable logging rather than to perpetuate the stand. If the two viewpoints can be reconciled and co-ordinated, they will interact to the benefit of our forests.

The South, notably the states from Virginia to Texas, inclusive, until recent years has probably furnished the best example of unintentional forestry, where a very haphazard, unscientific, and crude system of diameter cutting has been followed for 60 years or more. The significant point in this situation is that, although practically all the virgin timber in the South Atlantic States was removed before 1900 and the major production since has been from second- and third-growth stands, the peak of lumber production in this region was not reached until 1926, which indicates that, in general, growth has kept up with cutting. If selective logging had been followed, if many small trees had not been sacrificed for 2″ × 4″ or other small material, and if efficient fire protection had been assured, the yield could have been very much larger. North Carolina, one of the oldest settled states, is one of the leading lumber-producing states in the Nation.

The best and most economical method of cutting, from the viewpoint of permanent growth, has not until recently been generally understood and appreciated by the owners of timberlands. Silviculturists are not generally concerned with or interested in logging methods and costs, and loggers formerly cared little about the future of the stand. A reconciliation of the two viewpoints, a vital desideratum in the practice of American forestry, is gradually being achieved. Foresters generally are not sufficiently acquainted with the inhibitions and limitations of logging methods. However, much progress is being made. The U. S. Forest Service and private companies have contributed much toward better utilization practices.

Those concerned with the utilization phases of forestry should show a more sympathetic interest in the objects and demands of the silviculturists. But it is equally important that the latter be fully informed and sympathetically interested in logging problems and procedures and in the limitations and possibilities of marketing the various grades and sizes of the trees to be felled. The logger looks to the present, the silviculturist to the future; each should have an understanding of and interest in the other's viewpoint.

Until growth increment can be shown to equal or exceed capital accumulation, loggers cannot be expected to sacrifice current returns for prospective or deferred profits. The forester should be interested in decreasing logging costs, broadening the demand for forest products, and determining profitable outlets for woods and mill waste, for thereby he will be directly assisting in the better business management of our forests on a sustained-yield basis. Economical logging is an important keynote to successful forestry practice.

EXTENT OF THE UTILIZATION OF TREES FELLED IN AMERICAN FORESTS *

Great wastage has attended the conversion of our trees into usable forms. A large proportion of trees felled in our forests is still wasted. Wastage in both the woods and manufacturing operations may be classified as avoidable and unavoidable. Under our present economic conditions, most of it—probably 70% to 80% or more—has been unavoidable. Extensive wastage may occur in the woods because profitable means of utilization may be limited. Hodgson found that more than 6 million cords of sound wood, or 42 cords per acre, were left unused in logging in the Douglas fir region. In 1947, 27% of the volume of timber in this region was left to rot or burn.† Considerable

* For further information, see *Wood Waste in the United States,* U. S. Forest Service, Washington, D. C., 1947.

† From *Salvage Logging in the Douglas Fir Region of Oregon and Washington,* by E. E. Matson and J. B. Grantham, Oregon State College, Corvallis, Ore., 1947.

progress has been made in the last 10 years. Degree of utilization is now directly related to market prices, as described elsewhere. High prices mean more complete and efficient utilization; also the wastage may be less on low-cost than on high-cost logging operations in the same region.

On the basis of studies made in various regions it is estimated that, of all the trees felled for lumber in our forests, the wastage is approximately as shown in Table I, depending on local utilization conditions.

TABLE I

Woods Losses	*Percentage of Total Wood Volume Wasted*
Stumps	2– 3.0
Tops, limbs, and branches	10–12.5
Defective and decayed trunks and boles shattered in felling	4– 5.5
Miscellaneous, such as improper log lengths, transportation losses en route to destination, decay in storage	1– 2.0
Total	17–23
Manufacturing Losses	
Bark	7– 8
Saw kerf	8–10
Slabs	8– 9
Edgings and trimmings	6– 8
Seasoning	3– 4
Remanufacture	2– 3
Miscellaneous	1– 1
Total	35–43
Total woods and manufacturing losses	52–66

The total volume of usable rough lumber produced from the average tree is, therefore, only 34% to 48%, depending on species, region, and local conditions. It is apparent from the table that woods waste and losses are only about one half the actual manufacturing losses.

The greatest wastage in felling and conversion of our forests is found in the large timber of the Pacific Coast region, particularly in such species as Douglas fir, western hemlock, Sitka spruce, and redwood. The wastage in conversion of western white pine, northern white pine, and some of the more valuable hardwoods is generally very small. It is considerably limited also in ponderosa, sugar, and the southern pines.

It has been determined by the U. S. Forest Service that only 8% of the cubic volume of the average western white pine in northern Idaho measured from the ground line to a top diameter inside the bark of 4 in. is wasted in the woods operation. Only 20% of the trees altogether is wasted, not counting, however, sawdust, edgings, and other material used as fuel. Likewise only 11% of the ponderosa pine of the same region is wasted or lost in woods operations. Western larch is generally long butted; the wastage for this species owing to its shaky condition and accumulation of pitch amounts to 7.9% of the tree in addition to 5.7% of the tree lost in the stumps.

Woods waste and logging losses may be largely attributed to the following practices:

1. Improper measuring of log lengths; also, cutting to avoid crooks, crotches, large knots, and other defects; this is exclusive of trimming allowances.

2. The customary practice of cutting logs in even lengths, such as 10′, 12′, 14′, 16′ instead of odd and even lengths, such as 12′, 13′, 14′, 15′, 16′.

3. Carelessness in cutting high stumps and the failure to take all the possible merchantable length from the tops. Western white pine is generally utilized to top diameters of 6″ to 12″ or an average of 9″ and often less; ponderosa pine tops are utilized to a spread of 7″ to 12″, or an average of 10″ in the Northern Rocky Mountain region. Some western softwoods and southern hardwoods are cut to 15″ to 18″ or more in the tops. Felling trees from springboards in the Northwest results in much loss of good wood. Small and short top logs are frequently left in the woods because they cannot be handled profitably.

4. Unnecessary allowance for trimming regular lengths. Allowances of 2″ to 4″ are customarily used in the East and South, but frequently 3″ to 12″ or more are found. In the big timber of the Pacific Coast long lengths are sometimes cut with 10″ to 36″ or more allowed for trimming. Carelessness is responsible for much of this unnecessary allowance.

5. Breakage in felling due to carelessness, resulting in shattered trunks, broken tops, split logs, lodging in merchantable trees, splintered butt logs, and so on. Breakage is also due to rough terrain and "crossing" (that is, trees falling on other felled trees or on windfalls) especially in redwood and other West Coast timber.

6. Losses in transportation. Logs are sometimes lost from the cars enroute in railroad transportation. Floated logs are often left along stream banks after spring freshets or when flushed from splash dams, and many logs are lost as "sinkers" in floating, rafting, and stream

driving. Losses are also due to jams in flumes and chutes. The expense of retrieving these logs is often greater than the value to be secured from the product.

7. Losses due to insect and fungus damage where logs are left unduly long on rollways or in woods storage during summer or warm seasons. The losses from fungus damage in hardwoods, particularly sapwood, may be especially serious.

8. Damage to the residual stand, especially in the Northwest where cable-hauling systems are used.

9. Uncontrolled fires in slash burning after logging.

FORESTRY AND LUMBERING—COMMERCIAL FORESTRY PRACTICE

Definite progress has been made by private owners to perpetuate the growth of timber on their forest holdings. Earliest efforts were in the direction of fire protection to prevent both current and future losses of growing timber. Then several companies, notably in the Northeast, the redwood region, and the southern pine region, installed nurseries for planting trees. Concurrently with this progressive movement, many attempts were made to control insects and diseases, and improved practices were introduced in turpentine operations in the longleaf and slash pine in the Southeast. Still later, companies introduced cutting methods designed to promote natural regeneration or to increase the growth on their holdings on a sustained-yield basis. More recently, several companies, largely in co-operation with the forest experiment stations, made careful studies to determine profitable diameter limits for cutting which are discussed in the chapter on selective logging.

Many regional lumber manufacturers' associations have employed foresters to assist member companies and others to develop sustained-yield plans of forest management. Hundreds of foresters are now employed on logging operations. The tree-farm movement sponsored by the American Forest Products Industries has been of great assistance in the progress of private forestry.

The large holdings of the big lumber and pulpwood companies are generally managed according to very efficient and acceptable forestry principles. The smaller timber tracts, especially the farmers' wood lots and other small woodlands, are not usually managed along acceptable forestry principles.

EPOCHS IN THE DEVELOPMENT OF THE AMERICAN LUMBER INDUSTRY

From an historical viewpoint, the industry has passed through four distinct periods of development, each featured by gradual expansion and growth. No sharp lines of demarkation can be drawn for the course of the development and expansion of this industry. The division into these periods, therefore, has been made along broad lines, concurrently with the changing economic and social conditions of a rapidly growing country:

1. The colonial, pioneer, and early development period from the earliest settlements to about 1828. During this time the forest was largely an obstruction to settlements. There was great wastage in conversion and use because much of the forest was felled for the growing of agricultural crops. The forest had little if any value except for local home and farm construction, fuel, and a relatively small export trade in spars, ship timbers, and construction lumber. Maine quickly became the leader in the industry and so continued throughout this period, indeed, until the center of the industry shifted to New York about 1850. Small mills driven largely by waterpower constituted the chief activity. Many timbers and even lumber were hand-hewn. Lumber production did not exceed 1 billion ft. per year even at the end of this period.

2. Rapid expansion in population and development of the Middle and Far West from 1828 to 1880. This period was marked by a rapid increase in immigration, the expansion of the population over the Alleghanies to the Middle and Far West, and construction of canals and railroads to facilitate transportation. The great fertile plains of the Central States and prairies were a boon to the agricultural settlers. The prairies west of Illinois were largely treeless. In order to provide lumber for home and general construction, the lumber industry developed with rapid strides. Michigan and Wisconsin, soon after New York and Pennsylvania, became the great centers of lumber production. Chicago, St. Louis, and Detroit became great centers of consumption and distribution for a rapidly developing white pine industry. Mississippi River towns such as Keokuk, Dubuque, Clinton, St. Louis, and many other points became centers of white pine lumber production and later of fabrication supplied by logs from the Wisconsin and upper Mississippi rivers and their tributary streams.

This period was featured by the introduction and use of steam as a motive power, the development of the circular saw to replace the water-driven muley saw, the cutting of hemlock for the bark alone in Pennsylvania, and great wastage in cutting only the better logs in the pineries of the Lake States to provide cheap lumber for the upbuilding of the Middle West. Lumber production rose rapidly until it reached about 18 billion b.f. toward the end of this period.

3. Industrial development and expansion from 1880 to World War II. With the practical depletion of the best of the virgin timber in Michigan and Wisconsin in the decade from 1890 to 1900, the industry shifted rapidly to the southern pine of the South, to the white pine forests of northern Minnesota, and to the Far West. The year 1909 marked the peak of lumber production of the country, when 44 billion b.f. were sawed. White pine was quickly supplanted in supplying the great Middle West by the southern pines. Douglas fir and ponderosa pine from the Far West began to move east over the newly constructed transcontinental railroads. The great era of railroad expansion and settlement through the Middle West and to the Pacific Coast had slowed down. Immigration began to decline. Prices of lumber as well as of other commodities steadily rose. Competing structural materials, such as brick, concrete, fiberboards, and steel, began to supplant wood during the latter part of the period. The warning of a possible timber famine no doubt contributed to a change in the public mind regarding the uses of lumber. This was the great period of stumpage accumulation by the larger lumber and timber-holding organizations.

4. Post-war reconversion conditions (1945–). The serious housing shortage which was aggravated by (1) strikes, (2) decreased labor efficiency, and (3) lack of men both in the woods and at the sawmills was an outstanding feature of the postwar period. Lumber was classified as a critical material during the World War, and there was relatively little construction for dwellings and normal requirements. This condition created an increased demand for lumber after the war. Concurrently with it, however, the high cost of labor and resulting high costs of housing and other construction prevented a rapid return to normal building operations. Inventories were low in retail lumber yards.

Stumpage prices, logging costs, and lumber prices were 100% to 150% or more above the prewar levels. Although employment and demand for all classes of goods including forest products were at high levels, the high costs of construction were a definite deterrent to increased lumber usage. Woods labor had been attracted to the more remunerative war industries, and the readjustment to normal conditions was slow.

This period was also marked by:

1. The introduction and use of mechanical devices to meet high costs of woods labor.

2. Great progress in the development of forest management by private timber owners throughout the country. Federal attempts at legislative controls of private logging operations met with little favorable response or sympathy from private operators. State controls, however, made considerable progress in the North, South, and Far West. The necessity of growing timber for their own requirements became evident to most of the larger lumber and pulpwood companies.

3. The replacement of men responsible for logging operations by more skilled and better-trained men. In many logging operations, foresters assumed leadership, both for efficiency measures and for long-range timber-management plans. It became widely recognized and accepted that forests must be so logged that their growth would be continued and even increased. Simple and rudimentary forestry measures, such as fire prevention and protection, although important, were expanded into selective logging with variable-diameter limits and more efficient and complete utilization of the raw material (standing timber), due to higher prices and the fact that a much greater percentage of the bigger holdings of our commercial forests were being placed under sustained-yield forest management

RECENT DEVELOPMENTS IN UTILIZATION OF OUR FORESTS

During the years since World War II many changes have come about in the utilization of our American forests. Some of them have had a profound influence on logging methods and practices. Intensive competition on our lumber markets has forced the greatest economies in logging procedure. Among the outstanding changes that may be noted during this period are:

1. The small sawmill is very steadily but positively replacing the large lumber mill as a source of lumber production. This has been particularly noteworthy in the southern pine region which has been the most important source of lumber production during the past 45 years or more. The small sawmill is, at the present time, the chief

source of lumber production except in the forests of California and the Northwest (including the Inland Empire), where the large mill will continue to be the principal source of production until the virgin forests have been removed.

2. Good roads and improved motorized transportation are revolutionizing methods of logging and log transportation to the sawmill plant. Truck transportation has very steadily and positively replaced, to some extent, water transportation, the railroad haul, and even the flume and chute. It has also replaced several skidding devices formerly resorted to in certain sections, as in the South.

3. The tractor has replaced many skidding methods formerly in common use in the forests of all regions. The tractor has resulted in cheaper skidding as well as in leaving the woods in better condition for future growth.

4. Selective logging based on scientific studies or rough approximations is replacing the indiscriminate felling of trees in the woods. Studies have determined the minimum merchantable size limits of profitable logging and lumber manufacture. Notable advances along these lines have been made in the South, Pacific Northwest, Idaho, California, and the Lake States. Effective studies have been made in the southern pine and northeastern forests to determine the lowest profitable limits of tree sizes. This has resulted in more profitable conversion of the raw product as well as in better silvicultural practice to insure continuous and permanent yields in the future.

5. Mechanized felling, bucking, and loading devices have been widely introduced and used to offset the increased wages and decreasing efficiency of woods labor. These are described elsewhere in the text.

6. Efficiency measures and integrated utilization have been adopted by many of the larger timber-holding and operating companies. Undoubtedly many advances have been made through a more scientific approach to the utilization of timber, particularly the adoption of methods, formerly regarded as worthless or of little value, to eliminate waste in logging operations. The increasing tendency is toward more intensive fabrication of the primary product of the sawmill. Many wooden parts and articles formerly made at or near the centers of consumption are now being made at the sawmills. Economy in labor charges as well as in handling and transportation charges is being effected by this procedure. Many examples could be cited to illustrate this tendency: auto-body parts, refinements in various types of floor-

ing, boxes, shooks, crating materials, furniture, and machinery parts. Integrated utilization is developing rapidly, such as saw logs, pulpwood, veneer logs, and other products from the same logging operations.

UTILIZATION IN EUROPE AND THE UNITED STATES CONTRASTED

Forest utilization does not constitute a very serious economic problem in Europe because of the paucity of available forest material, the density of the population, and the relatively large demand for lumber and other forest products. Poland, Czechoslovakia, and Rumania, and particularly Sweden, Russia, and Finland, are important lumber-exporting countries. Countries of dense population in western and southern Europe are important lumber-importing countries.

The per capita consumption of forest products is much lower in Europe than in the United States. Tradition and custom are important factors in the consumption of wood. Practically no exterior construction, such as on dwellings, barns, or buildings of any kind, is fashioned from lumber, as is customary in the United States. Particularly is this true in central, western, and southern Europe. Economic conditions make possible almost complete utilization of forest products. It is estimated that probably 80% to 94% of all the wood grown in Europe outside of Russia, Finland, and Sweden is utilized, whereas in the United States only about 34% to 48% of the trees felled in our forests are ultimately used in one form or another. We are approaching European conditions in many ways, and, therefore, it is interesting to know the situation that prevails there.

Some outstanding features of western European utilization which may be contrasted with those prevailing in our country are:

1. Relatively few important species are produced in Europe. This situation tends to simplify utilization problems. Scotch pine and Norway spruce probably compose 70% to 80% of all the lumber and other forest products in Europe. Three other species are of considerable importance but not comparable with these two in volume, namely, silver fir, beech, and white oak. There are some ash, larch, birch, Corsican pine, and other minor species, but the total constitutes a very small percentage. In some specialized localities, as in southwestern France, maritime pine is the important tree, but it is limited in area although found in pure stands throughout the region known as the Landes.

2. Trees are small in contrast to the varied sizes prevailing in the United States and the very large sizes found in our virgin timber of the Pacific Coast states. Most of the trees in Europe are produced under some form of systematic forest management. Because of the limited possibility of rotation, it is not economical or profitable to wait until these trees reach large sizes. These small trees produce sound, good-quality, but exceedingly knotty logs. The knots, however, are the small tight variety, and lumber is readily converted to the various products in sawmills and woodworking establishments. Practically no clear boards are produced in Europe. (This would indicate that we have been excessively prejudiced against the use of knotty lumber in the United States. There has been a widespread fetish here in favor of clear boards, but with the increased cost and difficulty of producing them it is likely that we will use more and more knotty lumber.)

3. Lumber is produced from a large number of relatively small but permanently located sawmills. Most European sawmills are of the gang-saw type where the logs are "sawed alive." The periodic growth of standing timber is generally cut in these small permanent sawmills. (This is in marked contrast to the past nomadic character of our sawmill industry in the United States where mills set up for cutting a given tract are dismantled and towns are abandoned when the virgin timber has been removed. We will probably gradually approach European conditions in this respect. In some of the older settled communities this has already happened. The stabilization of our sawmill communities and the permanence of mill locations will contribute much to employment and social welfare.)

4. The per capita consumption of lumber and forest products in Europe is exceedingly low compared with conditions prevailing in America. Probably 40% of our lumber production normally goes into house and dwelling construction. The exteriors of wooden siding so commonly found in the United States are practically nonexistent in Europe. We have been a great nation of wood users; large available supplies and inexpensive forms of forest products have tended to make us so. In Europe, on the contrary, the scarcity of standing timber, the relatively slow rates of growth, and the heavy demand compared with the supply, resulting in high prices, have tended to keep the per capita consumption of lumber and forest products at a relatively low figure.

5. In spite of competition from other structural materials, wood finds a large utilitarian demand throughout Europe. It is still used for flooring, furniture, interior trim, boxes and crating, and for a multi-

tude of purposes where wood has proved a better and less expensive material than any other. Europeans build with greater sense of permanence than we. There is a reverence for old things which does not exist to the same degree in our younger country. Roofs, for example, are generally built of tile or slate, and, on account of their great weight, wooden rafters are employed largely for their support. The American tendency toward the introduction and extended use of steel in railway carriages, freight and motorcars, and the like does not exist to the same extent in Europe. European railways have returned to wooden cross ties after experimenting with steel and iron.

6. In spite of the intensive development of forestry, some countries such as Germany, Switzerland, France, Belgium, and Denmark are extensive importers of lumber and other forest materials. This demonstrates very clearly that, even though lumber is rather high in price and every effort is made to replace wood with other materials, still there is an important demand for wood in all those countries.

7. Europe has a very different economic and social structure founded on centuries of custom and tradition. Wasteful practices commonplace in this country are seldom seen in European countries. Even the bark, branches, twigs, and leaves are utilized in some of the European forests. Although there will be a tendency in this direction in America, it will probably be many years before we can apply the extreme utilization practices which are commonly found in Europe. Custom and tradition are powerful factors in this situation. Unusual customs which date back to the feudal system and medieval times, giving the peasants certain rights to use forest products, still prevail in parts of Europe.

8. A good, complete, and permanent system of roads throughout the European forests simplifies and reduces logging costs to a minimum. Transportation facilities are the heart of successful logging and forestry practice. A complete network of woods roads, some installed centuries ago, offers marked contrast to most of our forest areas, especially under virgin conditions where many and expensive means of transportation must be constructed before the forests may be utilized.

FACTORS WHICH DETERMINE DEGREE OF UTILIZATION

As described elsewhere, only a small portion of the trees actually felled in the American forests is ultimately used. The American nation is still a relatively young and pioneer one. It is, moreover, passing through experimental stages of development. Economic and social influences which have had such an important bearing on stabilization in the use of forest products in Europe do not obtain to the same degree in the United States. The progress of the American lumber and associated industries has been marked by great wastage. Even now a vast amount of good material is not utilized.

The factors which determine the degree of utilization may be briefly summarized as follows:

1. Demand for material is reflected in prices. The extent of utilization follows price levels very closely. Increased prices mean more complete utilization, and, conversely, low prices mean increasing wastage. When prices are attractive, softwood trees in the East are often cut to 6″ in the tops, low stump heights prevail, and utilization is excellent. When prices are depressed, eastern hardwoods are only cut to 12″ to 14″ in the tops and often to 8″ to 10″ when the market is sufficiently attractive. There is a striking disparity between prices received for the upper and lower grades of lumber. The latter are generally sold below cost of production and are difficult to dispose of even at sacrifice prices. Only the bigger and better logs are used in times of depressed prices. Demand has been the greatest single factor in the utilization of American forests. In Europe this demand is sufficiently high to utilize profitably nearly all the forest produce grown.

2. Products must be salable. Even if a demand apparently does not exist for certain materials, a potential market may exist. Merchandising efficiency must be developed to furnish a profitable outlet for nonutilized material. For example, for many years tupelo, black gum, and aspen were regarded as "weed trees" and found practically no market. The trade regarded them as inferior woods because of difficult seasoning and other qualities. Now they enter very substantially into the hardwood markets. For many years white, red, and other firs and western larch were nonutilized trees in the Far West. Now local, statewide, and even national markets have been developed for these woods. Similarly, beech was considered almost worthless, but through efficient preservative treatment to extend its life and service as a cross tie and better seasoning and manufacturing methods a good market for it has been developed.

3. Transportation costs must not be excessive. Freight rates on common carriers, or the costs of transporting material from the woods to its ultimate market, are often the greatest barrier to the use of any material. Wood is a commodity of great bulk and weight in proportion to its value. There may be a very definite demand or outlet for a product, and it may be salable, but the transportation charges for getting it to

the market where it is usable may be prohibitive. Until the Panama Canal was developed, West Coast woods did not enter the eastern markets to any large extent. After World War I, water rates from the West to the East were reduced to as low as one half to one third the prevailing rail rates. This was an important factor in placing West Coast woods on the large consuming markets of the Atlantic Seaboard. For many years, a virgin silver fir forest in the Apennine Mountains of Italy was not utilized because the high transportation cost to market was prohibitive, even in a country of high wood values and large lumber imports. In Transylvania, near densely populated centers, there were beech forests that were not marketable because of excessive transportation costs. Many thousand cords of good fuel wood are decaying annually in the Adirondacks and other sections because the cost of handling and trucking or shipment by railroad to a nearby consuming market prevents profitable utilization.

4. Form, size, and condition must lend themselves to utilization. Much sawdust is wasted because its form and condition make it unadaptable for any large commercial use. One tenth to one fifth of the total volume of some of our logs is turned into sawdust. This sawdust is occasionally used but not to any large extent. Only a small number of slabs from sawmills are generally consumed, except for fuel, lath, and pulp, because few products can be made from them profitably. Large quantities of material are left in the woods because of broken tops or shattered logs, or because disease and insects render their condition or form unavailable for utilization. There is heavy wastage in the process of conversion of the raw material to the ultimate form of utilization. In these various manufacturing processes large quantities of small forms of wood are developed for which no profitable market may be available. Several hundred thousand cords of slabwood, however, are converted every year into paper pulp. More than 6000 carloads of waste materials are developed in the New York woodworking and manufacturing plants every year. Definite progress is being made to eliminate this enormous wastage.

5. Cost of handling must not be too high. The product of the forest may be in demand, and cost of transportation to the final market may be reasonable, but logging waste may be so scattered and the resultant cost of assembling so high that the material cannot be marketed profitably. Frequently this situation may be closely related to the transportation costs previously referred to, but it is largely the labor item that is excessive and prevents the profitable ultimate use of the material. For example, fuel wood, woods waste, tops, defective logs,

and broken trees developed in logging operations and odds and ends of different manufacturing plants may be available in large quantities, but the cost of assembling them in sufficient quantity to justify a commercial enterprise to utilize them or transport them to market may be unduly prohibitive. Separating this waste by species or sizes to suit consumers' demands may result in excessive costs.

6. Custom, prejudice, and tradition may prevent adequate utilization. The American people have for many years been unwilling to use knotty, wormy, or stained lumber for many purposes. Frequently only the butt logs of the large virgin white pine were used in the early lumber days of the United States, notably in the harvesting of the great virgin forests of Michigan, Wisconsin, and Minnesota, because the lumbermen could sell only clear boards. Now, however, the public is discovering the desirability of using knotty panels, pecky cypress, and "sound and wormy" chestnut and oak. Knots do not necessarily weaken structural timbers when used in some positions. Stained lumber may serve certain structural purposes as well as lumber not so affected, but it has been generally considered defective and undesirable heretofore. One often observes knotty lumber in large doors and interior trim, as well as in flooring, in various public and private buildings in Europe. With the passing of the virgin forests and the necessity for using second-growth timber, the American people must be gradually educated to the use of lumber and other products formerly considered defective, undesirable, or nonusable. Because of ill-founded prejudice and tradition, much disastrous competition from other materials, as explained elsewhere, has often resulted. This has had an important influence on the intensity, completeness, and nature of the utilization of American forests. Numerous substitute materials have replaced wooden shingles. In many sections of the country, the wooden shingle is again replacing some of these substitute materials, because shingles properly manufactured and conditioned have proved to be a superior article for the purpose. Fiber boxes have replaced the wooden box for the shipment of many commodities. This has adversely affected the utilization of second-growth white pine in New England and many other sections of the country. Competition from steel, brick, cement, and many other materials has also seriously interfered with the extent of the utilization of lumber in the United States.

7. Laws affecting the use of lumber and other wood materials may hamper utilization. In many municipalities, ordinances to terminate or restrict the use of wood and other inflammable materials, in the more heavily congested sections of these cities, have very definitely inter-

fered with the use of wood for framing, siding, and the usual building operations, as well as in various structural forms. They have also seriously interfered with the use of shingles. With the growth of population in the United States and as living conditions become more stabilized, it is only to be expected that wood will be used less and less in our larger cities for structural purposes. Brick, cement, and stone are more permanent forms of exterior construction, and the expression of popular opinion in favor of them through legal enactment is a natural development. Many crosscurrents of competition for business between lumber and substitute materials have no doubt had a part in periodically influencing the degree of utilization of lumber and other materials in our larger communities. In some municipalities many unfair practices and restrictions have been corrected in recent years. In spite of these legal inhibitions, however, a vast amount of lumber is still used in all our great cities for sewers, subways, and other underground construction, for docks, wharves, and railroad terminals, and temporary construction in connection with all building operations. The use of wood very closely follows intensity of population. Legal barriers and influence, however, may play a very important part in the utilization of lumber and other forest products.

PRINCIPAL CHARACTERISTICS IN THE ECONOMIC STRUCTURE OF THE LUMBER INDUSTRY

In analyzing the present and future utilization possibilities of our American forests, it is appropriate to point out some of the outstanding difficulties and vulnerable points in the economic position of the industry. Though it is a large and important national industry, with many strong and notable features, it has also some detrimental elements. Because of the nomadic and migratory character of the occupation, the rapid changes in utilization of species, the crosscurrents of competition and distribution by rail and coastwise ocean traffic, it is only natural that some outstanding weaknesses have developed. In fact, they have been present for a number of years as the result of unorganized control and disorderly production which led to excessive competition among the various species of wood, and between wood and other materials. With 50,000 or more widely scattered producing units in normal times, it has been exceedingly difficult to organize the industry to combat and correct these chronic evils. Among these evils may be mentioned:

1. Instability of price levels. Other commodities produced from a relatively small number of large manufacturing units can be controlled in output much more readily than lumber, which is produced from so many species in widely distributed sections of the country and from so many small and large mills. Largely as a consequence of this situation, instability of price levels has placed in a most unsatisfactory position the logger, the manufacturer, and the wholesale and retailing distributing agencies, as well as the ultimate consumer. Price changes in the lumber industry have been generally rapid and subject to frequent and abrupt fluctuations.

2. Failure to standardize the product properly as to recognized quality, size, and condition. Lack of dependability of the product has been most disconcerting and unsatisfactory to those determining building and construction specifications. Methods of standardization of grades, trade and grade marking, labeling, and the observance of recognized and uniform specifications of size and condition are doing much, however, to improve and overcome this situation.

3. Excessive overhead and carrying charges, such as interest, insurance, taxes, and depreciation on large timber properties acquired at relatively high stumpage levels. This condition has frequently resulted in hasty and disorderly liquidation of standing timber assets and the flooding of a market already unstable in price levels and chronically oversupplied with material. It has been indicated by experience that on some large investments for sawmills and logging equipment, in addition to standing timber, depreciation charges may be disproportionately excessive if the life of the operation is less than 7 years, whereas the overhead charges, such as taxation, interest, protection from trespass and fire, may be excessive if the operation runs over a period of 12 years.

4. Abnormally high production costs caused by large-sized sawmills cutting 150,000 to 500,000 b.f. or more per day in a single shift, together with long log hauls from comparatively inaccessible timber. There is an exceedingly wide variation in production costs among logging and lumber manufacturers in all forest regions. Only those enjoying low costs for stumpage, logging, transportation, saw milling, and overhead are in a satisfactory position to compete successfully in the Nation's markets for the available business. The day of the successfully operated big mill is practically past except on the Pacific Coast where vast timber supplies of large trees justify the installation and operation of large producing units.

COST FACTORS

The lumber business has three distinct phases: (1) logging, (2) lumber manufacture, and (3) sales and distribution. The latter two have been generally standardized on a fairly uniform and efficient operating basis. Therefore, costs may be approximately determined in advance. Factors of cost are generally well under control, so that it is possible to estimate the cost of lumber from the log pond to the pile with some degree of accuracy. The big variable, however, is logging costs, which differ with every "chance" and local condition. It is the logging phase of the lumber business, therefore, where profits are likely to be made or lost. Good judgment in the determination of the best methods, equipment, and personnel to be used is of very great importance. Various systems may be employed in determining the cheapest method of transportation. It is often a serious problem whether tractors, trucks, railroads, or steam skidders may be used to best advantage. The correct solution of the transportation problem is the key to successful logging. Money is rapidly made or lost through the selection of the method of bringing out the logs. Frequently combinations of transportation systems must be employed and each one used where it will handle the logs on the cheapest basis. Probably in few other industries are there so many variable factors as in logging.

Costs of logging may be classified in two broad categories—those of fixed and of direct expenditures. In the former are such cost items as taxes, interest, amortization of indebtedness, and insurance (both fire and compensation). Direct costs include labor, equipment, repairs, camp maintenance, salaries, and general overhead.

The principal factors that comprise the cost variables on every logging job are:

1. Timber—its size, density, weight, and merchantable stand per acre.

2. Distances of (*a*) minor and (*b*) major transportation.

3. Topography as related to transportation and felling conditions, with degrees of gradient, changes in elevation, drainage, and relation of natural drainage basins to each other.

4. Labor supply and efficiency.

5. Federal or state restrictions as to hours or conditions of labor, compensation insurance, forest-fire regulations, and regulation of the weight and size of loads on public highways.

6. Cost and efficiency of equipment and supplies.

DETERMINATION OF THE LOGGING PLAN

For every logging operation, the factors governing the selection of the plan to be pursued must be carefully reviewed. The special features and characteristics of the tract as a whole and in relation to the method of log transport must be thoroughly studied. After the timber estimate, topographic maps, and other engineering preliminaries have been completed and the financial policies determined, a careful appraisal of the probable costs involved must be made. Transportation methods and improvements are too frequently considered on their own merits without reference to the entire operation. On some logging chances, steam driving, flumes, or chutes have been used to transport logs from the landings to the sawmill or to further means of transportation where truck transportation should have been used. It is estimated that the saving possible in reduced costs of minor transportation alone, had trucks been installed as a means of major transportation, would have been sufficient to turn a deficit into a profit.

Several important factors must be carefully reviewed and considered in the selection of methods to be pursued on every logging operation, whether small or large. Costs are always considered on a basis of charges per thousand feet. Among the more important factors to be studied and considered are:

1. Costs of construction for major transportation or the use of methods already installed.

2. The cost of operation and maintenance of this form of major transportation.

3. Camp construction together with operation and maintenance in relation to the conduct of cutting and transport operations from each camp site. Local facilities such as ranches, farms, trailer camps, and communities must be carefully considered as they may serve the purpose better from the viewpoint of cost than would the construction of separate camps. For the larger operations, separate camps must generally be installed.

4. The costs of construction for minor transportation. There may be two or three methods of skidding logs to the major method of transportation.

5. The cost of operating and maintaining minor transportation facilities.

6. Loading and unloading costs involved in these methods of major and minor transportation.

7. The cost and operation of landings, log decks, and the like, under these methods of loading and unloading.

8. The costs of felling, log making, and swamping under the aforementioned method of minor transportation.

These factors must be considered in the light of all the others influencing the equipment and labor service. None of them must be considered or treated separately. They are intimately related and integrated. The difficulty in estimating the advantages and costs of each part of the operation is accentuated with the increase in the number of methods that hold out promising possibilities, especially in such regions as northern Idaho where practically every known North American method of logging has been used. Frequently the only possible method to be installed is readily discernible after a preliminary examination. In eastern Canada, the usual method is the sled haul in winter and stream driving in spring. With the increasing number of small logging units throughout the South and East, however, the question of the most economical method of bringing out the logs is a most important one to be decided.

LOGGING CONTRACTS

Logging contracts, known also as cutting contracts, timber-sales contracts, or as timber-sales agreements as they are designated by the U. S. Forest Service, are used to explain the conditions and protect the buyer and seller in the sale of stumpage. When a lumber company cuts its own timber, such a contract is not necessary, but sawmill operators and loggers frequently purchase timber from private individuals, companies, and Government agencies. The contracts are legal agreements between the buyer and seller to cover the conditions under which timber is cut and removed from a given area. They may be documents of 1 to 35 or more pages. They should be properly witnessed and should clearly set forth all the conditions which have generally been discussed, explained, and agreed on verbally before the signing of contracts.

Among the principal provisions are:

1. Names and addresses of parties (seller and buyer) to the contract, including guarantee of ownership by the seller.

2. Location and area involved, including description by legal subdivisions, reference to deeds, maps, or well-known and established marks such as roads, streams, fences, blazed lines, marked corners. Included is generally an estimate of the probable volume of standing saw tim-

ber, cords of pulpwood, poles, cross ties, or other products to be removed.

3. Conditions of felling and bucking, such as stump heights, top diameters, cutting only marked trees or those designated by size, species, and location; also breakage, lodging, use of windfalls, and so on.

4. Conditions of skidding and log haul, such as machinery or methods permitted, use of timber for skidways, corduroying, cabins, or other purposes in connection with logging, or with road, skidway, and landing construction.

5. Observance of state compensation laws.

6. Protection of owner against claims, suits, or liability arising from personal injury or property damage incident to the felling and removal of timber from the premises of the owner.

7. Conditions governing the prevention, control, and suppression of forest fires, disposal of brush, felling of snags, defective or undesirable trees; also suspension of operation during periods of low humidity or extreme fire hazard.

8. Advance payments prior to felling, covering partial or full value of the timber to be cut.

9. Basis of payment, such as agreed-on log scale, estimate of marked timber, or lump-sum settlement. Large sales are generally made on the basis of actual log scale at landings prior to loading for transport. Small sales are made on the basis of an estimate or a lump sum. This should also include the terms of merchantability of logs, such as 50% or 70% sound. Payment may also be on the basis of logs classified according to log grades.

10. Penalties for infractions of rules, such as cutting trees not marked or designated, unnecessary breakage or destruction in felling and hauling operations, incomplete utilization of the stems due to high stumps, wastage in tops, or improperly measured log lengths and overrun. A bond may be required for the proper fulfillment of contract conditions.

11. Conditions of bidding. Government sales require advertising for a stated period of time. Some sales are made after advertising in local papers or sending out letters to prospective buyers.

12. Duration of sale. The time of beginning and termination of the cutting operation should be explicitly stated. Provisions may be made for extension of time with readjustment of stumpage prices, penalties for extensions, or the recognition of climatic factors preventing operations, or "Acts of God" known as "force majeure."

INTEGRATED LOGGING—PRELOGGING AND RELOGGING *

Within recent years, largely as a result of changing conditions and circumstances, several new developments have occurred. These are briefly described as follows:

Integrated logging consists of selecting those species and sizes best adapted to particular products with the objective of securing greater

FIGURE 1. Good example of integrated logging on a felled area of Norway spruce in Germany. In order to effect complete utilization of all available material, the largest trees are cut for saw logs, and the smaller sizes and tops are used for poles, piling, posts, and pulpwood. Tops and even branches are piled for fuel wood. The entire stem is stripped of all bark for tanning purposes. Stumps are cut low and even the smallest limbs and twigs and sometimes stumps are grubbed out and utilized for charcoal or fuel wood. Much fuel wood is required to drive automobile trucks in many parts of Europe.

revenues by taking advantage of changing market conditions. For many years integrated logging has been practiced in northern Idaho where red cedar poles and some other products were taken from the stand before the main logging job was started for saw logs. This was

* For a detailed discussion of several aspects of these subjects in the Northwest see *Salvage Logging in the Douglas Fir Region of Oregon and Washington*, by E. E. Matson and J. B. Grantham, Oregon State College, Corvallis, Ore., 1947.

done to avoid unnecessary breakage. This was a form of both integrated logging and prelogging. An illustration of integrated logging was described by R. R. Reynolds * showing that a southern pine tree 15″ in diameter under certain conditions was estimated to be worth 88 cents as pulpwood, $2.64 when converted into saw logs, and $5.50 as a pole. This indicates that with changing market conditions standing timber should be converted and sold in those sizes, shapes, and products which bring the highest prices. Another good example of integrated logging is in the Anaconda district of Montana where the forests are cut for a great variety of products, such as saw logs, posts, lagging, fuel wood, converter poles, mine props, cross ties and telephone poles. An excellent example of integrated logging is found in the forests of Crosset Lumber Company near Crosset, Ark. At Brewton, Ala., the T. R. Miller Lumber Company operates a large sawmill, box factory, veneer plant, timber-treating plant, turpentine still, and cooperage plant. The various sizes and species in the forests are logged and directed to these various operations so that the best prices will be obtained. For example, in the treating plant timbers, cross ties, poles, and posts are treated.

A broad example of integrated logging on the several southern National Forests is shown in Table II which includes all of the products cut for the fiscal year of 1947.

TABLE II

Items	*Per Cent*
Saw timber for lumber, veneer logs, and cooperage stock	59
Chestnut acid wood	18
Pine and hardwood pulpwood	11
Miscellaneous, including poles, piling, cross ties, posts, and distillate wood	12
Total	100

Prelogging has recently been developed on the West Coast and is designed to remove those trees which ordinarily would be destroyed or reduced in grade during the felling and yarding operations of the larger and heavier trees. It is generally conducted on those areas where there are two distinct sizes of timber, namely, the large old-growth overstory, and the small pulpwood sizes which often form an understory. The old growth is much more commonly found. Species removed in prelogging are white fir (*Abies amabilis*), western red cedar, and West

* See Southern Lumberman, Nashville, Tenn., December 1947.

Coast hemlock. The upper-diameter limit is generally 20″ outside bark on the stump. Long logs and frequently "package logging" are common. Minimum diameter limits are governed by the manufacturing processes; usually a top diameter of 4″ is the minimum limit. Where there are both an overstory and understory, the removal of timber in prelogging consists of taking out as much as 40 cords of wood per acre.

Relogging or recovery logging varies widely throughout the Pacific Northwest where it is largely developed. The amount recovered is dependent on the degree of utilization practiced in the original logging operations. Some of the larger companies take out a comparatively low recovery of 2 to 10 cords per acre of pulpwood. The size limit of the material to be recovered from former logged areas is generally governed by the processing equipment at the conversion mills. The minimum size is generally a log with a 4″ top diameter, 16′ long, and the maximum diameter is 20″. Package logging is often used with relogging, and long lengths are strapped together in packages and loaded on trucks in this way. In one case a company made considerable profits from cutting high stumps of western red cedar and converting these into bolts for shingles. The use of chain saws is very helpful in cutting off stumps formerly cut to a height of 6′ to 15′ or more.

DISTRIBUTION OF LOGGING AND OTHER COSTS FOR PRODUCING LUMBER

Tables III and IV present the distribution of costs starting with logging expense and stumpage values through to the shipment of the finished product f.o.b. sawmills.

Table III shows the costs in percentages of the total according to the various items involved in producing lumber.

Table IV is based on the identically same costs divided into direct, overhead, and stumpage costs. Both are indicated in percentages and are based on several hundred million feet produced in 65 operations in various parts of the South and Southeast, including both southern pine and hardwoods.

These tables show certain very significant facts: Logging is by far the most important phase of producing lumber from an expense viewpoint. The costs of sawmilling and stumpage are other important items. From the viewpoint of direct costs, labor is the largest single item involving approximately one half at least of the total cost of producing lumber.

TABLE III

COSTS OF PRODUCING, SELLING AND SHIPPING LUMBER

Average of 65 Mills. According to Principal Phases of Operations

Items	*Per Cent*
Logging	33.5
Sawmilling	16.0
Yards and sheds	5.2
Dry kiln	3.8
Planing mill	6.9
Total manufacturing	65.4
Shipping	3.0
Selling	2.8
General administration	12.7
Depreciation	2.9
Manufacturing and conversion	86.8
Stumpage, including purchased logs	13.2
Total	100.0

TABLE IV

DISTRIBUTION OF COSTS FOR 65 MILLS

According to Direct, Overhead, and Stumpage Items

Direct Costs	*Per Cent*
Labor	41.7
Materials and supplies	15.6
Contract costs, including the above 2 items	8.4
Other—miscellaneous	2.7
Depreciation	2.9
Total direct costs	71.3
Overhead	
Wholesale commissions	1.6
Advertising and trade promotion	0.2
Salaries	3.7
Taxes	2.6
Insurance	2.7
General overhead	4.7
Total overhead	15.5
Timber	
Timber—stumpage and purchased logs	13.2
Total	100.0

These two forms of accounting for costs on the same operations are shown in Tables III and IV.

DISTRIBUTION OF LOGGING COSTS

The principal item of logging costs from a national viewpoint has been the main log haul or major transportation. The next most important was the skidding or minor transportation cost. The present most important item is falling and bucking, owing to high labor wages. This varies, however, with local conditions such as the size of timber involved, climatic conditions, topography, cost and efficiency of labor and mechanized methods, length of haul, and other factors which may vary widely even within restricted areas. For example, in northern hardwoods and in Canadian and northeastern pulpwood operations falling and bucking comprise the largest item of cost in logging.

Table V gives one example of the distribution of logging costs from stump to the sawmill or pulp mill in eastern Canada and the northeastern section of the United States. In Canada falling and bucking are done with manual labor to a large extent.

TABLE V

Items of Logging Costs *	*Cost per Cord in Per Cent (80 cu. ft. of Solid Wood per Cord)*
Cutting (falling and bucking)	32.0
Yarding (skidding)	11.8
Hauling to drivable stream	14.7
Stream driving	9.0
Camp construction, maintenance, and overhead	13.0
Stumpage	13.5
Workmen's compensation	2.0
Forest protection (and crown ground rent in Canada)	4.0
Total	100.0

* Data supplied by L. A. Nix, consulting forester of Toronto, Canada.

Four companies alone in eastern Canada annually produce about 1,500,000 cords of wood. The largest one, on the Gatineau River, has a capacity of 1500 tons of paper per day.

LOG GRADES

During the past 20 years there has developed a demand for the separation of logs into grades. These grades are based largely on the prospective grade outturn in lumber. The most effective and universally adopted plan of log grades was developed in the Northwest in Douglas fir and associated species. There is no universally adopted plan of log grades in the South and North, but in local regions and particularly with certain mills, there is a fixed plan of separation of logs into grades, generally three in number. Very different prices are paid for the different grade logs. Therefore, loggers must give special attention to proper lengths and elimination of defects in crosscutting so that the highest quality logs are produced.

In some regions there has been a separation of the standing timber into tree grades based largely on the quality of the first or butt log. Thus there may be a very different stumpage value for various trees, even of the same size and species. This means that very careful selection and marketing of trees to be cut must be employed in the woods, especially by those experienced in lumber grades or the eventual outturn of these trees and logs expressed in value.

Rising stumpage values and especially the high prices paid for all grades of lumber during and since World War II have given increased emphasis to the subject of log and tree grades.

Lists of hardwood log grades devised by the U. S. Forest Products Laboratory and of northern hardwood tree grades based largely on the character of the first log found in the trees as cut * are given here.

HARDWOOD LOG GRADES

By U. S. Forest Products Laboratory

No. 1 Log

Lengths 10′ or longer, butt logs 12″ or larger; logs other than butts 15″ or larger.

Logs under 16″ in diameter must have not less than five sixths of each of three faces clear in not more than two cuttings not less than 7′ long, and in logs 16″ and over not more than two cuttings not less than 5′ long.

Sweep limited to 15% deduction in gross scale.

No logs admitted with a net scale less than 60% of gross scale.

No. 2 Log

Lengths 8′ or longer; diameter 10″ or larger; logs under 10′ long must be 12″ or larger.

Logs 10′ long, 10″ or larger, must have not less than two thirds of each of three faces clear in not over two cuttings 3′ or longer.

Logs 12′ or longer will admit not over three cuttings per face.

* The subject of log grades is so extensive that it cannot be given adequate treatment in an elementary text of this kind. For further information see various pamphlets as published by groups and individual companies and articles as found in the lumber trade journals.

Logs under 10′ long, 12″ or over, must cut not less than three fourths clear on three faces in not over two cuttings 3′ or longer.

Sweep limited to 30% deduction in gross scale.

No logs admitted with a net scale less than 50% of gross scale.

No. 3 Log

Lengths 8′ or longer; diameters 8″ or larger. Logs must scale not less than 40% of their gross volume.

No. 4 Log

Tie cuts. Must have sound centers. No rot. Sweep or crook not more than 4″ in 8′.

NORTHERN HARDWOOD TREE GRADES

Tree Grades	*Value Groups*
No. 1	Trees, the first log of which are grade 1 logs
No. 2	Trees, the first log of which are grade 2 logs
No. 3	Trees, the first log of which are grade 3 logs
No. 4	Trees containing a net of less than 40% of their gross log scale

RELATIVE VALUES OF PRIMARY FOREST PRODUCTS

No special study has been made by Government or private authorities to determine accurately the annual value of primary forest prod-

TABLE VI

Primary Products	*Value (in Million Dollars)*
Logs for lumber, sawn cross ties, and other sawn products	$1200
Fuel wood	600
Fence posts	200
Pulpwood	250
Veneer logs	100
Hewn ties	50
Poles	30
Piling	20
Distillation wood	8
Naval stores	120
Tanning materials	10
Nuts	10
Maple sirup and sugar	10
Christmas trees and greenery	10
Pharmaceuticals and oils	3
Miscellaneous bolts, logs and similar primary products	10
Total	$2631

ucts, that is, the logs used for lumber and other purposes as felled, skidded, and hauled out of the woods to some processing plant. Logs and bolts of 2′ to 4′ in length, up to 100′ or more in length are taken out of the woods for a great many different purposes. Primary products, other than logs, are also included.

Table VI shows the approximate value of the principal primary products of our logging and associated woods operations as delivered to sawmills and other processing plants. The total value is estimated to be $2631 million.

CHAPTER II

Forest Resources and Stumpage Valuation

The forests of the United States now comprise about 25% of the total area of the country or about 76% of the primeval forest area. Originally they occupied 822,238,000 acres or nearly one half the total land surface of the country.

FIGURE 2. Aerial view of large logging operation in the Big Cypress Swamp near the Everglades of Florida. In the foreground is skidder bringing logs to the railroad. Logs are hauled to Perry, Florida, several hundred miles away for sawing into lumber. This is the largest body of virgin cypress left in the country. *Courtesy Southern Cypress Manufacturers Association.*

Much of our forest is found at high elevations in the mountainous regions, conspicuously in the western states where the timber is relatively inaccessible or of insufficient size or of too inferior quality to be of commercial importance for saw logs, pulpwood, or other forms of forest products. Forests in these locations are chiefly valuable for watershed protection and seldom enter commercial markets.

Many of our forests have been cleared for agricultural purposes or burned or are located in swamps. Under these conditions they are of little or no value in furnishing usable commodities.

The U. S. Forest Service has separated the American forests into eight broad major regions. They occur in natural geographic and topographic units and generally exhibit distinctive logging and utilization problems. In some regions, notably in the East, similar types of forests and logging methods are found in two or more regions. On the other hand, hardwood logging is concentrated largely in the Mississippi Delta, the southern Appalachians, and adjacent sections of the Southeast region which is largely devoted to southern pine production.

AREAS

The total forest area of the country is 623,828,000 acres, of which only 461,044,000 acres are classified as commercial forest areas.

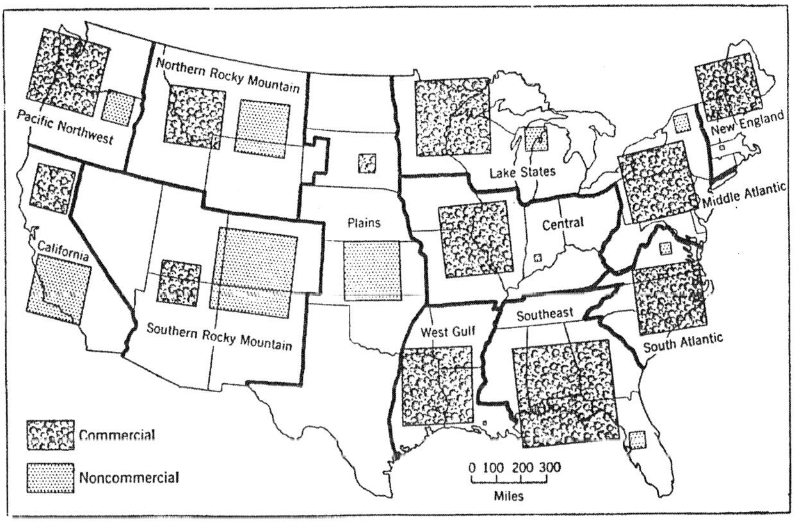

FIGURE 3. Location of commercial and noncommercial forest areas in the United States. This demonstrates that more than 75% of the potential growing areas for our future timber supplies are found in the North and South. Less than 25% are found on the West Coast and Rocky Mountains. *Courtesy U. S. Forest Service.*

Only about one half the saw-timber area is virgin or old growth, and most of this is in the West. The second-growth timber, especially in the South, however, is rapidly growing to commercial size. Second-growth timber already plays an important part in the lumber markets of the United States, as described elsewhere.

About 40% of all the saw-timber areas are found in the South. Only 23% of our saw-timber area occurs west of the Great Plains; yet this region contains the bulk or 65% of the volume of standing timber. About 37% of the saw-timber area of the country embraced within the North contains only 14% of all the saw-timber left in the country.

Of the total area devoted to forests only about one fifth or 98,855,000 acres is virgin or old growth. About 75% of this area is found west of the Great Plains.

OWNERSHIP

Ownership of our commercial forests has been subdivided into three classes: (1) private, (2) Federal, and (3) state, county, and municipal.

1. *Private companies or individuals own 344,973,000 acres or 75% of our commercial forest area.* These are chiefly lumbermen, farmers who own large areas of wood lots in small and scattered tracts, pulp and paper companies, and others interested in owning and holding forest properties. Most of the larger holdings controlled by private companies are found in the Pacific Coast region and in the South. Farmers own about 40% of all the privately owned commercial forest area.

2. Federal agencies own or manage 19% of our commercial forest area. A large share of this area is in National Forests which are operated by the U. S. Forest Service; a relatively small percentage is controlled by the U. S. Indian Service and the National Park Service, War Department. A very high percentage of these federally controlled forests is found in the Rocky Mountains and Pacific Coast States. In some states such as California, Idaho, Montana, and Oregon, a large share of the forest area is found within the National Forest boundaries. A small percentage of the National Forests is located in Michigan and Wisconsin, New England, Pennsylvania, the Southern Appalachian Mountains, the Southeast, and the South.

3. States, counties, and municipalities own 6%. There has been a pronounced forward move in the acquisition of forest areas by these agencies. More than 30 states now own or control forests. New York and Pennsylvania each have more than 2 million acres which are set aside in their state forests and parks.

Of the 461 million acres of commercial forests, 345 million acres are owned by private individuals or interests; 89 million acres are owned

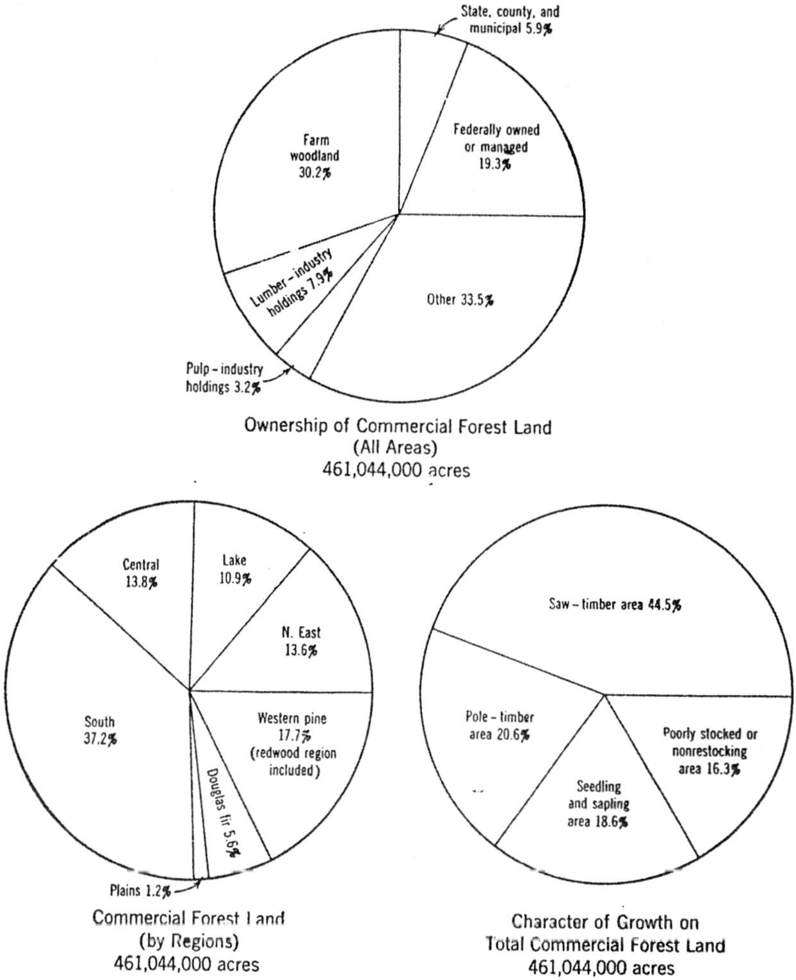

FIGURE 4. Commercial forest areas of the United States. The top circle shows ownership by classes. This indicates that the lumber and pulp industries own a very small share of the total. Most of the areas are owned by farmers and other small woodland owners. The lower left circle shows the commercial forest by regions. The circle on the right shows the same forest according to character of growth. This indicates that saw timber occupies a very large per cent, namely 44.5% of the total present commercial forest area. *Courtesy U. S. Forest Service.*

and managed by Federal agencies; and the remainder, or 27 million acres, is owned by the states and local political units. It is apparent, therefore, that the private companies own the bulk of the standing timber, both in area and in volume.

As a large share of our forests is privately owned, logging methods have been pursued without any Government supervision. However, several states have passed laws affecting certain aspects of logging.

VOLUME OF STANDING TIMBER

The volume of standing saw timber in the United States is estimated to be 1601 billion b.f. or about one third the original stand. Fifty per cent of the commercial saw timber is old or virgin growth. The U. S. Forest Service estimated the standing timber in 1920 to be 2,214,893 million b.f.

The region west of the Great Plains contains 65% of all the remaining saw timber far removed from the great consuming markets of the Nation. The South has 21% and the North 14% of the standing saw timber.

The entire eastern forests (including North and South) contain only 35% of the standing saw timber. Lumbering and clearing for agricultural purposes have removed a very large portion of the once rich and varied forests of the region lying east of the Mississippi River. In spite of the large areas devoted to forests in the South, those states contain only 21% of the saw timber, but they possess an important and rapid-growing second-growth forest which has great possibilities for the future.

Practically no hardwoods of national importance are found in the Rocky Mountain or Pacific Coast States.

LEADING COMMERCIAL SPECIES

The leading species found in the saw-timber forests of the country are as shown in Table VII.

Altogether about 60 commercially prominent species are found in the lumber markets of the United States. Two western species, Douglas fir and ponderosa pine, comprise about 38% of all the standing timber. For more than 50 years, southern pine has been the principal tree cut for lumber. Many kinds of wood formerly regarded as of little value or even worthless have entered the lumber markets within recent years.

Among these now assuming some prominence which formerly were considered either of inferior quality or rather difficult to season, in

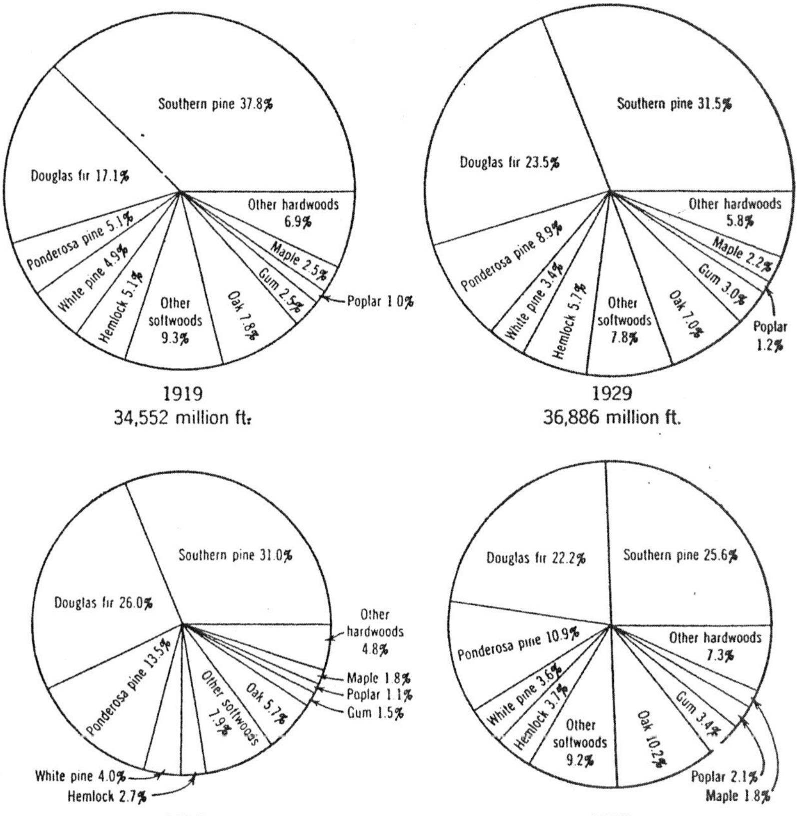

Figure 5. Trend of lumber production by important species. This shows that southern pine has been the leading species for a good many years. Douglas fir has been second in production and has advanced in relative volume in the past. Ponderosa pine has increased in importance, and white pine and hemlock have decreased in production. Oak has also increased in importance. *Courtesy U. S. Forest Service.*

comparison to the more valuable species, are western hemlock, tupelo and black gum, western larch and beech. Most of these woods have important intrinsic values and are now more favorably regarded in the lumber markets. Some of the true firs and aspens have had more recent recognition and acceptance on our markets. It is apparent from

TABLE VII

Species	*Million feet Board Measure*
Douglas fir	430,028
Southern pine	188,327
Ponderosa pine	185,025
True firs	113,541
Oak	101,381
Western hemlock	97,207
Beech, birch, and maple	67,887
Sugar pine and western white pine	40,224
Redwood	38,114
Spruce (western)	35,880
Spruce and fir (eastern)	27,007
Red gum	26,476
Larch	26,606
Lodgepole pine	22,060
Tupelo and black gum	18,957
Cottonwood and aspen	15,662
White and Norway pine	15,385
Hemlock (eastern)	14,921
Yellow poplar	12,347
Miscellaneous	123,965
Total	1,601,000

Table VII that one species, namely Douglas fir, comprises more than one fourth of all the saw timber in the country, and nearly one half of all the western softwoods. Oak is the leading hardwood species available in the American forests.

About 25 species (or groups of species, as in the case of southern pine and the true firs) comprise 93.6% of the total stand of saw timber. Three species of softwood, namely, Douglas fir, ponderosa pine, and southern yellow pine, form about 50% of the total standing timber of the United States.

SUMMARY OF FOREST RESOURCES *

The following is a summary of the statistical position of forest resources of the United States and the available supply of timber for the lumber and other industries.

1. The total forest land area of the United States is 623,828,000 acres, of which 461,044,000 acres are classified as commercial forest.

* Supplied by the U. S. Forest Service.

2. Of the commercial forest area of the United States, 205,176,000 acres are classified as saw timber. Of this, 59 million acres are in the West, 98 million in the South, and 48 million in the North.

3. The total volume of saw timber in the United States is 1,600,972 million b.f. Of this, 1,296,477 million b.f. is softwoods, and the remainder or 304,595 million b.f. is hardwoods. Of the remaining saw timber available, 840,340 million b.f. is classified as on virgin areas. Virgin timber, mostly in the West, represents about one half the volume

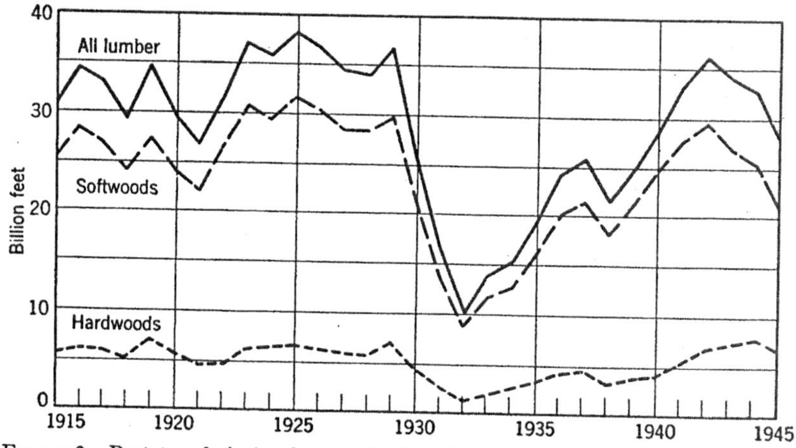

FIGURE 6. Past trends in lumber production showing cut of all species as well as softwoods and hardwoods. *Courtesy U. S. Census Bureau.*

of standing saw timber. Of all the volume of saw timber, 1,042,556 million b.f. is in the West, and the remainder is in the East.

4. The ownership of commercial forest in the United States is largely private. The 461 million acres of commercial forest are owned as follows:

	Million Acres	*Per Cent*
Private	344,973	75
Federal	88,957	19
State, county, and municipal	27,114	6
	461,044	100

5. The volume of saw timber on the commercial forest land (914,272 million b.f.) is largely under private ownership. Of this, more than two thirds are industrially owned and somewhat less than one third farm-owned. A total of 622 billion b.f. is Federal owned, of which the National Forests contain 518 billion b.f.

TABLE VIII

LUMBER PRODUCTION BY IMPORTANT SPECIES

	1909 *M ft.*	*1919* *M ft.*	*1929* *M ft.*	*1939* *M ft.*	*1945* *M ft.*	*1945* *% of* *Total*
Southern pine	16,277,135	13,062,938	11,629,689	7,749,188	7,210,066	25.64
Douglas fir	4,856,378	5,902,169	8,688,700	6,494,301	6,237,312	22.18
Ponderosa pine	1,499,985	1,755,015	3,288,237	3,360,004	3,056,281	10.87
Hemlock	3,051,399	1,754,998	2,098,668	665,259	1,038,999	3.69
Eastern and northern white pine *	3,900,034	1,723,642	1,247,878	1,004,262	1,022,625	3.64
White fir	89,318	223,422	307,241	97,712	499,266	1.78
Redwood	521,630	410,442	485,606	345,003	443,807	1.58
Spruce	1,748,547	979,968	563,953	346,159	357,819	1.27
Cedar	346,008	332,234	308,839	263,693	302,586	1.08
Idaho white pine	§	§	§	§	246,128	0.87
Cypress	955,685	656,212	532,022	421,584	228,388	0.81
Sugar pine	97,191	133,658	349,294	308,929	228,103	0.81
Larch	421,214	388,121	244,719	111,488	190,722	0.68
Lodgepole pine	23,733	16,281	30,897	54,803	65,358	0.23
Balsam fir	108,702	68,030	37,602	20,002	‖	
Other softwoods †	....	....	....	....	12,412	0.04
Total softwoods	33,896,959	27,407,130	29,813,345	21,242,387	21,139,872	75.17
Oak	4,414,457	2,708,280	2,574,495	1,432,119	2,858,862	10.17
Red gum	706,945	851,431	1,103,703	382,693	970,678	3.45
Poplar	858,500	328,538	435,953	276,383	578,463	2.06
Maple	1,106,604	857,489	823,552	445,163	522,100	1.86
Tupelo	96,676	143,730	306,634	271,486	438,703	1.56
Beech	511,244	358,985	186,558	119,564	352,140	1.25
Cottonwood and aspen	265,600	144,155	165,306	129,858	209,038	0.74
Birch	452,370	375,079	374,377	140,738	151,869	0.54
Other hardwoods ‡	2,200,406	1,377,259	1,102,109	535,083	900,619	3.20
Total hardwoods	10,612,802	7,144,946	7,072,687	3,733,087	6,982,472	24.83
All Species	44,509,761	34,552,076	36,886,032	24,975,474	28,122,344	100.00

* Includes jack pine.
† Species not specified
‡ Other hardwoods include ash, basswood, chestnut, elm, walnut, and other lesser species.
§ Included under Ponderosa Pine.
‖ Included in other softwoods.
Source: U. S. Census Bureau.

STUMPAGE AND STUMPAGE APPRAISAL *

DEFINITION

Stumpage is the value of standing timber. It is generally expressed in dollars per unit of production such as thousand board feet, cross ties, cords, posts. It may, however, be expressed in terms of value per

* Although stumpage and stumpage values come properly under the subject of forest valuation, they are treated briefly here because logging studies and appraisals are often made to determine stumpage values. Also stumpage values may determine if logging operations are likely to be profitable or not.

acre. The word stumpage may also be used in a general sense to mean the standing timber itself. In this case there is a distinction between stumpage and stumpage value.

STUMPAGE TREND

Stumpage values have steadily increased for many years in the United States. Fifty to a hundred years ago, stumpage values of $1

FIGURE 7. Good stand of white pine, hemlock and spruce in northeastern region. Practically all of the white pine cut in the Northeast and Lake States is from second-growth timber.

to $5 per acre were common. Since there may have been 5000 to 40,000 b.f. per acre, these values represented from a few cents to a dollar or more per thousand board feet. Generally speaking, they have gradually risen during the past three decades, and much more sharply since 1940 than before that time.

The highest peaks for lumber prices were in 1920 and again in 1948. Stumpage prices reflected the increases in lumber prices. Generally, stumpage values do not fluctuate so sensitively nor so violently as do the prices of lumber or other forest products. Probably the widest fluctuations have been found in northern white pine. Fifty to seventy-five years ago much of this stumpage was sold for as little as 25 cents per thousand board feet. For well-selected timber, stumpage of this species sold during recent years for as much as $20 to $25 per thousand board feet.

FACTORS WHICH DETERMINE STUMPAGE VALUES

About 75% of the commercial timber areas in the United States is privately owned. In periods of active business many sales are consummated.

Many factors enter into the determination of stumpage values. Each tree literally has a different stumpage value; in fact, each log within the individual trees varies in its stumpage value. There has for many years been a very wide range of stumpage values within each species, depending on the factors of location, accessibility, size, form, condition, demand on the market, and so on.

Some of the more important factors in the determination of stumpage values may be briefly summarized here.

1. Species, quality, size, and density of standing timber. White pine, white ash, and black walnut are much more valuable species than tupelo gum, incense cedar, balsam fir, western hemlock, and such species. Large-sized trees generally produce higher percentages of upper grades because they have fewer defects and therefore command a higher stumpage value. Where timber stands are dense and there is a high yield per acre, the cost of logging will thereby be reduced. Stumpage values within a species are always higher for dense stands of timber than for those of sparse growth. If 100 million ft. of timber is to be hauled over a truck road built for this purpose, the cost of hauling per thousand feet will be less than if a smaller amount is to be hauled, and, therefore, the stumpage value is higher.

2. Accessibility, distance from the market, and general location. The cost of logging, manufacture, and transportation of the product to the market has a very direct bearing on stumpage values. In difficult logging chances where the costs of felling, skidding, and the like are excessively high, stumpage values are obviously lower than in those where more favorable logging conditions exist, and where skidding, hauling, and so on can be done at a relatively low cost.

3. The competitive demand for timber, together with the quantity sold and quantity of other adjacent timber available. The cost of logging may be exceedingly high on a relatively small tract because of excessive charges for depreciation and the like, whereas on a tract of large size where the period of logging is spread over a number of years, the stumpage value, when the carrying charges such as interest, and insurance are taken into account, may be reduced to some extent.

4. Demands of the market. The kind of product, its salability, and so on may have a definite bearing on stumpage value.

FIGURE 8. An unusually heavy and valuable stand of cove-type hardwoods in the southern Appalachian Mountains. Many of these stands contain up to 20,000 b.f. or more per acre but generally only 5000 to 9000 ft. per acre.

5. Form of sale. Such factors as a flat price for the entire tract or a fixed rate per thousand board feet for the entire tract or for each different species of timber cut may have an important bearing on the stumpage value. Timber in wood lots in the East is frequently sold by the lot rather than on the basis of thousand board feet.

6. Terms of sale. Required time and form of payment, cutting restrictions, requirements of diameter and top limits, fire protection clauses, close utilization, and similar conditions influence stumpage values, as evidenced in National Forest timber sales. Stumpage for immediate cutting has ordinarily a higher value than that held for long-term liquidation.

METHODS OF ARRIVING AT STUMPAGE VALUES

Most of the sales of stumpage in the United States have been the result of bargaining or an approximation as to value, depending on other sales in the same region. In the early days of the American lumber industry many sales were made on the basis of a flat price per acre or lump sum for the entire tract. With the rise in stumpage values, it became necessary to make a more accurate determination of values per thousand board feet rather than set a per-acre or lump-sum basis for the transaction.

Two rough methods are commonly applied in determining stumpage values, as follows:

1. The cost of logging, log transportation, and lumber manufacture, and the cost of sales, depreciation, and overhead, such as insurance, interest, and taxes, are added together. To this sum is added 20% or some theoretical estimate for profit and risk. This total is subtracted from the actual sales return of the lumber, the difference being the stumpage value of the timber. For example, the following formula is used in calculating stumpage by this method, all costs and values being expressed in terms of dollars and cents per thousand feet board measure:

$$S - \left((L + M) + \frac{L + M}{5}\right) = \text{stumpage value}$$

or

$$S - \tfrac{6}{5}(L + M) = \text{stumpage value}$$

where L = logging cost
M = manufacturing cost including sales expense, overhead, etc.
S = sales return or f.o.b. mill selling price

For example, following out this formula in actual practice, assuming the following costs, we have

		Cost per M.b.f.
L = logging cost	Felling and bucking	$ 7
	Hauling logs to mill	8
M = manufacturing cost	Sawing and piling	15
	Hauling to market or loading on cars at railroad	5
	Overhead expense, depreciation, incidentals, etc.	3
	Total	$38

If a margin of 20% of this total is added for profit and risk, or $7.60, the total cost of putting the lumber on the market, plus this additional 20% is $45.60. If the average price received for the lumber f.o.b. mill is $60 per M.b.f., the total cost of $45.60 including profit and risk margin is subtracted from the selling price, $60, which leaves $14.40 for the stumpage price.

2. Another method is to calculate the total cost of production without any allowance for profit and risk. In the foregoing instance, this total would be $38. Subtract this total from the sales price of $60, which would leave a balance of $22. This balance is then equally divided between the seller or timber owner and the purchaser or operator, resulting in a stumpage value of one half of this difference or $11. Although the costs and returns are the same by these two illustrations, there is considerable difference in the actual stumpage value. Some regard the realization value or the net proceeds after the costs of logging and manufacture have been deducted from the sales price as the stumpage value.

Forest service method of determining stumpage value. The Forest Service, after an exhaustive study of conditions which determine stumpage values on National Forest timber sales, has arrived at two methods of determining stumpage values.* These are known as the investment method and the overturn method. The calculation under the former method is as follows:

The average amount of money employed in the operation including working capital and fixed investment must be first determined. A definite percentage of the average investment gives an annual sum which must be set aside as a profit margin. This sum divided by the yearly cut is the profit margin per thousand board feet. The sum of the

* *Instructions for Appraising Stumpage on National Forests,* U. S. Forest Service, U. S. Department of Agriculture, Washington, D. C.

margin and the depreciation per thousand board feet and current logging and milling costs deducted from selling and milling prices gives the stumpage rate. Expressed as a formula this calculation is

$$X = S - \left(Lc + Mc + D + \frac{\text{per cent of } (A + W)}{\text{annual cut}}\right)$$

where X = stumpage price
S = average selling price
Lc = logging costs
Mc = manufacturing costs
D = depreciation of fixed investment
A = average fixed investment
W = working capital

All except A and W are expressed as amounts per thousand board feet log scale.

Illustrating the application of this method is the following:

Selling price (1000 b.f. at $60 per M)		$60.00
Logging costs	$15.00	
Milling costs	15.00	
Depreciation ($24,000 divided by 10,000 M)	2.40	
Margin for profit and risk (15% of $235,000 divided by 10,000 M)	4.03	
Total charges		36.43
Stumpage price		$23.57

Calculation under the Overturn Method. Profit under this method is the percentage of the overturn, or the entire production cost per thousand board feet at date of sale, including operation charges and depreciation but not stumpage price. The calculation may be expressed by the following formula:

$$X = S - (Lc + Mc + D) + \text{per cent of } (Lc + Mc + D)$$

or transposing

$$X = S - (1 + \text{per cent}) \times (Lc + Mc + D)$$

Lc, Mc, and D represent logging costs, milling costs, and depreciation, all in amounts of thousand board feet log scale. X is the stumpage price and S the average selling price log scale. Per cent represents the rate of profit margin allowed on the overturn.

If we take the operation used to illustrate the investment method, the following result is obtained from the formula:

Depreciation is $2.40 as before, logging costs $15, milling costs $15, and selling price is $60 f.o.b. mill, all in terms of log scale. If 25% on the overturn is regarded as fair, the calculation becomes:

	Per M.b.f.
Logging costs, log scale	$15.00
Milling costs, log scale	15.00
Depreciation costs, log scale	2.40
Total	$32.40
Margin for profit and risk at 25%	8.10
Total cost and margin	$40.50
Selling value log scale	60.00
Stumpage indicated	$19.50

It should be noted that the margin for profit and risk in this calculation is $8.10 compared with $4.03 under the investment method with a rate of but 15%.

The overturn method is chiefly applicable to smaller operations or at least where the investment is very small in comparison with current operation costs. It is used largely in railroad transaction operations. It is of especial value in small sales where the investment is low. The investment method, on the other hand, is applicable to larger operations, is in accordance with usual business practice, and permits ready comparison with practice followed in other industries.

PRINCIPAL PRODUCTS OF THE FOREST

Of the many products of the forest, lumber is the most important, comprising 55% of the total and 70% of the saw timber cut. About two fifths of our lumber comes from the South and another two fifths from the Pacific Coast. Oregon and Washington produce almost one third of all the lumber produced.

The second most important item produced from our forests is fuel wood. Much fuel wood comes from tops and limbs of trees cut for saw logs, pulpwood, and other products as well as from sawmill waste, such as slabs, edgings, and trimmings. Fuel wood makes up about 18% of all the products of the forest, and most of it, or two thirds, is hardwoods, the remainder being softwoods.

Pulpwood is the third most important product of the forest, constituting 11% of all timber cut. There has been a rapid increase in recent years in the production of pulpwood, notably in the South and in the Pacific Northwest. In the South, more than 50 mills have been constructed within recent years. The South is now the leader in pulpwood production, exceeding the northern states by 40% and the West by 80%. Southern pines are the principal species used for pulpwood in the South. Spruce, fir, hemlock, and pine are used principally in the North; in the Pacific Northwest, western hemlock, Sitka spruce and white fir are the principal species used for pulpwood.

Lumber, fuel wood, and pulpwood comprise about 84% of the annual cut of forest produce, the remaining 16% being distributed among a great many items. In order of importance, these are:

1. Veneer logs.
2. Hewn ties.
3. Round mine timbers.
4. Fence posts.
5. Cooperage materials.
6. Shingles.

Other less important items are poles, distillation wood, export logs and hewn timbers, piling, tanning extract wood, wood for excelsior, and many logs, bolts, and pieces of special specification. Among these may be mentioned laundry stock, baseball bats, handle stock, and furniture stock.

CHAPTER III

Forest Labor and Housing

GENERAL

Labor cost is the major item in both the logging and manufacturing branches of the lumber industry. It includes practically the entire cost of felling and bucking and is a large component part of major and minor transportation. The financial success, therefore, of any logging enterprise depends very largely on the quality and efficiency as well as the amount of labor used. There is generally a wide variation in wages paid, efficiency, and methods of remuneration in different parts of the country.

Forest labor has traditionally been recruited from men familiar with rough and arduous conditions existing in our forest regions. Usually the work has been characterized by long hours and hazardous labor, requiring considerable strength, skill, and endurance. In the early logging days of the American colonies, and notably in Maine, New York, and Pennsylvania, men frequently worked in the woods during the fall and winter from dawn to dark. Manual labor was used very largely in many phases of the operations. The timber was located near drivable streams and was of such size that it was readily handled by manual labor or horse skidding. Only a few laborsaving or mechanical devices were in existence until the large timber of the Far West was exploited. Mechanical skidding devices were used in logging the Lake States forests, but they were not developed to replace manual labor to any large extent until the Pacific Coast forests began prominently to supply the lumber demands of the country. The nomadic character of the industry has resulted in the rapid building of camps and forest communities, the concentration of labor in temporary quarters, and the subsequent abandonment of woods settlements and some sawmill towns. With second-growth forests increasing in importance in the lumber industry, however, stable forest communities have already been established and will become more significant.

ORIGIN, CHARACTER, AND NATIONALITY OF FOREST LABOR

The early logger was the American settler. His first occupation was to clear the forest for a home. With the rapid rise in lumber production during the nineteenth century, the extension of agricultural development westward, and the concentration of woods activities in the Northeast and later in the Lake States, it was natural that French Canadian labor, attracted by the wages and accustomed to the rigorous work of the woods, dominated the woods labor in the early days. Political events in Canada and the difficulty of earning a substantial living there no doubt contributed to the important part played by this type of labor in the exploitation of the northern and northeastern forests. When the industry became centered in Pennsylvania, many of these French Canadians moved there and subsequently to the Lake States. With the expansion in lumber production and the development of the Lake States as a center of production concurrently with a rapid increase in immigrants of Scandinavian origin, the Scandinavians rapidly replaced the French Canadians, who preferred to remain near their homeland and to continue in the woods work of the Northeast. Scandinavians were generally familiar with woods work in Europe. It was natural, therefore, that the Swedes and Norwegians should dominate the woods labor of the Lake States.

The woodsmen of the South have always been predominantly Negroes. In the Northwest and Pacific Coast States the lumber industry was developed largely by men with capital and with organizations from the Lake States. The prominent type of labor found in the Northern Rockies and on the West Coast, therefore, is of Scandinavian origin, with a strong admixture of Finns and representatives of other nationalities.

In the Southwest Mexican labor was used in the woods for several years. Then Scandinavians came in from the Northern Rockies. The type has again shifted back to Mexican and mixed American.

Throughout the woods labor camps of the country, local ranch and farm labor are often used. In many sections, especially on small sawmill operations, men work in the woods in the late fall and winter, and on the farms in the spring and summer seasons. Many men engaged on small logging jobs live at home, board with local farmers, or "batch" in trailers.

DURATION OF EMPLOYMENT

Woods labor generally continues the year round. Until about 1900, woods work in the Northeast depended on snow for winter logging, with stream driving or floating in the spring. In the Far West and in the South where a large portion of our lumber is produced, logging continues the year round. In fact, the only seasonal logging commonly found is in parts of northern Idaho, in the high Sierra Nevadas where deep snows interfere with logging during the wintertime, and in limited sections of the northeastern and Lake states where log transportation is dependent on the spring drive.

Year-round logging operations continue generally from 275 to 300 days per year. The rest of the time is accounted for by Sundays, holidays, shutdowns for repairs or bad weather, inventories, or other delays. Heavy rains and high water in the South and low humidity (fire hazard) in the Far West often cause "shut downs" during logging.

COMPENSATION

PERIODIC

Labor is paid on the basis of hourly, daily, weekly, and monthly wages. Truck drivers, hook tenders, woods bosses, and such are generally paid on a daily, weekly, or monthly rate. Truck drivers are often paid on an hourly basis.

Highest wages have generally been paid on the Pacific Coast where labor is usually the most efficient. The wages in the South are probably the lowest, but labor in this region is the least efficient. Much more supervision is required in the South than on the West Coast, and this factor adds to the labor expense in the former region. However, when small crews are used in the South on the many small operations, less supervision is required.

CONTRACT

Felling and bucking, skidding, railroad construction, truck haul, and other forms of woods operations are often let out by contract for either a flat price for the job or, in case of felling and bucking and often transportation, on a per thousand board feet basis. The plan known as the gyppo * system has become very prominent in the Northern

* Also spelled jippo. In the East and Northeast, the same type of subcontractor is known as a coonskinner.

Rocky Mountains and on the West Coast where contractors or subcontractors take over many of the woods operations on a contract basis. Piecework is the prevailing practice in felling and bucking in practically every forest region. This is a form of the contract plan of compensation. There is a wide variation in the efficiency of men em-

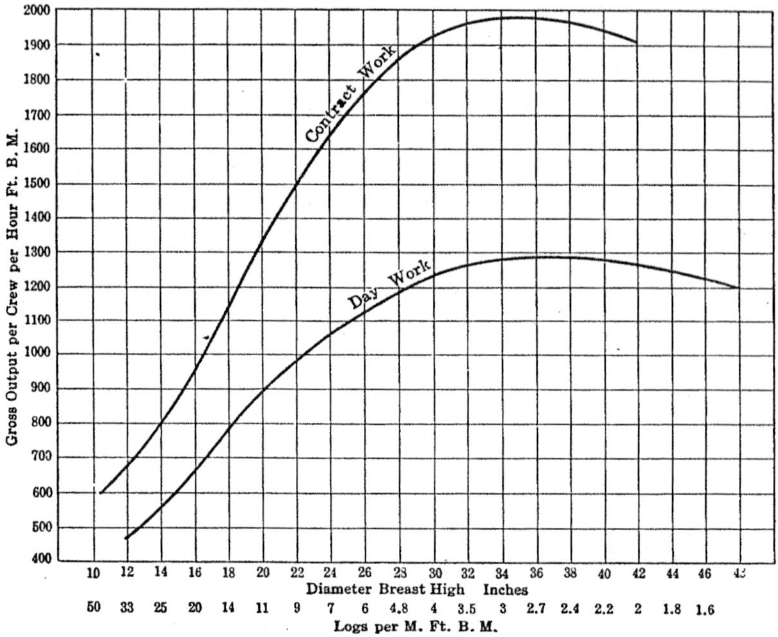

FIGURE 9. Comparison of felling and bucking output by contract and day labor (crews of 2 men) in ponderosa pine cutting 8 to 12 M.b.f. per acre on slopes up to 30%, light brush and windfall and negligible cull in the Inland Empire. *After M. Bradner et al.*

ployed on a piecework basis. Efficiency in these operations depends on the character of the timber and ground, the experience of the worker and his willingness to work, and the number of hours employed on the job, together with the efficiency of the tools used. A survey of 126 of the larger southern operations showed that 85 logged on contract, 41 did not, and 44 logged both on contract and with their own crews.

BONUS SYSTEMS

Bonuses are sometimes paid on logging operations. The bonus is generally based on production or output in thousand board feet above a stated minimum. Bonus payments are also based on efficiency, service and attendance plus production. The bonus system is probably applied more commonly in sawmill operations than in logging jobs. Sometimes a bonus is paid on the basis of low stumps, efficiency in felling and bucking, and similar factors.

On the Pacific Coast, the base for a bonus paid by some companies for felling was 60 sq. ft. of stump area per felling crew. A bonus per square foot above this base was paid.

In California pine logging, men in a skidding crew received a bonus, depending on the type of work, over and above the schedule output per day. There was a definite rate per hour for less than the standard schedule, a base rate for the standard schedule, and a rate per thousand board feet for all above the schedule. The bonus may be figured for every week, month, or the time spent on a setting.

COMPARISON OF MAN-HOUR PRODUCTIVITY

Although the hourly wages paid in logging and other phases of the lumber industry are very much higher on the West Coast than in other parts of the country, the *cost* to the employer of wages paid in the South, when expressed in productivity, is actually the highest in the lumber industry. This is due, as previously explained, to the much smaller-size timber and the unavoidable use of much manual labor in the South. High-powered machinery which is required to handle and move huge logs on the West Coast is not adaptable to use in southern small-size timber. The operating, depreciation, and other charges incident to this high-powered and expensive machinery would prevent its profitable use under southern conditions, and, furthermore, it is not adapted for use in small-size timber. Thus, wages in logging must be related to productivity.

Table IX shows the man-hour production for the West, North, and South and the effect thereof, on respective wage costs, on a straight-

TABLE IX

Lumber-Producing Region	*Productivity, b.f. per Man-Hour*	*Straight-Time Avg. Hourly Earnings*	*Cost per M ft. of Production Based on Straight-Time Avg. Hourly Earnings*
West	89.7	$1.18	$13.21
North	72.7	0.73	10.32
South	33.4	0.52	15.52

time hourly earnings basis. The figures are taken from an analysis of the U. S. Bureau of Labor Statistics study of wages in the lumber industry. These figures, announced recently, apply to 1944 conditions.

The significant phase of Table IX is that the average Douglas fir log will yield nearly 11 times as much lumber as the average southern pine log, as indicated previously. Furthermore, Douglas fir grows in stands yielding 35,000 to 60,000 b.f. per acre and often much more, whereas southern logging operations deal with second growth or regrowth which is widely scattered over large areas with merchantable stands of from 1000 to about 10,000 b.f. per acre.

TABLE X

MAN-HOURS REQUIRED FOR LOGGING 1000 BOARD FEET BY REGIONS *

All regions	8.1
Southern pine	9.9
Western pine	6.6
Douglas fir	4.9
Redwood	7.8
Southern hardwood	13.6

* From *Labor Requirements in Lumber Production*, by B. H. Topkis, U. S. Bureau of Labor Statistics, 1937.

Since 1937, man-hours required for logging, manufacture, selling and administration, and transportation have advanced very markedly, but with increased mechanization of logging there is a tendency to decrease the number of man-hours to produce 1000 b.f., especially in Douglas fir and other big timber on the West Coast.

LABOR ORGANIZATIONS AND WAGES

Labor is most highly organized in the lumber industry in the Pacific Northwest Douglas fir region. It is also well organized in the western pine industry, principally in Oregon, California, Washington, and Idaho.

For some years, the American Federation of Labor has been the principal union organization in sawmills in these regions, whereas the Congress of Industrial Organizations has been the principal union in the logging phases of the industry in the same regions. The former union is represented by the Lumber and Sawmill Workers of America of the United Brotherhood of Carpenters and Joiners, whereas the unit among the loggers is the International Woodworkers of America, affiliated with the Congress of Industrial Organizations.

The cost of labor has advanced materially in all sections of the country. It is an important reason for the increase of prices for stumpage and saw logs as delivered at the sawmills. The increase of freight rates has sharply increased the cost of transportation of lumber from the sawmills to the retail yards or other destination. However, the advancing costs of labor have affected the costs of logging, particularly in those phases where manual effort is concerned. Thus, the cost of felling and bucking with or without mechanical felling devices has advanced relatively more than any other phase of logging because of high wages.

The minimum wage now paid in the Pacific Northwest as applied to Douglas fir operations is $1.40 per hour. The actual average pay is $1.61 per hour. These rates represent about 100 per cent increase above those prevailing a few years ago.

In the southern region where the important southern pine and hardwoods are produced and there is more manual effort used in logging than in other regions (because of the relatively scattered and small-size trees), sporadic attempts were made at labor organization ever since World War I. There was, however, no recognizable labor organization in the lumber industry until 1938 when the American Federation of Labor organized six plants in Arkansas. By 1943, the number of unionized mills rose to 40 and in 1946 to 150 mills. Since then no substantial increase has occurred in the number of mills unionized.

The Bureau of Labor Statistics of the U. S. Department of Labor has shown that in October 1946 an average wage of $0.64½ per hour was paid in the southern pine logging industry. The minimum rate in early 1948 was $0.60 per hour, and the actual average wage paid was $0.66 per hour.

The average Douglas fir logs yield about 1100 b.f., whereas the average southern pine log averages 100 b.f. Mechanization is obviously the result of the huge size of the timber logged as well as the ingenuity of the logging industry in devising methods of handling trees and logs of such large size.

Thus, the big timber plus methods of mechanization means a much higher production per man-hour on the West Coast than in the South. Therefore, the labor cost expressed in production per thousand board feet in the South is much greater than on the West Coast.

The present minimum wage scale of $0.40 per hour became effective in October 24, 1945, and the 40-hour week went into effect on October 24, 1940. However, this minimum wage has never been put into effect in the lumber industry. The actual wages paid have always exceeded the minimum legal permissible scale. Many sawmills operate 48 hours per week and pay time-and-a-half for the extra 8 hours.

There have been sporadic but unsuccessful attempts to unionize logging operations and sawmill units in the Northeast, eastern Canada, the Lake States, Southern Appalachians, and Central States. However, because of the widely scattered locations and the small size of individual logging and sawmilling operations, the attempts at unionization have not been successful. Many of these small logging operations are conducted for relative short periods of time. Farm labor which devotes at least a part of its time to farm or other work, as well as to logging, is not easily unionized. Furthermore, the workers themselves are often the bosses or managers of the logging and sawmill companies involved. Some of these sawmills operate only for a few weeks or months per year, and the crews work principally in farm or other work. Under these widely scattered and small unit organizations, attempts at the formation of unions have not been successful. Logging crews of from one and two men to five to ten men constitute the entire woods crew for many small mills. These men often live at home with their families and drive daily several miles to and from work by auto.

CAMPS

HISTORICAL

Until recent years logging camps, notably in the Northeast and Lake States, were of rough crude construction, generally built of logs and characterized by few comforts or sanitary facilities. Remote and inaccessible locations and the necessity for low costs were responsible for these conditions. Men often worked daily 12 to 14 hours or more. Tiers of wooden bunks with straw for bedding, a large stove in the center of the bunkhouse, and wires strung above to dry wet clothing were the prevailing practice. Rough coarse food with emphasis on beans was provided in the cook shack. These two main buildings, together with the wanigan or commissary (camp store), a horse barn, root cellar, quarters for the logging boss and camp clerk, and blacksmith shop, generally composed the series of buildings which were erected in summer or fall and occupied for one or more seasons.

Sometimes the bunkhouse and cook shack were combined in one building. Food and supplies for winter use were toted in by team, boat, or both. The 60-man camp was the prevailing size. The southern camp consisted largely of crude shacks or tents for the Negro woodsmen.

The coming of the railroad changed conditions radically. With quick easy access to the outside world, portable houses and family camps replaced the old log camp. There was greater permanency of location. Men could be moved greater distances to and from the felling and skidding operations as they were transported on trucks or on log trains.

The heavy timber of the West Coast also meant more permanent locations on account of the large yields per acre. A much smaller area

FIGURE 10. A typical Adirondack pulpwood logging camp, housing 50 to 60 men, which produced 4000 to 5000 cords in a year. Logs are of spruce, which will later be used for pulpwood. The cook shack is in the left foreground, bunkhouse is on the right, blacksmith shop and stables are in the background, and the camp office is beyond the bunkhouse. Where railway or highway transportation is available, camps are generally made of boards. On pulpwood driving operations camps are generally constructed of logs.

was cut over each year than in the white pine and spruce operations of the North and Northeast.

The most notable advances in camp construction, sanitation, recreation, and food supplies have been made in the Pacific Northwest. This has been especially true since World War I. The camps are now characterized by shower and tub baths, separate iron beds and mattresses, electric lights, recreation facilities, and the best quality and variety of food, well prepared and served by competent chefs. The old rough bull cook and cookee methods have been replaced in some of the larger and better types of logging camps by trained and highly paid experts in the culinary field. Large camps have virtually disappeared in the southern pine region with the passing of the large concentrated holdings. Scattered stands, good roads, and improved trucks permitted logging crews to live in their own homes.

FACTORS OF LOCATION

The following are the governing factors in camp location:

1. Convenience to work. This is of the greatest importance. Railroads and motor trucks are often used to transport men quickly to and from camp, thus saving time and increasing efficiency and general satisfaction. If men walk to and from work, the distance one way should not exceed 1 mile or preferably ½ mile.

2. Dry ground, for health and sanitary reasons. A good sandy soil with favorable drainage is desirable.

3. Good pure water supply from a spring or well. Many operations have abandoned the use of streams because of contamination.

4. Convenience for toting in men and supplies. This is of relatively little importance because of the accessibility by railroad or by motor truck on highways. In the Northeast and Canada where toting is important, this item may be considerable and is often 7% to 10% of the total cost of the logging operation.

The factors which control the removal of camps are:

1. Distance to work. In the Northeast and Lake States, the maximum is generally considered to be 1 to 1¼ miles by foot. By rail or motor truck it has not been considered advisable to require more than 20 to 30 minutes to go to work; a much shorter time is preferred.

2. Footage cut at each camp. On the West Coast some camps are moved after cutting about 100 to 200 million ft. or, in many cases, every 3 to 5 years. The northern and eastern camps are generally used for 1 to 3 years, generally 2 years. In pulpwood operations, camps are moved after cutting approximately 5000 to 30,000 cords of pulpwood. In hardwoods and small timber operations, camps are moved after cutting 2½ to 3½ million ft. of saw logs.

TYPES

Camps may be classified as to (1) method of construction; (2) size or number of men housed; (3) character and period of occupancy, such as family camps, summer camps; (4) portability; (5) ownership—that is, company or contractors' camps; and (6) land or floating camps.

Logging camps are generally of board construction. Log camps are found only in the Northeast and eastern Canada and occasionally in the Lake States and Northern Rocky Mountains. Log and tent camps are the prevailing rule on all driving operations where the cost of toting in lumber would be prohibitively expensive. Generally rough, tight, knotted boards of No. 1 or No. 2 common grade are used. Frequently they are double boarded, with felt or building paper between the boards for additional warmth. Neither board nor log camps except portable board camps are generally moved to a new location when the adjacent timber has been cut. However, in board camps the doors, window frames, and so on are generally dismantled and removed to the new camp location. Some floating camps have been used in the South, in Alaska, and in British Columbia. Trailers are often used on small logging jobs for portable and small stationary mills in the North, Northeast, and South. Small houses such as tourist-camp cabins and built-on skids are used by some southern mills. They remain from 2 to 6 months at a "set." Much logging is done by farm labor for small mills, and the workmen live in trailers or at home. Men often drive 30 miles or more to go to work from their own homes in many parts of the country.* Thus, the large camps of 50 to 300 men or more are found only on the largest operations in the West and a few in the South, the Northeast, the Appalachians, and eastern Canada.

Logging camps usually vary in size from 10 men up to 300 or more. The largest camps are generally found on the Pacific Coast. The usual pulpwood camp in the Northeast generally has 50 to 100 men. The size of camp depends on the amount of timber to be cut, accessibility to work, and the method of transportation employed. There is a distinct tendency toward using semipermanent camps and transporting the men out to the woods and back on cars or trucks. This provides greater stability and satisfaction for the men. They can have their families with them, which contributes to the development of community life with schools, churches, home gardens, and similar advantages. Family camps are also found in the Negro logging camps of the South and Southeast. Many large companies in the West and South have established camps that are models of sanitation and comfort with libraries, hospitals, reading rooms, and recreational facilities.

Most logging camps are kept in operation throughout the year. However, in the Northern Rocky Mountains where deep snows are prevalent during the winter, camps are generally located from May to November. This is also true of logging camps in the higher Sierra Mountains of California.

* A study of 90 large southern pine operations revealed that on 39 of them the men went to work in their own conveyances, indicating that they lived in their own homes and did not require camps.

On railroad operations, many of the camp buildings are portable. They may rest on logging cars or may be set down with the log hoist on temporary foundations along the tracks. Car camps, that is, those

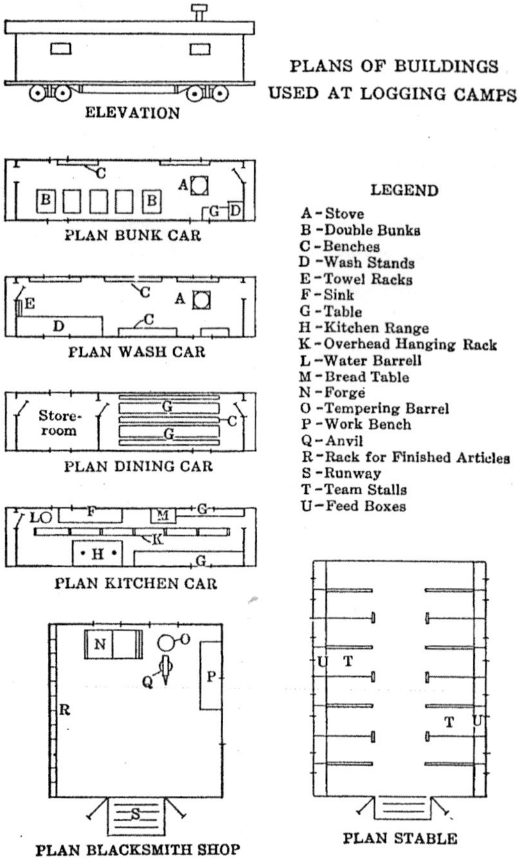

FIGURE 11. Elevation plan of a typical portable car at the top and floor plans of six of the most important buildings used in a logging camp in the Southern Appalachian Mountains.

resting on old logging cars or composed of old cars, are sometimes used in the Northwest and in the Northern Rocky Mountains.

Portable board camps when transported from place to place are estimated to last 7 to 10 years, whereas car camps are estimated to last 10 to 15 years or more. In some sections, as in the Northeast, there is a trend toward small and separate bunkhouses, sometimes of the portable "knock down" plan. They have several advantages, such as: (1) they can be used in rough and steep locations; (2) occupants can regulate temperature, lighting, and air conditions to suit themselves; (3) groups can select their own company; (4) greater pride is taken in cleanliness and neatness; (5) the likelihood of contagion and infection is less; and (6) they can be readily cleaned and

FIGURE 12. Typical logging camp located in an attractive valley on an old Douglas fir cut-over area below hillside of seed blocks left after logging to regenerate the harvested area. These modern camps include bathing, recreational and reading facilities, comfortable living quarters, bunkhouse for single men and quarters for married men with mess hall, repair shop, general offices, fire-fighting equipment, etc. *Courtesy Weyerhaeuser Timber Company.*

fumigated. Camps of this type, however, are more expensive to construct. On the small-mill operations of the East and South, loggers and woodsmen generally live in nearby farmhouses. Frequently the loggers are the farmers and their sons working in the woods in the winter and on their farms during the spring, summer, and fall. In eastern Canada many farmers are employed in the woods from November 1 to May 1. Tent camps are used in the Northern Rocky Mountains during the summer and also for driving camps along the streams of the Northeast and Canada. They are likewise used in the South on both pine and hardwood operations as well as for small sawmill operations.

The log production per man-day in the logging camps varies widely. In the heavy sugar and ponderosa pine stands of the central Sierra Mountains in California, a camp of 125 men produces about 165 M.b.f. per day with four Diesel tractors and one high lead donkey. In the Northeast and South, a general figure of 1000 b.f. per man-day should be produced by each man in camp. In the heavy timber of the Pacific Northwest the relationship is frequently 1500 to 2500 b.f. per man-day in camp. This is obviously due to the large amount of machinery and other labor-saving devices.

Log camps in the Northeast getting out 4000 to 5000 cords of pulpwood yearly and accommodating 50 men generally cost $2000 to $3000 without equipment. These camps can be built by 10 men in 15 days, or 150 man-days.* A logging camp of this type has the size buildings shown in Table XI.

TABLE XI

Buildings	*Size—Inside Dimensions*
1 bunkhouse	30′ × 40′
1 cookhouse	30′ × 40′
1 stable for 10 teams	30′ × 30′
1 camp office for log boss	18′ × 20′
1 blacksmith and woodworking shop	18′ × 20′
1 root cellar in hillside	10′ × 20′

The equipment for a logging camp of this type generally costs $1000 to $4000 additional, depending on its quality, permanence, and completeness. The cost of the large and more elaborate camps on the West Coast may be several thousand dollars and sometimes as much as $50,000 or more for the more substantial buildings and permanent types of camps.

The duties of the men vary in nearly every camp. In a camp of 60 men employed in winter sled hauling, draying, and chaining 5000 cords of spruce pulpwood in the Northeast in rough topography to a drivable stream, the camp personnel was employed as follows:

1 camp boss or foreman.
1 camp clerk.
1 barn boss or stabler.
1 cook.
2 chore boys.

* This is assuming that logs and other materials are delivered on the site ready for assembling.

1 blacksmith.
5 men unloading sleds at stream landings.
10 men loading 4′ pulpwood bolts on sleds at skid piles.
10 teamsters who, in addition to driving teams, place the bolts on the sled loads during loading and assist in unloading.
15 road men (locally known as road monkeys or chickadees) who do miscellaneous work connected with road operation such as graveling and haying on steep slopes, operating sprinklers at night, operating Barienger brakes, plowing, and shoveling snow and general road repair and upkeep.
13 men draying (bobbing) and chaining. About 2500 cords of pulpwood are drayed and chained on the steep upper slopes (the remainder or 50% having been horse-skidded direct to the haul roads during the summer, and, after crosscutting to 4′ lengths, the bolts were piled on skids along the winter haul roads awaiting loading on sleds as previously described).

The 20 men required for a skyline Diesel skidder weighing 240 tons and of 450-hp. capacity with a 105′ steel tower (the tail spar being located at maximum distances of 2000′) in Douglas fir timber, cutting 40,000 to 60,000 b.f. per acre were as follows: (Logs up to 64′ in length were taken, and from 1000 to 8000 b.f. were skidded on each load):

1 skidder foreman in charge of crew.
1 skidder operator.
1 loader who loads logs from landing to log cars.
1 engineer or third operator.
1 head or top loader on log cars.
1 tong shaker who sets the tongs for the loader.
1 chaser who unhooks the chokers at the railroad landing.
1 hooker, in the woods.
4 choker setters in the woods.
1 whistle punk who signals the starting and stopping of operations.
1 assistant foreman or second rigger who rigs the tail trees.
4 third riggers who assist the second rigger.
1 powder monkey to shoot the choker holes.
1 block swamper who chops the slash away from the guy line stumps where the tail spar blocks are attached.

The previous two lists illustrate the types of labor required on two operations, the first using largely unskilled labor whereas the latter used men of great skill, ability, and experience. Every operation differs in the number of men employed, depending on the local conditions of timber, topography, and method.

IMPROVED SANITATION, FOOD, EQUIPMENT, AND EFFICIENCY

Camps have been vastly improved in sanitation, equipment, and efficiency during recent years. The most notable advances have been made on the Pacific Coast where electricity, shower baths, improved heating devices, drying rooms, separate beds and mattresses, ample light, ventilation and drying conditions prevail, and many camps pro-

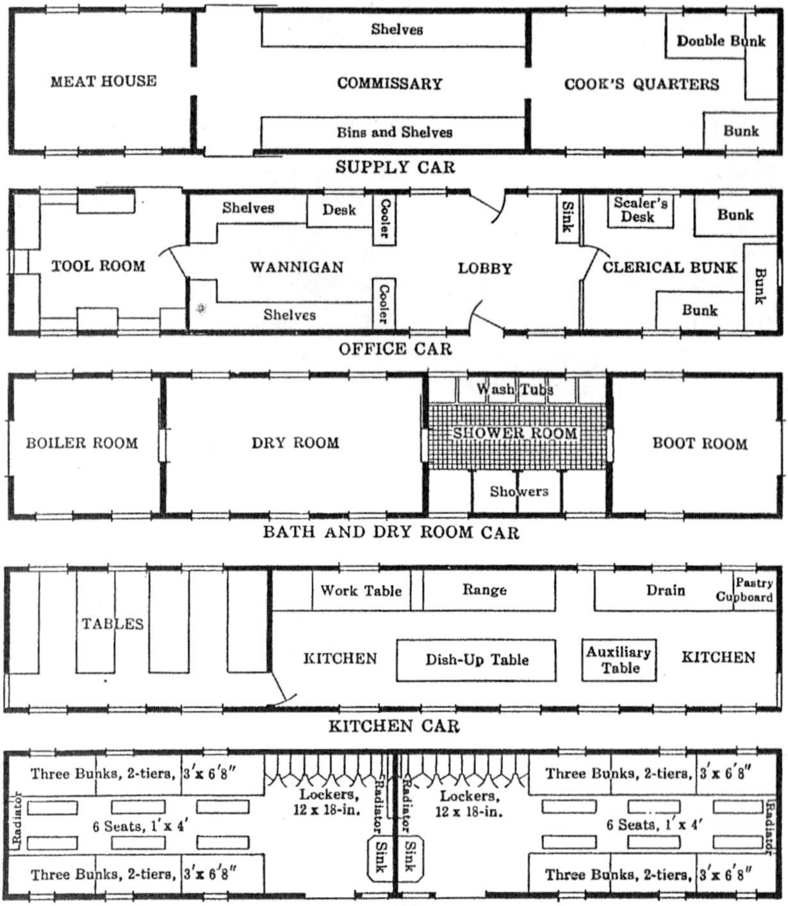

FIGURE 13. Floor plans of five cars used in one of the largest and better-equipped car camps in the Northwest. Excellent facilities, such as shower rooms, drying rooms, opportunities for recreation and reading, are provided in many of these camps. In matters of sanitation great advances have been made during recent years in all regions.

vide recreation and writing rooms. These modern conveniences have contributed in a very substantial way to the improvement of the morale as well as to the operating efficiency and general satisfaction of the men employed.

The preparation, storage, and care of food have received wide attention within recent years. Great improvements have been made in all parts of the United States.

ORGANIZATION OF LOGGING OPERATION WITH TRACTORS AND TRUCKS FOR SMALL MILL OR CONCENTRATION PLANT PRODUCING 3 TO 5 MILLION B.F. PER YEAR

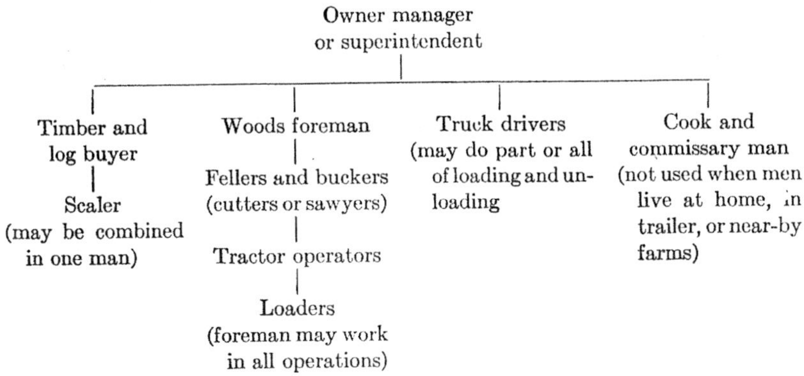

FIGURE. 14. Organization chart for small operation in South or Northeast.

The cost of feeding the men varies widely with the prevailing local customs in each region. A study in the Lake States reported the approximate distribution of charges for feeding men in the average logging camp as shown in Table XII.

TABLE XII

Items	*Percentage of Total Food Costs*
Labor in and about cookhouse	20
Meat, including lard, oleo, sausage	50
Flour	3
Sugar	3
Potatoes and fresh vegetables	4
Canned goods, including milk and fruits	15
Miscellaneous	5
Total	100

Most of the states in the forest regions have enacted laws affecting the construction and operation of all forms of camps, including those devoted to logging. These regulations have to do with the sanitation, location, light and air space, disposal of wastes, water supply, and so on. Some states specify the distance camps must be located from a lake, stream, or watercourse. Generally, permits must be secured for the establishment of camps. Definite regulations apply to the use of fire extinguishers, the location and source of water supply for the camp, location and drainage for horse stables, ventilation and flyproofing, the disposal of wastes and garbage, and the amount of cubic feet of air space in the sleeping quarters. These and other regulations are established by public health or similar state departments in practically every state in the important forest regions.

Among examples of state regulations affecting the housing of men in logging camps, the following may be cited as typical regulations:

1. Camps shall not be located at a distance less than 200 ft. from a river, brook, or lake.

2. The source of water supply shall be a stated distance, generally 100′ to 200′ from and above, in elevation, any possible pollution.

3. Every structure serving as a habitation shall be furnished with ventilation flues and windows of sufficient number and dimensions to ventilate and light the interior properly. In some states, a minimum of 600 or more cu. ft. of air space must be provided for each occupant of sleeping quarters.

4. Floors must be elevated a minimum distance above the ground; side walls must be a minimum height; beds must be elevated above the floor a stated distance and must be separated at least 2′ apart.

5. Adequate facilities must be provided for excluding flies from the kitchen and dining rooms.

6. There must be a system for proper disposal of garbage and refuse, such as burning or burying a minimum distance from the nearest camp building.

7. Latrines shall be located, constructed, and cleaned in accordance with stated specifications.

8. Provision must be made for isolating men with contagious diseases.

9. Penalties are usually provided for infraction of all rules by a properly constituted state official.

COMPENSATION AND OTHER LAWS AFFECTING LOGGING COSTS

There are many state and Federal regulations regarding the prevention and suppression of forest fires, the construction and operation of camps, and other phases of woods operations. Companies engaged in logging, in some states, are required to bear heavy charges for accident compensation. These requirements vary with different states. In some they are exceedingly burdensome and in others correspondingly light. The Pacific Northwest has taken the leadership, generally, not only in the matter of compensation laws affecting the protection of the employees against accident but also in the matter of fire protection, housing, sanitation, and such. The accident-compensation insurance requirements have been an important factor in forcing the adoption of the contract form of logging in practically all parts of the country. Compensation insurance generally comprises 2% to 11% or more of the total logging costs in the various states. These costs may be based on the number of employees or on the total wages paid, and may vary with the risks involved, duration of employment, or previous experience record of accidents. There is generally a minimum charge for logging operations. In some states, a yearly minimum of about $200 is charged for the lowest form of compensation insurance. This may vary up to many thousands of dollars for a company employing large numbers of men.

The first workmen's compensation law of its kind to be declared constitutional was that in Washington which became effective in 1911. State compensation legislation was first attempted in New York in 1898. Insurance of the employer's liability to pay compensation is an essential feature in nearly every state and may be made effective in three ways, as follows:

1. Private insurance in stock or mutual companies.
2. Self-insurance with proof of ability to pay.
3. Insurance in state funds.

In Oregon, the employers' rates vary from 10 cents for office employees to several dollars for special work for each $100 of payroll. Rate reductions up to 20% are made on operations that demonstrate a favorable accident record. State records indicate that on an average, in Oregon, the employers pay 87 cents and workers 9 cents, the remainder coming from interest and miscellaneous sources.

Logging is generally classified as a hazardous form of employment. In Washington, logging companies pay 6% of the payroll for compensation insurance.

According to A. S. Cook, the percentages of accidents by classes of work in felling (cutting) and skidding in Quebec are shown in Table XIII.

TABLE XIII.*

Phases of Work	*Per Cent of Accidents*
Swamping out roads	4.5
Felling	26.3
Branching or limbing	30.6
Bucking	9.4
Skidding and piling	29.2
Total	100.0

* From *Pulpwood Cutting—Efficiency of Technique*, by A. Koroleff, Canadian Pulp and Paper Association, Montreal, 1941.

CHAPTER IV

Partial Cutting (Selective Logging)

DEFINITION AND SCOPE

Partial cutting, sometimes referred to or known as selective logging, consists generally of the removal of some of the larger, mature, or overmature trees with widely flexible diameter limits to obtain current profits from the logging operation and at the same time to improve and perpetuate the forest for later cuttings. In young stands the smaller trees may be cut for pulpwood and similar products and the larger trees left for further growth, as in the South. Partial cutting is not strictly a silvicultural system as it does not conform to the exact principles of the selection system, the shelterwood system or one of the clear-cutting systems of silviculture. Moreover, it does not follow the previous practices of harvesting our American forests which in the past have generally been based on securing immediate profits and the exploitation of the forest with little or no thought of the future. Partial cutting has been widely applied both to private and publicly owned forests. The theory underlying the ultimate objectives is the leaving of thrifty-growing well-formed trees that have the ability and room to grow more rapidly in the years to come as the result of releasing the crown cover and permitting the remaining trees to grow in size and, therefore, in value. It is in some ways an American method of logging our forests with definite improvement over previous practice and yet not conforming to any one of the approved and standard forms of silvicultural cuttings so long practiced in Europe. In some respects it may be regarded as a temporary expedient to insure the permanence and stability of our forest growth and thus serve the lumber and other forest-products industries in maintaining sustained yield on small farmers' woodlands as well as on large timber holdings.

Partial cutting in even-aged stands may begin at an early age, namely, 10 to 15 years in southern pine stands or 25 to 35 years in the Lake States and in the Northeast. The plan is to take out a profitable crop every few years and at the same time leave the best-formed and rapid-growing trees to grow into larger and more valuable sizes. In each stand, the plan is to maintain the forest in good, thrifty, and healthy condition and at the same time secure the maximum permanent production and yields from a given tract. For example, in the Lake States, the 6″ trees of jack, red, and northern white pine and spruce on favorable sites will double in volume and value in approximately 10

FIGURE 15. Selective logging of middle-aged mixed hardwood stand on the Allegheny National Forest in Pennsylvania. About 12 cords of distillation wood per acre were cut from a stand averaging 30 cords per acre. Photograph taken about 5 miles southwest of Kane. *Courtesy U. S. Forest Service.*

years. Still later, the volume will continue to increase rapidly, while the value goes up even more rapidly with the demand for rural electrical poles, cabin logs in resort areas, piling, railroad ties, and other products that have a value higher than pulpwood and can be made from trees not yet sufficiently large to yield saw logs. If some of the trees in a good and even dense stand are harvested and utilized, the entire remaining stand will benefit from the removal of these trees. If a partial cutting is not made, the entire stand may stagnate, and the weaker and suppressed trees may die out in the competition and thus be lost for profitable harvesting.

Partial cutting has gradually evolved in American logging practice during the past 25 years or more, and much knowledge and experience have been gained through this practice. Additional information will doubtless be gained from its further application and the results of its use. Therefore, in many ways it is in a transitional stage of development.

Selectivity may apply to individual trees, groups, or even areas. Thus from a technical forestry viewpoint, the choice of the individual trees to be cut may involve silvicultural selection, economic selection, group selection, or other factors, depending on the condition and age of the stand and individual trees as well as the demands of the market.

Partial cutting as used in the American forests is not to be confused with the selection system. The latter applies to uneven-aged stands and tolerant species and is widely used in Europe and, to a less extent, in the United States. Thus, the name selective logging has been misconstrued by some of the laymen as well as by some foresters as a practical application of the selective system.* The R. B. Goodman operations in mixed hardwoods and hemlock in Wisconsin are one of the best applications and illustrations.

Partial cutting is both a silvicultural and a utilization procedure and is now widely used on private logging operations as well as on National Forests. On the latter, emphasis is placed on stand improvement rather than current income. Selective logging is a term that has been used for rather crude and unscientific procedures as applied to our woods for many years. Unfortunately, under the guise of selective logging, many of our forests have been abused by the more valuable, larger, and better specimens being taken out without reference to future conditions and rates of growth. This has resulted in "high-grading" and deterioration of the stand for future growth and cutting operations.

However, partial cutting recognizes the well-established principle that logging costs increase with decreased tree sizes. Conversely, the larger the trees, the lower the costs of logging expressed in the unit of value, namely, per thousand board feet. Thus, there is a distinct but flexible relationship between tree sizes and profits. Small tree sizes may be profitably logged when lumber prices are attractive or whenever small sizes may be marketed. Frequently, however, many trees are felled during the most rapid rate of accelerated growth and when lumber markets are glutted with trees cut below commercially profitable sizes. Furthermore, the grade outturn from small-sized logs is disappointing, and in some cases logs are brought to sawmills and manufactured at a loss rather than at a profit.

* For further review of the various silvicultural systems, see *Practice of Silviculture*, by R. C. Hawley, John Wiley & Sons, New York, 1947.

FIGURE 16. Selective logging in the mixed-oak type in the Hudson Valley, New York. About 2000 b.f. per acre was removed from merchantable stands of about 5000 b.f. per acre. Some of the trees from 16″ to 32″ d.b.h. were marked, depending on density of the crown cover, condition, and rapidity of growth. The tree being felled was a white oak 26″ d.b.h. *By Life Photographer Dmitri Kessel. Copyright Time, Inc.*

Within the past 30 years, many studies have been made to determine the sizes of trees of various species which may be profitably logged and also when trees have reached their maximum of *financial growth*. Unfortunately some individuals and companies have used selective logging as a means of determining the minimum-sized trees which should be cut as a strictly business procedure. Still others have introduced and used it both as a means of securing current profits and to insure regular, continuous, and profitable future cuts. The adherence to a fixed or rigid diameter limit usually results in a definitely gradual deterioration of the stand. It generally leaves unprofitable trees which occupy favorable growing space. Ultimately these trees either die or gradually reach the minimum diameter size.

CONDITIONS UNDER WHICH PARTIAL CUTTING MAY BE PRACTICED

Partial cutting has been widely introduced and used in all parts of the United States with the exception of large mature and overmature stands of Douglas fir as well as those of mature lodgepole pine, aspen, and other types of even-aged forests which do not readily lend themselves to a system of partial cutting.

The change within recent years from railroad logging to truck haul has meant an opportunity for frequent cuts, that is, cutting cycles more frequent than has been the custom in the past. Also taking a smaller volume per acre has been practiced with success. That is, the results have been currently profitable as well as maintaining the growing capacity and condition of the forest for future cutting. The tractor with its ready flexibility in taking out individual trees here and there has also contributed to a better application of partial cutting. Tractor logging combined with truck haul have changed methods of logging in all sections of the country and, where the logging has been done with care, the forest has been left in a better condition for future growth. Thus, improved procedures in logging have tended to increase the application of the theory of partial cuttings. In many operations, test plots have been established, and trees are counted and girth measurements taken. The results of these formal studies have demonstrated the practicability of logging areas by selective processing and the improved growth rates for future logging. Individual trees are marked by blazing or paint spots or otherwise designated for cutting. Forest owners, both public and private, are learning a great deal about the practicability of applying various phases of partial cutting to virgin conditions and especially to regrowth forests, notably throughout the South, the North, and the East. In more recent years, experimental areas have been cut in the heavy timber of the Northwest as well as in the Rocky Mountains to determine the best methods of partial cuttings.

FIGURE 17. Selectively cut southern pine stand 35 years old in eastern Virginia. An average of 10.7 cords of pulpwood were cut per acre, and an average of 20 cords per acre are left for increased rate of growth and later removal. Note the large number of stumps and the dense stand left for later periodic logging. This is an excellent example of sustained yield logging and management. *Courtesy U. S. Forest Service.*

Therefore, selective cutting in its true silvicultural sense is especially adapted to rapid-growing and uneven-aged stands and theoretically to tolerant species. Growth increment in selectively cut stands has proved to be definitely greater. The removal of the large-crowned and dominant trees that are mature or overmature definitely stimulates the growth of the trees that have been left.

In the South, lumbermen have learned that trees 8″ to 12″ in diameter offer generally the minimum limits to which pine saw timber should be cut and 5″ to 7″ the minimum for pulpwood cuttings. This is, in effect, partial cutting in its true application to young and growing stands. Because of the many improved highways and good woods roads, together with the open stands and level topography and excellent markets for saw logs, pulpwood, and other products, integrated logging with the tractor and motor truck has proved highly successful from a forestry standpoint as well as profitable.

Where stands are heavily cut for pulpwood, cross ties, and other products, promiscuously and recklessly and with no thought of the future, there has been a definite deterioration of the condition of the stands as well as a slower growing rate per acre for the future. In the pulpwood and hardwood stands of the North and Northeast, partial cutting has been used for many years. Generally, trees cut for saw logs are taken down to 14″ to 18″ in diameter in northern hardwoods and hemlock. However, all trees should not be cut down to any definite diameter limit. Some trees up to 20″ to 25″ in diameter in such stands may be left under special conditions. The diameter limits depend on the density of the stand, growing conditions, tree diameters, the number of trees in the various diameter classes, the prevalence of valuable or low-value species, current lumber and stumpage prices, and the management plans of the owner.

In the old-growth and dense timber of the Douglas fir region, clear cutting is almost imperative because of the even-aged stands, the large-sized virgin trees, and the density of the forest. In that region many of the stands, as indicated elsewhere, are clear-cut. The ponderosa pine type found throughout most of the West is generally logged by some method of partial cutting more than any other type in the American forests. This is explained later in connection with the discussion of the application of partial cutting on National Forests.

ADVANTAGES OF PARTIAL CUTTING

These may be summarized as follows:

1. Only trees yielding a satisfactory current profit and those whose future growth rate appears uncertain are cut. Higher grades of lumber which command the best prices on the market are obtained, as only the largest and most profitable trees are removed. Thus, the minimum limits of profitable tree sizes are determined. A smaller volume is cut with a greater profit per thousand board foot or per cord on the operation.

2. Felling and bucking and, generally, other logging costs are reduced. Although fewer trees per acre than under previous cutting conditions are felled, the costs are distributed among the larger trees, and, therefore, the cost per thousand board feet is lower.

FIGURE 18. Example of selectively logged lodgepole pine stand in the Rocky Mountains after felling and before skidding. Some excellent examples of integrated utilization are found in lodgepole pine stands, especially when near mining and smelting operations where saw logs, poles, posts, mine props, lagging, converter poles, cross ties and fuel wood are sometimes cut from the same operation. Low stumps and complete utilization feature timber sales on the National Forests. A good residual stand left for increased growth and cutting is evident. The diameter of lodgepole pine is seldom more than 16″–18″ d.b.h., and this species often has more merchantable-size trees per acre than any other species.

3. Partial cutting improves the condition of the forest by favoring the rapid-growing and valuable species to be left and thus insures a better-quality and continuing product for the future.

4. Young timber on the area is left in good growing condition for the future, thus putting the forest on a sustained-yield basis. Sometimes trees large enough to give promise of another cut in 15 to 20 years are purposely left in the process. Thus partial cutting may, in effect, be an improvement cutting.

5. Partial cutting on a shelterwood basis is a distinct advantage in some stands in assuring satisfactory regeneration. The "shelterwood trees" will add favorable growth, if thrifty full-crowned trees are reserved, and may be harvested when reproduction is well established.

6. Better fire protection is assured by leaving a smaller amount of slash; the proportion of small trees is reduced and, thereby, the amount of undesirable low grades of lumber is reduced; improvements for woods operations can be used in future cutting, thus reducing the cost of future installations. Partial cutting prolongs the life of the lumber operation and contributes to the land values for resale. It may also definitely contribute to the stability and permanence of the communities in which sawmills and other forest industries are located.

OBJECTIONS AND DISADVANTAGES

Some disadvantages have been adduced to selective and partial logging. These are:

1. For some forms of products such as cross ties, poles, piling, pulpwood, and chemical wood, there may be no marked advantage in cutting only the larger trees. However, all these products may be taken in many forms of application.

2. Greater care must be exercised in removing the timber. Supervision, marking, and similar operations must be done by skilled and efficient employees to secure the best results. The area must be carefully cruised, mapped, and studied before the plan is put into operation—then alterations may be made as the management plan is revised to fit changing conditions.

3. Immature timber must be protected and taxes paid on it while the forest is gradually improving in value. These costs as well as interest and expenses of similar nature must be assumed in any plan of forest management. These accumulated costs plus the value of timber uncut may be serious deterrents to the plan unless the charges are reasonably low. The value growth may be equal to or exceed the volume growth. Certainly, the holding of young and immature stands for the past several decades has proved to be profitable during recent years.

4. Partial cutting is not applicable to certain overmature stands such as the 80- to 120-year-old stands in the Lake States. These stands produce little if any seed, and volume loss is excessive. However, full-crowned mature trees may often be left. The disadvantage is that partial cutting requires a careful study of the stand in the light of knowledge of the seed habits of the species.

5. Partial cutting cannot be practiced on areas cut under enforced liquidation of standing timber assets. This has been a powerful factor in the Northwest region and in other sections where the burdens of interest, taxation, and other carrying charges have induced exploitation without any thought of the future. Financial restrictions may be the

FIGURE 19. Selective cutting in redwood where the mature and larger trees are felled and the others are left for later cutting. The tractor lends itself admirably to selective logging because it is a mobile unit, readily adapting itself to skidding and other woods work. Here it is supplying power for an electric chain saw. On the upper right part of the tractor is shown a spare saw.

motivating factor in preventing the practice of partial logging, especially where the policy of the company does not include the holding of property for future yields. Some forests are cleared for agricultural or real-estate-development purposes—under these conditions continuous forest management is impossible.

6. Unless partial cutting is carefully done, it may result in a deteriorated stand of poor trees unable to grow into something valuable on the market.

These disadvantages are not entirely insurmountable and in some cases may not be very important if our forests are to be continued on a system of forest management.

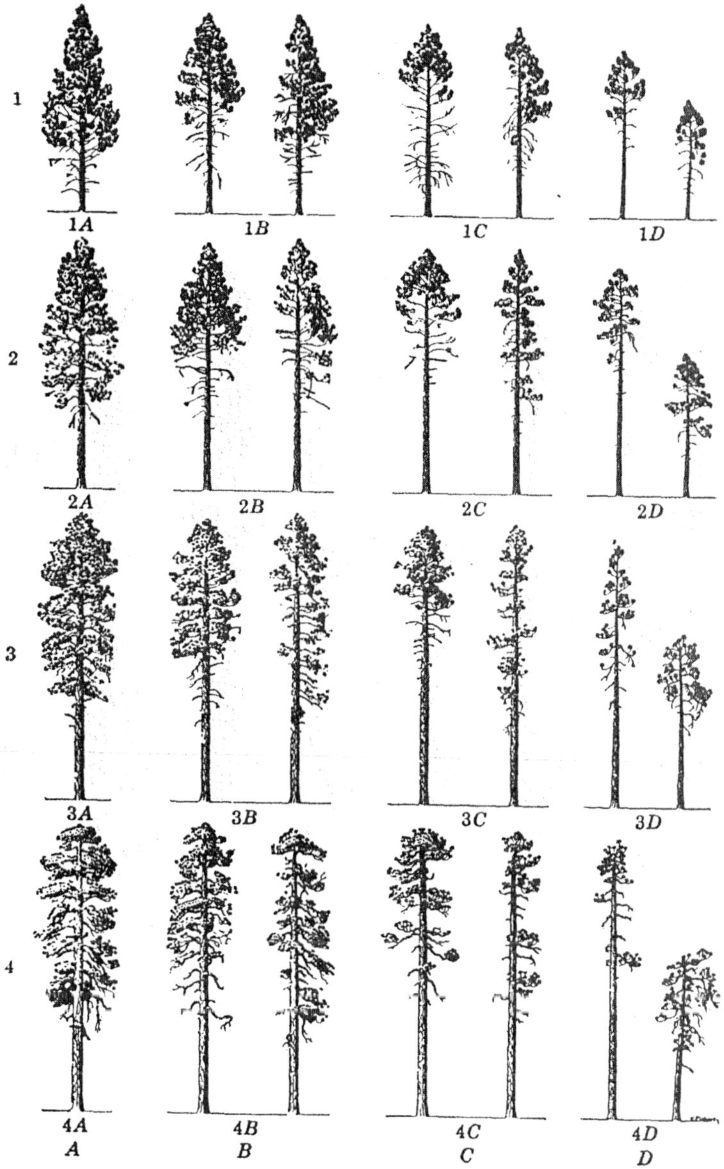

FIGURE 20. The Keen Ponderosa Pine Tree Classification based on age and vigor. The numbers at the left refer to age classes as follows: (1) young trees rarely over 20″ d.b.h.; (2) immature trees rarely over 30″ d.b.h.; (3) mature trees rarely under 40″ d.b.h.; and (4) overmature trees with wide latitude in diameters but usually large sizes in dominant trees. The numerals *A, B, C,* and *D* refer to crown vigor classes as follows: *A*—full vigor; *B*—good to fair vigor; *C*—fair to poor vigor; and *D*—very poor vigor. *Courtesy U. S. Forest Service.*

APPLICATION OF PARTIAL CUTTING ON NATIONAL FORESTS *

Partial cutting in some form has been the principal method of harvesting timber on the National Forests for some years.

FIGURE 21. Typical selective logging of ponderosa pine. The harvesting program called for removal of 50% of the merchantable stand by volume. The residual stand will grow much more rapidly after the opening of the crown cover. Many age classes are represented. The 11 stumps in the foreground were cut very low resulting in efficient utilization.

The following are leading examples:

1. *Ponderosa pine type* (the most widespread example geographically), found in all the western states from Canada to Mexico and from the West Coast east to South Dakota. Selective or partial cutting is used wherever this tree is found in the National Forests in merchantable quantities. Illustration of its application in this type is shown in Figure 20, known as the Keen tree classification which is used

* From information supplied by Ira J. Mason, Chief, Division of Forest Management, U. S. Forest Service, Washington, D. C.

TABLE XIV

Condensed Table of Marking Instructions for Western White Pine

Stand Description					Cutting Practices				
Age Class	*Ribes Potential*	*Vigor*	*Condition of Mixed*	*Stand Class*	*Method of Cutting*	*Maximum Volume to Remove in First Cutting*	*Slash Disposal*	*Broadcast Burning Practice*	*Regeneration*
Immature (*A*)				*A*	Light partial	25% of basal area	Lop and scatter. Pile and burn concentrations	None	Natural
Mature (*B*)	Light (1)	Good (*a*)		*B* 1a	Partial	40% white pine 50% all species	Lop and scatter. Pile and burn concentrations	None	Natural
		Poor	Sound (*b*)	*B* 1b	Seed tree	90%	Pile and burn	None	Natural
					Clear-cut	100%	Broadcast	Single or double	Planting
			Defective (*c*)	*B* 1c	Clear-cut	100%	Broadcast	Single or double	Planting
	Heavy (2)	Good	Sound (*a*)	*B* 2a	Partial	40% white pine 50% all species	Lop and scatter. Pile and burn concentrations	None	Natural
			Defective (*b*)	*B* 2b	Partial	40% white pine 50% all species	Lop and scatter. Pile and burn concentrations. Finally broadcast	Single or double	Planting
		Poor	Sound (*c*)	*B* 2c	Clear-cut	100%	Broadcast	Single	Planting
			Defective (*d*)	*B* 2d	Clear-cut	100%	Broadcast	Single or double	Planting
Overmature (*C*)				*C*	Partial or clear-cut	100%	Lop and scatter. Pile and burn. Finally broadcast	Single or double	Planting
Uneven-aged (*D*)				*D*	Very light partial		Lop and scatter. Pile and burn concentrations	None	Natural

All letters and numbers used in this table refer to the following types or stands:

A. Class *A*, merchantable immature stands
B. Class *B*, mature stands
1. Light potential ribes population
   (*a*) Stand of good vigor
   (*b*) Stand of poor vigor with mixed sound
   (*c*) Stand of poor vigor with mixed defective
2. Heavy potential ribes population
   (*a*) Stand of good vigor with mixed sound
   (*b*) Stand of good vigor with mixed defective
   (*c*) Stand of poor vigor with mixed sound
   (*d*) Stand of poor vigor with mixed defective

C. Class *C*, overmature stands
D. Mature or overmature, uneven-aged open stands

as the basis of marking prior to logging. Live timber is marked on the principle of removing the poor-thrift mature trees, a part of the good-thrift trees of larger diameter, and the overmature trees.

One of the excerpts taken from timber sale contracts for ponderosa pine in eastern Oregon as used in 1948 is as follows:

> Using the Keen tree classification as a basis, Class 3C and 3D trees, Class 3B trees 36 inches d.b.h. and over, Class 4 trees except such 4A trees as may be needed as seed trees, and any trees which, in the judgment of the forest officer, are undesirable as part of the future stand shall be marked for cutting; *provided,* that a marked living tree of any classification must have, in the judgment of the forest officer in charge, a sufficient realization value to equal at least the direct logging and milling costs of its merchantable logs, exclusive of all charges for stumpage, for depreciation, and for margin for profit and risk, including interest.

2. *Mature but not overmature western white pine* in northern Idaho and western Montana. See Table XIV for marking instructions as devised by the U. S. Forest Service at Missoula, Mont.

3. *Southern pine,* throughout the region with very few exceptions.

4. *Hardwoods,* both northern and southern, except aspen in the Lake States.

The important exceptions to the use of the principles of selective logging on the National Forests is in the harvesting of the following types which generally lend themselves to some form of clear cutting:

1. *Mature and overmature Douglas fir* in the Pacific Northwest. This holds true generally except in about one fourth to one third of the stands where a residual growth can be developed for future cutting and where the topography is favorable to tractor logging (a skidding machine peculiarly well adapted to selective logging).

2. *Mature and overmature lodgepole pine* except thrifty stands in early maturity as found in the Northern and Central Rocky Mountains.

3. *Overmature stands of jack pine* in the Lake States.

The application of the principles of partial cutting is very largely found in individual tree selection, rather than in group or area selection.

The approximate minimum diameter limits of merchantable species and percentage of merchantable stand cut are very variable, depending on local circumstances, especially the conditions found in the stands to be cut, as, for example, insect attacks. The officials lay very little stress on minimum diameter limits. They do, however, stress the following considerations:

1. Individual tree thrift.

FIGURE 22. Selectively cut Douglas fir in the Olympic National Forest, Wash. Note stumps of larger trees felled. Douglas fir is generally cut clear except as noted in the text. *Courtesy U. S. Forest Service.*

2. Evaluation of insect susceptibility—largely bark beetles.
3. Removal of individual tree specimens incapable of additional *financial* growth.
4. Spacing of trees left for release and accelerated future growth.

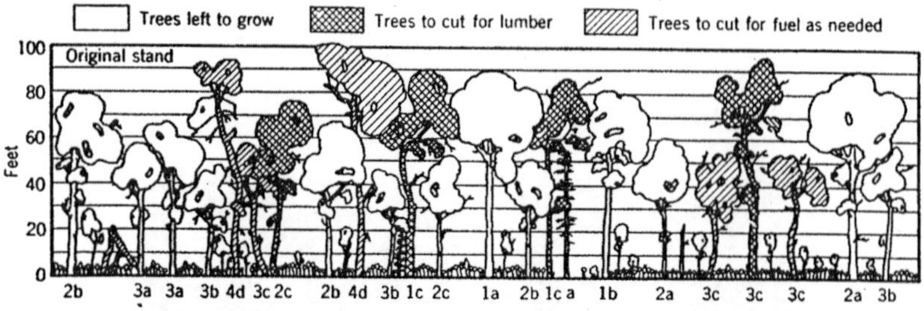

Before cutting, this forest averages 8000 b.f. per acre. Of this, 3200 b.f. is in 20 trees of declining vigor but of fair to good wood quality. These trees are commercial. They are marked for cutting. 4800 b.f. per acre is in trees of fair to good vigor and quality. These are left to grow. Also growing into commercial size are more than 40 trees per acre 8″ to 12″ d.b.h. which are carefully protected during logging. Over 200 trees 2″ to 8″ d.b.h. are likewise protected.

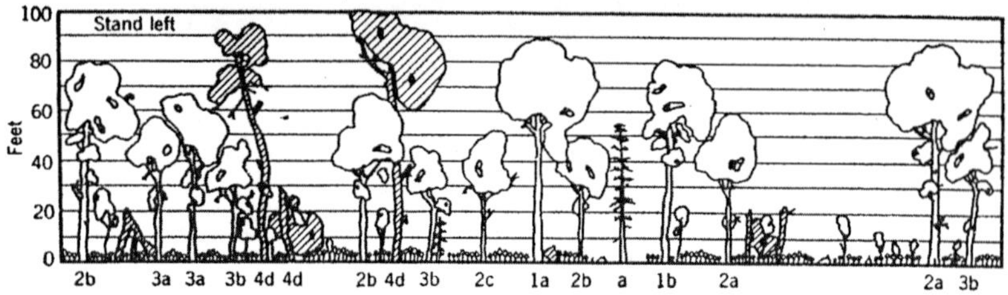

The stand which is left now contains 4800 b.f. in 40 commercial trees of fair to good vigor. There are few or no trees of declining vigor or quality. In 20 years this stand of trees will grow. The larger trees will increase in size and improve in quality. Some of the trees under commercial size will reach commercial size.

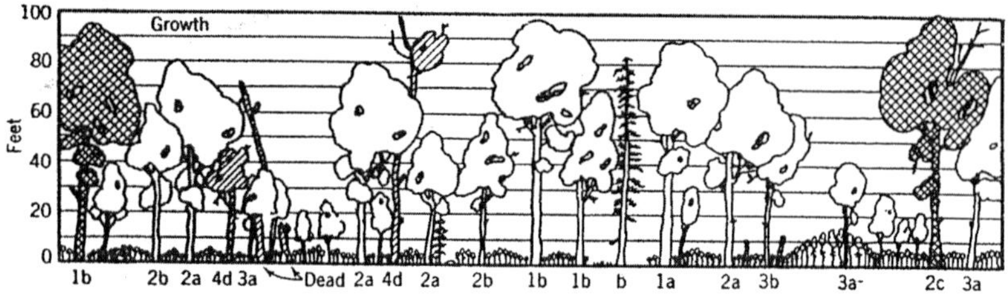

Twenty years after selective cutting the trees which were left to grow will have added 3000 b.f. of high-quality commercial wood to this average acre. The total stand will have almost reached its original volume. It will contain 7800 b.f. of commercial saw logs. Individual trees will increase in size, vigor, and quality.

There still remain in this stand 15 to 20 trees of lower than average vigor and quality. These should now be selected for the second cut. Again a stand of 4000 to 5000 b.f. should remain to grow.

FIGURE 23. Relationship of tree classes to good selective cutting in hardwood forests (Lakes States). *Courtesy U. S. Forest Service.* Tree quality and vigor class symbols are taken from *Comparison of Investment and Earnings of Hardwood Trees in the Northern Lake States,* U. S. Forest Service, Milwaukee, Wis.

In ponderosa pine stands, the general rule is to cut about one third to one half of the merchantable volume of the stand. The constant emphasis is to reduce the proportion of the initial cutting.

In southern National Forests, logging consists largely of sanitation and improvement cuttings. Currently, there are few sales for cutting growing stock of importance because there are very few mature or overmature elements in the stands. Cuttings are largely for pulp-

FIGURE 24. Typical selectively cut pulpwood stand where 16 cords of southern pine have been cut and at least an equivalent amount left for increased growth and later cutting. *Courtesy U. S. Forest Service.*

wood and some saw logs. The southern National Forests consist largely of young and immature stands because they were acquired after the merchantable timber was removed. Where the principles of partial cutting are employed, minimum diameters,* if used, vary widely by species, products, and localities. For example, on National Forests in Arkansas, shortleaf pine 8″ d.b.h. is the minimum for saw timber and 5″ d.b.h. for pulpwood. On the Coastal Plain, loblolly, slash, and longleaf may be cut to 10″ for saw logs and 5″ for pulpwood. High-grade hardwoods (yellow poplar, white ash, white oak, and magnolia) are cut to about 14″ d.b.h., and low-grade hardwoods (post,

* Where minimum diameters are stated, it should be understood that not all trees of those sizes and larger are cut. The diameter is really a method of defining a merchantable size rather than an authorization to cut every tree at and above the stated minimum limit.

TABLE XV

SUMMARY OF SPECIES AND DIAMETER SIZES CUT BY SELECTIVE LOGGING ON TRACT OF 32.3 ACRES IN CENTRAL NEW YORK

Diam. B.H.	*Sugar Maple*	*Bass-wood*	*Red Oak*	*Ash*	*Beech*	*Hem-lock*	*Tulip Poplar*	*Black Cherry*	*Elm*	*Butter-nut*	*Yellow Birch*	*Total*
15	18	14	3	10	5		3	2			1	56
16	22	16	4	11	7		1	2		1		64
17	24	18	12	8	9	4	3	3	1	1	1	84
18	28	18	9	9	7	4	4	2	1			82
19	20	15	4	7	7	6	4	1				64
20	12	5	7	3	5	5	4					41
21	9	5	5		3	1	2		1	1		27
22	8	2	5		3	4	1					23
23	6		6	1	2	6		2				23
24	3		8		2	4	1					18
25	3	2	4	1		1	2					13
26	3		6			3						12
27	2	1	4		1							8
28	1		3	1		2						7
29			4									4
30			2									2
31			1			2						3
32			1									1
Total	159	96	88	51	51	42	25	12	3	3	2	532

black and water oaks, hickory, beech, and the like) throughout the South are cut to 16″ d.b.h. In the Piedmont areas, shortleaf and loblolly pines are utilized to 8″ minimum top d.i.b. for saw logs and 5″ minimum top d.i.b. for pulpwood. Both silvicultural and utilization practices in the South have changed rapidly with higher stumpage values. Therefore marking rules and management plans have been altered.

EXAMPLE OF SELECTIVE LOGGING IN CENTRAL NEW YORK

In 1947 a typical small tract of 32.3 acres of Central New York hardwoods and hemlock was cut 31 years after thinning to eliminate the poorly formed trees, low thrift, and overcrowded specimens that were not growing well. Species partly or wholly eliminated from the stand were beech, soft maple, hemlock, ironwood, elm, popple, and poorer specimens (unthrifty, poorly formed, and defective) of the better species. The favored were white ash, tulip poplar, basswood, black cherry, and red oak, all rapid-growing species. The principal volume of the stand was composed of sugar maple with red oak second. The average volume cut per acre was 3832.7 b.f. The average number of trees cut per acre was 16.5. About one half the volume of the stand was cut, leaving trees up to 23″ in diameter to form the basis of later cutting. There were 53 trees per acre 8″ and up d.b.h. left in the stand after cutting and also many more between 2″ and 8″ d.b.h. Most of the trees had dominant crowns and had reached financial maturity. Trees left in the stand after cutting, as tallied, were 75 from 2″ to 8″, 31 from 8″ to 12″, and 22 from 12″ to 23″ d.b.h.; they were distributed as to species about the same as those cut, with a somewhat larger proportion of sugar maple, especially in the smaller sizes, due to tolerance. The average price for stumpage was $25 per thousand board feet.

Table XV shows the number of trees cut for each species and diameter. The diameters of trees marked for cutting ranged from 15″ to 32″ d.b.h. This is typical of a good many small wood lots of the region that have been successively cut for about 100 years and are still growing at a favorable rate.

PART II

PREPARATION OF LOGS FOR TRANSPORT

CHAPTER V

Felling and Bucking*

WOODS TOOLS

Woods tools and equipment are characterized by strength, durability, and efficiency for the work to be done. With the exception of axes, saws, and some metal parts, the tools and handles used in felling, bucking, and general woods work are made in the camp blacksmith shop.

AXES

Axes are used principally for chopping undercuts in felling, limbing, topping (small timber), and swamping. They are also used for hewing, splitting, and general woods work. An axe has two parts: (1) the head, consisting of the cutting edge or bit, and the poll or body with an eye or hole in it; and (2) the handle or helve which is inserted in the eye. Axes may be single or double bitted, the double-bitted type being generally employed in woods operations throughout the United States.

Single-bitted axes are used on small operations in second-growth timber in the Northeast, the Central States, the Lake States, and the South, especially where they may be used for other purposes such as a maul for driving stakes, spikes, and splitting. Some of both types are used in many parts of the South and the East. Where a lightweight axe is desired and used for dual purposes, single-bitted axes are preferred.

Axe handles are made preferably of second-growth sap hickory but sometimes of red or heart hickory and frequently of hard maple or other species. Occasionally they are made in the camp blacksmith shop. Generally they are oval in cross section, but in eastern Texas

* For excellent and detailed coverage of conditions in eastern Canada, see Pulpwood Cutting—Efficiency of Technique, by A. Koroleff, *Forester*, Canadian Pulp and Paper Association, Montreal, 1941.

and Arkansas octagonal handles are often used. Double-bitted axe handles are straight, whereas single-bitted ones are generally curved.

The standard length of axe handles is 36″ before being driven into the head. In redwood timber and other parts of the Pacific Coast, handles of 42″ to 48″ in length are sometimes used. In the Lake, Central, and Eastern States 34″ handles are ocasionally used, and in New England and New York 28″, 30″, and 32″ lengths are common.

The bit is generally forged with crucible steel welded on a mild-steel body and carefully tempered and tested for toughness and strength. The sides of the blade should be slightly convex and proportioned to enter the wood easily, burst the chip, and emerge clear without wedging or binding. Head weights vary with the character of the timber to be cut and the work to be done. Generally they are 3½ to 5 lb. or more. The heavier types up to 6 lb. are generally used in the large timber of the Pacific Coast. The heaviest head weighs 7½ lb.

Broadaxes used for such tasks as hewing cross ties, squaring logs, and facing chute logs, have bits 10″ to 12″ or more in width with heads weighing 5 to 7 lb. and handles 24″ to 38″ long.

Special axes are made for making cuts in which to insert gutters for turpentining in longleaf and slash pine in the South. Turpentine axes for cutting boxes at the base of these trees are seldom used, because of improvements and changes in the method of producing resin in the South. Correct grinding of axes to sharpen them is of great importance because the ease and efficiency in use and the length of life in service are greatly increased thereby. The following rules should be observed:

1. Grind slowly on a wheel kept very wet. Frictional heat destroys the temper.

2. The edge must not be so thin that an indirect blow will easily break it.

3. Start to grind the axe 2″ to 3″ back from the cutting edge, and grind to ½″ from the edge. Then in grinding the cutting edge maintain a convex bevel shape. A concave edge leaves insufficient support and breaks easily. A straight bevel is better than a concave one, but it is not so efficient.

4. Maintain a fan-shaped edge.

Where much wedging or driving is done, a maul should be used instead of an axe. Axes should not be used on metal wedges. Frost crystallizes tempered steel; therefore, in cold weather axes should be warmed before use.

SAWS

Saws are used principally in making the crosscut in felling trees and in bisecting (bucking) the fallen stem into logs. They are also widely used in making logging improvements. They may be classified as:

1. Two-man crosscut saw generally used in felling and bucking.

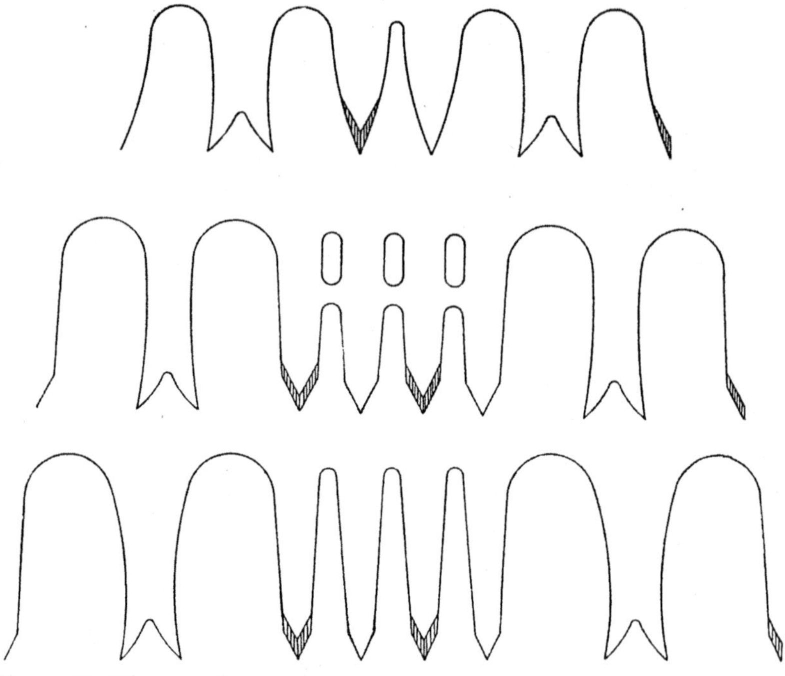

FIGURE 25. Three usual types of crosscut saws. The upper one with 2 cutter teeth and 1 raker is used for crosscutting oak and similar hardwoods. The center one illustrates the type with 4 cutter teeth and raker used in southern yellow pine. The lower one shows 4 cutter teeth and raker with wide, deep gullets and long lancelike teeth, used in Pacific Coast timber. The last type is of 13 gage on the tooth edge and the other two are 14 gage on the tooth edge.

2. One-man crosscut saw used in bucking the largest trees on the Pacific Coast and in cross-tie and other small-sized products.
3. One-man bucksaws used in both felling and bucking (principally bucking) spruce and balsam fir in the Northeast, Lake States, eastern Canada, and Newfoundland. The bucksaw is used especially for the smaller trees, 10″ or less in diameter. It is occasionally used both for felling and bucking trees up to 18″ in diameter.

Sizes. Lengths of two-man crosscut saws vary from 4½′ to 8′ in the small timber of the East and South. Saws 6′ to 12′ in length are generally necessary for the large timber of the Pacific Coast. On some large operations in Douglas fir timber the same 8′ crosscut saw is used as in felling longleaf pine in the South. Special lengths may be required in felling redwood, but few trees which are over 6′ in diameter at the stump are felled. The longest saw for felling redwood was 22′.

Saws for bucking Douglas fir are generally 7′, 7½′, or 8′ in length.

In width, saws vary from 3″ to 9″ in the center. They are generally wider at the center than at the ends. Usually the cutting edge is

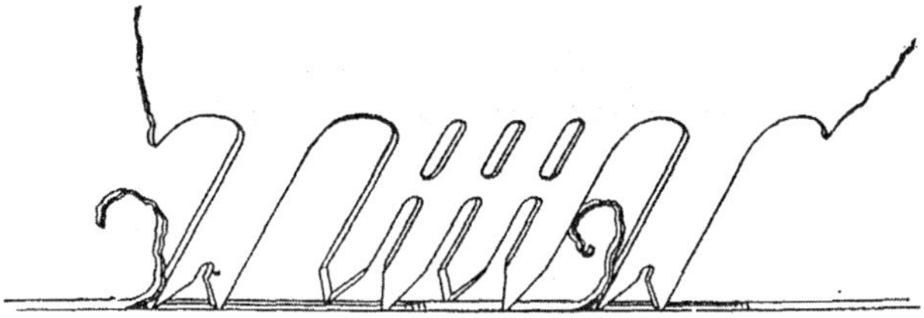

FIGURE 26. Sketch illustrating operation of the cutting and raker teeth on a crosscut saw. The four cutting teeth shown in the center are slightly longer than the raker teeth.

curved to facilitate elimination of sawdust. One-man crosscut saws are generally thicker and stiffer than two-man saws to prevent bucking on the forward or pushing movement. With two-man crosscut saws, each man pulls the saw through the wood, letting the weight of the saw do the work. The gage of the saw is generally 15 to 17 at the back and 14 at the cutting edge for western use, and 17 to 20 at the back and 14 at the tooth edge for the East and South. The saw kerf is generally 3/32″ thick or somewhat more.

Saw handles are generally removable because the saws frequently become wedged. Handles are generally 1½″ in diameter, 14″ to 20″ in length, and are fastened by clamps or screws at each end. Handles are made of ash, hard maple, beech, birch, and oak.

The teeth of crosscut saws are generally of two types, namely, the cutter and raker or cleaver teeth. The cutting teeth are used to sever the wood fibers on each side, as shown in Figure 26. The raker teeth are used to remove the sawdust. Various arrangements of width and number of cutter and raker teeth are used as illustrated. Long-fibered and soft-textured hardwoods are sawed with more difficulty, because the cutting teeth tear rather than actually cut the fibers, and a larger quantity of dust is removed. The conifers are sawed more easily than the hardwoods because of their peculiar texture and structure.

Saws must overcome both the resistance of the wood fibers and the friction of the saw passing through the cut. Kerosene is used to facilitate sawing, especially with resinous softwoods such as southern pine, ponderosa pine, and Douglas fir. Green wood is more easily sawn than seasoned wood because moisture diminishes the hardness of the fiber. Hence pine, spruce, cypress, and other softwoods are more easily sawn when green. Wood of poplar, aspen, and willow is more readily sawn when seasoned than when green, owing to the hardwood texture.

For oak and similar hardwood species, it is customary to have two cutting teeth for each raker tooth. For southern pine, the ratio is four cutting teeth for each raker tooth. For Douglas fir, the practice is to use four lance-shaped cutting teeth for each raker with extra large deep tooth and raker gullets. The saw blade should taper from the cutting edge to the back to prevent binding and to facilitate the movement of the saw. The gage is, therefore, thinner at the back than at the tooth edge.

PEAVIES AND CANTHOOKS*

Both these tools are used as levers in rolling, stopping, and turning logs, especially on skidways, landings, and the like. They are especially useful in:

1. Rolling logs over or up skids in loading trucks, sleds, chutes, railway cars, and so on.
2. Unloading trucks, sleds, and similar vehicles at landings.
3. Rolling logs on sawmill log decks.
4. Stream-driving operations, especially as in breaking jams.

Handles for both are made of hickory, hard maple, and ironwood. They may be 2′ to 6′ but generally are 4′ to 5′ in length. Both tools have hooks of various types, depending on their use with soft, hard, or frozen logs; thick-barked species; large or small logs, and so on.

SPRINGBOARDS AND GUNSTICKS

Springboards are generally 2″ × 6″, 2″ × 8″, and 3″ × 8″ pieces of rough Douglas fir 4′ to 5′ long, with an iron lip or spur screwed on one end. The springboard is inserted into horizontal notches out into the tree. The weight of the feller causes the lip to "bite" into the tree.

* The peavy has a sharp pike at the lower end, whereas the canthook has no pike but a lip or toe on an iron band at the lower end. The canthook is used chiefly around sawmills in handling sawn timbers as well as logs.

Fellers stand on these boards to chop the undercut and pull the saw. Springboards are only used on the Pacific Coast to afford a place from which to work. Large bell-butted trees on uneven ground and steep slopes, dead and down timber, and so on generally require springboards when the stumps are to be cut 3′ to 10′ or more above the level

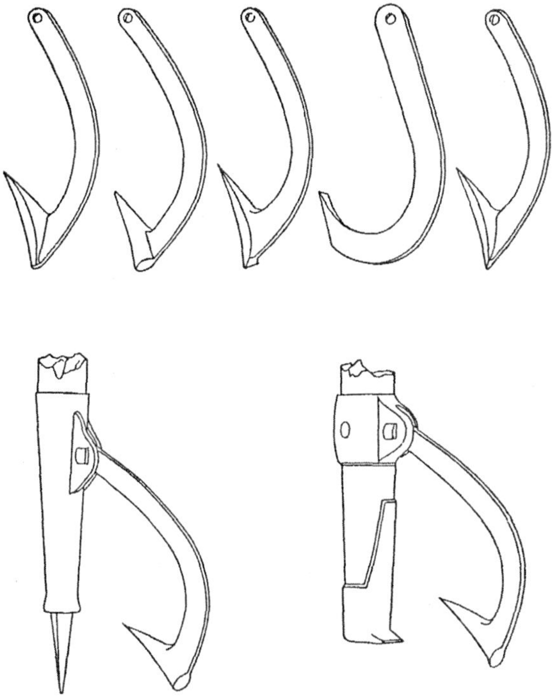

FIGURE 27. Peavy on lower left; canthook on lower right. Above are shown various types of hooks—left to right, they are duck-bill, round, diamond, chisel, and top-loader types.

of the ground. They are being employed less than formerly because trees are felled with lower stumps.* Also chain saws are more widely used.

Boards are placed between springboards on the largest trees, especially in redwood timber, thus forming a continuous platform around the tree. Men working on these springboards must wear calked shoes.

* The Long Bell Lumber Company, Ryderwood, Wash., formerly cut stumps from springboards at levels 6′ to 10′ above the ground. Recently they cut the same-size trees 18″ to 40″ above the ground, which is exceptionally low for large-sized virgin Douglas fir timber.

Gunsticks are scissor-like poles used in determining the direction of the fall of redwoods, especially when bedding is used to break the impact. They are generally made in the camp blacksmith shop of small board strips 6′ to 10′ in length. The feller sights down the two points at the end of the stick while the other ends rest at each end of the notch.

MISCELLANEOUS *

Wedges. Wedges are used in felling trees and to prevent binding in log-making operations in all parts of the United States. They are

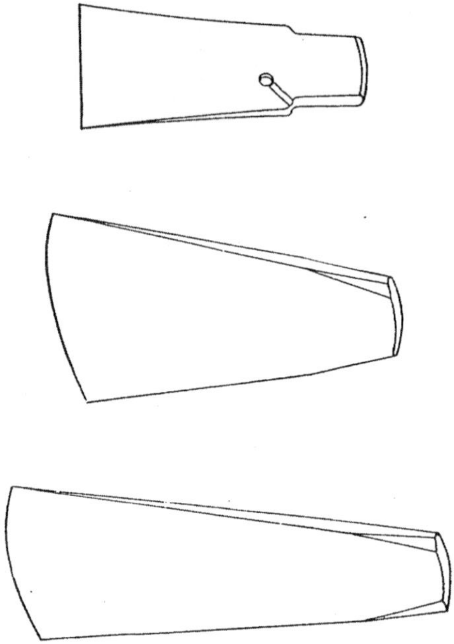

FIGURE 28. Three types of metal wedges. Above, a saw wedge; below, a felling wedge; and, in the middle, a bucking wedge. These wedges are made of various weights, generally 2 lb. or more, and in various lengths. In logging operations on the Pacific Coast, wedges play an important part in both felling and crosscutting operations.

generally purchased from equipment dealers but also are made in camp blacksmith shops. Various types are used, depending on the size and species of timber. Metal wedges of steel or iron up to 12″ long, 3″ wide, and weighing up to 9 lb. are used in the big timber of the Pacific Coast. Small metal wedges are satisfactory in other parts

* This includes only a partial list of woods tools and equipment commonly used.

of the country. The felling wedge has a much sharper (smaller) angle than the crosscutting wedge. Hardwood wedges are used on small logging operations throughout the East, the Northeast, the Lake States, and the South, but not exclusively, since metal wedges are also used in all these sections. In the northern hardwood forest, hard maple, beech, ironwood, and oak are frequently found. In the Central States and the South as well as the East, hickory, persimmon, and dogwood wedges are much preferred. Because of their common occurrence, however, oak, beech, and hard maple are the most widely used species. Wedges may be either sawn or chopped, and of varying sizes, depending on the local requirements and conditions.

Tongs. Tongs are used in skidding and loading logs. They vary in size and weight with the logs to be handled. Tongs up to 160 lb. or more are used in loading log lengths up tc 64′ or more in virgin West Coast timber.

Chains. Chains are used in skidding, trailing, and chaining logs with horses and tractors; in rafting and booming logs; in impeding the progress of logs on chutes; in binding logs loaded on sleds, trucks, and similar conveyances; in connecting two or three sleds in winter logging; and for many miscellaneous jobs about a logging operation, particularly in the forest east of the Great Plains.

Dogs, Grabs, Etc. Dogs, grabs, etc., are short heavy-pointed pieces of steel generally made by the camp blacksmith and driven into the logs. They are used in connection with chains and cables for such operations as skidding, rafting, and booming. They may be straight or curved and may be flat topped with a hole or ring (ring dog) attached to the top. Steel cables, chains, or ropes may be passed through the rings or holes to hold the logs together in rafting. Dogs are often coupled together (chain dogs) for trailing logs one behind the other or for skidding two or more logs together. They are widely used in pulpwood and hardwood logging and especially in saw-log chuting and similar practices.

Chokers. Chokers are nooses of wire cable placed around the end of one or more logs and are widely used in power and tractor skidding. They are generally 15′ to 45′ long with a choker hook on one end and a braided eye on the other. Chokers are generally used only in the largest timber and in skidding long lengths. They may bring in 1000 to 8000 b.f. on each haul. Maximum-sized logs up to 8000 b.f. or more are handled only with chokers in skidding operations.

Pike Poles. Pike poles are 10′ to 20′ long, generally made of ash, with or without a hook, and are used in stream driving, breaking log

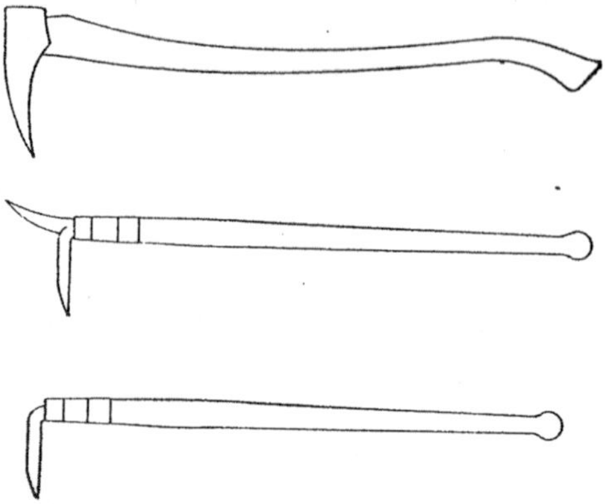

FIGURE 29. Three types of pickaroons. The type shown on the top is sometimes known as a hookaroon, which is used for handling poles, posts, or pulpwood. The lower type is used for handling various kinds of poles and cross ties. The middle type is used mainly for sorting lumber and miscellaneous small-sized products of the forest. Handles are generally made of ash and vary from 18″ to 36″ in length.

FIGURE 30. Variety of tools used on a hewn railway cross-tie operation. The tie maker's tools are, left to right, broad axe, chopping axe, measuring stick for length, pickaroon, one-man crosscut saw, kerosene bottle, and bark spud. A hewn cross tie is on the right. The former practice of hewing cross ties has decreased because of the high cost of labor compared with the cost of producing sawn ties from small sawmills where side cut lumber may be profitably sold.

jams, and handling logs in booms. They are also used in connection with log ponds at the mills. They generally have sharp pikes on the end to hold logs and poles and to move them about as desired.

Pickaroons. Pickaroons are short poles 35″ to 40″ long with a recurved pike or hook used in drawing or pulling small products such as cross ties, 4′ pulpwood, fuel wood, chemical wood, cooperage, and bolts down steep slopes or for short distances. They are tools primarily for hand logging as a substitute for skidding products short distances to haul roads.

MECHANIZED FELLING, BUCKING, AND LIMBING DEVICES

Concurrently with the increasing pay and often decreased efficiency of manual labor used for felling, bucking, and limbing, there have been more than 32 kinds of power saws of chain, circular, drag, and other types developed to counteract these high costs. One of the first large companies to use the chain saw widely was the Eastman Gardiner Company at Laurel, Miss.* This was in 1930. Pneumatic-powered saws have been used experimentally for years. World War II ending in 1945, helped to develop chain saws, as they were widely used for clearing areas for landing fields and for many other purposes as well as on logging operations to supply much needed lumber for the war. Increasing mechanization has already reduced the man-power requirements for 1000 b.f. of logs from about 15 to 22 man-hours in 1922–25 to only 5 to 7 hours at present, depending on local conditions.

An important advantage of the power saw is the lower stumps that may be cut with these machines and especially the greatly increased board footage per unit of time as contrasted with the customary manual labor. Records show that two to four times as much production on a man-hour basis have been demonstrated with these mechanical saws in several parts of the country.

There are, in general, three separate types that are used for power felling and bucking. There are 13 different types of circular power saws, 8 different types of drag saws, and 11 different types of chain saws. Most of them are gasoline-driven. Some are powered only by electricity. Some may be operated by gasoline, electric, or pneumatic power. They are generally two- or three-man saws. However, the small circular and some chain-saw types are operated by one man.

There is also a type known as mammoth tree shears which is mounted on crawler tractor and has hydraulic shear blades.

* See A New Mechanical Tree Feller, *Journal of Forestry,* Vol. 31, No. 6, October 1933.

FIGURE 31. Making the wedge cut or undercut on hardwood in the South with a Lombard 36″ gasoline-powered chain saw which is also made in 24″ and 48″ lengths. *Courtesy Lombard Saw Co.*

FIGURE 32. Usual method of felling tree with one of several types of two-man chain saws. Note low stump.

It is important that these machines be kept in good running order and that the teeth be kept constantly sharpened. In many cases, logging companies or individuals have loaned these machines to the fallers and buckers at no charge for the use of the machine, including gasoline and oil.

As felling and bucking are generally paid for on a piecework basis, the costs of these operations have been materially lowered, especially where it was exceedingly difficult to obtain labor at all, or where labor was inefficient as well as expensive (in production per man-hour).

FIGURE 33. Disston chain saw used in felling hardwood in the Northeast. The rotten heart defect was not disclosed by the apparent sound condition of the tree before felling. *Courtesy Disston Saw Co.*

In most sections of the country, very few mechanical felling saws were used before the last war, whereas there are thousands now being used in all the more heavily timbered sections. In felling pulpwood in eastern Canada, manual felling and bucking are still used, but the trend is toward mechanized operations. In brushy and swampy locations the manual methods are also still used. Some companies, after trying out the power saws, have returned to manual methods.

Power saws are widely used on landings where logs have been skidded in long or tree lengths by tractors and then crosscut before being loaded on trucks or railroad cars into shorter lengths suitable for use in the sawmills.

The lower costs and advantages enjoyed by these various power saws are due to the following reasons:

1. Greater production of board feet per unit of time, than with manual labor.

FIGURE 34. Lowther C-saw at work showing method of felling and crosscutting with same machine. This is particularly adapted to cutting small-size timber such as pulpwood in the open and flat "piney woods" of the South. *Courtesy Lowther Saw Co.*

2. Light weight for easy portability.
3. Durable machinery that stands up under vigorous and strenuous operation.
4. Low stumps, which increase the volume of trees felled, especially in the butt logs where the best lumber grades occur.
5. Less breakage and shattering of boles in felling.
6. Greater precision and, therefore, less loss of wood in log lengths.
7. All wearing parts of saws quickly and easily replaceable.
8. Saws readily repaired, replaced, or sharpened.
9. Relatively low initial cost which means low overhead and maintenance costs.
10. Nearly all types may be used to saw both horizontally and vertically with minor and quick adjustments.

Although considerable mechanical difficulties were encountered in the initial operation of some of these types and often the services of expert mechanics were required, these difficulties have been largely overcome. The speeds, saw kerf, power required; number of saws; reversibility and operating details vary widely with the different types. For pulpwood felling, small one- and two-man operated circular and chain saws are commonly used. For large timber involving trees up to 6′ in diameter or more exceptionally long chain saws are used. Power is generally supplied by a small gasoline engine attached to the machine. In case of electrically operated power saws, the power is supplied by tractor. Most of the machines are carried from tree to tree by hand. However, some types are operated from wheels. In open types of timber, such as southern pine, ponderosa pine, and some other stands where there is very little underbrush and the ground is fairly level, there is very little difficulty in the portability and operation of these power saws. However, in swampy locations and in regions of heavy brush, windfalls, and much dead and down timber, considerable difficulty may be encountered. Actual records show that two to five times the production is possible with these power saws as compared with manual labor.

The following are some of the types used for felling aside from the usual chain saws driven by gasoline.

EUREKA DRAG SAW

The Eureka air-cooled pony drag saw is used principally in felling redwood and Douglas fir. It weighs 120 lb. and falls trees 2′ to 20′ in diameter. It has a double-cycle air-cooled 4-hp. portable gas-driven

FIGURE 35. At Poison Lake, Calif., this D4 tractor equipped with an Atkins electric saw is felling 15 M.b.f. of ponderosa pine per hour. A generator mounted on a tractor supplies the current.

motor. It uses from 1 to 1½ gal. of gasoline per day and is used for undercutting, felling, and bucking.

It is used with two men, but in emergencies one man can operate it. These machines have produced 30,000 to 40,000 b.f. per day in Douglas fir and associated species in timber running 24″ in d.b.h. and up. Production figures for redwood are somewhat less.

ATKINS ELECTRIC-POWERED CHAIN SAW

Power for this saw is supplied by tractor. It is particularly used in big timber in Douglas fir, redwood, ponderosa pine, and species that

FIGURE 36. Large timber on the West Coast requires long chain saws. Power saws operate much more successfully in big timber than in small timber, once they have been set up, because a much larger footage is involved and less time lost in moving the saw per unit of production.

occur with them. Irregular terrain and brushy ground conditions found in Douglas fir and redwood timber require low cost operation under difficult conditions. The Atkins electric saw has been successfully operated under such circumstances. One method developed with this saw is the Humboldt undercut which is taken from the stump rather than from the butt log as has been the previous custom. This leaves the first log felled free of a sniped undercut which has been chopped with the axe in manual felling or cut out with one of the chain saws.

The advantages of this saw are:

1. It reduces the cost of manual labor 48% to 76% in felling and bucking operations.

2. It provides for lower stumps, thus increasing the volume of timber felled by 3% to 8%. Low stumps also mean somewhat cheaper skidding costs and decreased fire hazards.

3. An electric cable permits its use up to 1200 ft. from the power source. This gives the machine a wide latitude of action especially for large trees when once it has been installed. It can be used effectively for large footages.

4. Limbing can be done flush with the bole, making loading and trucking conditions simpler and cheaper. Limbing is often done with the one-man electric chain saw.

5. There are fewer accidents with this saw than with handsaws and axes, and, consequently, lower insurance rates are available when it is used.

ONE-MAN POWER SAWS

For felling, crosscutting, and limbing small-size timber generally up to 20″ d.b.h., several one-man power saws have been produced. Some are the chain-saw type, and one is the circular-saw type. They weigh 38 to 69 lb. each and vary in horsepower from 1½ to 4 per machine. They are especially adapted for use in pulpwood, fuel wood, and other small-size trees. They are also suitable for making thinnings in densely crowded stands or in plantations requiring thinning, either for profit or for silvicultural improvement of the stand.

A summary of four types of portable one-man power saws is shown in Table XVI. All are driven by gasoline engines.

TABLE XVI

ONE-MAN POWER SAWS *

	Trade Name			
	Universal	*Precision*	*Timber Hog*	*Sally Saw*
Horsepower	3	3½	4	1½
Weight	38 lb.	48 lb.	52–57 lb.	69 lb.
Standard blade	20″, 26″	20″, 26″, 32″	20″, 24″, 30″	12″ limit

* These saws are made by the following companies:
"Universal"—Power Machinery Company, Seattle, Wash.
"Precision"—Precision Parts Ltd., Montreal, Canada.
"Timber Hog"—Reed-Prentice Corporation, Worcester, Mass.
"Sally Saw"—Cummings Machine Works, Boston, Mass.

DISADVANTAGES OF POWER-FELLING SAWS

Although the many types of power felling and bucking saws have increased in use throughout the United States and Canada, there are

FIGURE 37. The chain saw is frequently used for making an "undercut" in bucking tree trunks. This is difficult with manual bucking. It also prevents frequent binding unless the heavy trunk is properly supported. *Courtesy E. C. Atkins Saw Co.*

many reasons why they are not more widely used. Among the reasons are:

1. Rugged and precipitous terrain, dense underbrush, and difficult ground cover, including windfalls, sometimes make it exceedingly slow and expensive to move power saws from tree to tree and even to and from the felling areas.

2. Cutters (fellers and buckers) are not generally adapted to the use, adjustment, and repair of these machines. Some models require constant attention, replacement of parts, sharpening, and adjustments. In remote locations, repairs and parts cannot be readily secured. Woodsmen are generally not good mechanics.

3. Owing to delays and lost time incident to setting up and operating the machines, together with the unwillingness of the cutters to change deep-rooted customs, cutting in small timber, as in eastern Canada and in other sections, is still done largely by manual labor with axes and saws. Table XVII shows an analysis of productive, semiproductive, and nonproductive time in northern hardwoods in New Hampshire.

TABLE XVII

DISTRIBUTION OF TIME IN FELLING AND BUCKING NORTHERN HARDWOOD SAW LOGS WITH A ONE-MAN ELECTRIC CHAIN SAW (HOMELITE) AT GORHAM, N. H., BASED ON BANGOR RULE *

Operation	*Time, %*
Felling	17.98
Butt off cut	2.04
Bucking into logs	9.98
Limbing with saw	0.56
Total productive time	30.56
Intermediate movements	17.04
Moving generator	10.65
Equipment, maintenance	3.18
Total semiproductive time	30.87
Saw pinch	1.19
Delays—mechanical	16.26
Travel	11.59
Eat and drink	9.53
Total nonproductive time	38.57

Total of productive, semiproductive, and nonproductive time—100%.

* Supplied by American Pulpwood Association.

4. In deep swamps, as in the South, power saws are not used because of their weight and the difficulty of moving them from tree to tree as well as of transporting them to the felling areas. The difficulties in deep-water swamps in both hardwoods and cypress have militated against their use.

Some of these difficulties are not insurmountable, and it is likely that the use of these power-felling saws will increase, especially as their mechanical efficiency is improved.

FIGURE 38. Bucking ponderosa pine log with power saw in California. Tractor at left furnishes electric power. The man on the log is wedging the saw cut to prevent binding.

GIRDLING

Girdling, or the incision of a continuous ring about the tree, under the bark, within the cambium layer, which deadens the tree, is practiced to a limited extent in some of the hardwood forests of the East and South.

Cypress is girdled in the winter months and left standing 3, 6, or occasionally 12 months. It is done to eliminate moisture from the tree, thus decreasing its weight, facilitating logging operations, and making it possible to float the logs. The primary reason, however, is to hasten the drying process, as cypress lumber is generally air-seasoned. Girdling permits better seasoning, with less degrades, especially in the sap grades. Also girdling means that the logs will float better in the log ponds at the sawmills.

Hardwoods in the East and South are girdled to a limited extent to eliminate them from the stand rather than as a phase of logging operations. One company in the Adirondacks has girdled 40,000 acres of hardwood, chiefly beech, on mixed hardwood and spruce stands which primarily are managed for spruce pulpwood production. As the spruce pulpwood is floated to market, and there are no facilities for transporting the low-value hardwoods profitably, they are girdled to make room in the stand for the more profitable spruce. During the recent advances in stumpage prices, hardwoods which would formerly have been girdled were sold at favorable prices, and so girdling is not so frequently practiced as formerly.

Teak is girdled in India and Java and left standing approximately 2 years before being felled.

FELLING *

SEASON

Winter felling best meets the requirements for conditioning lumber which is to be seasoned during the following early spring and summer

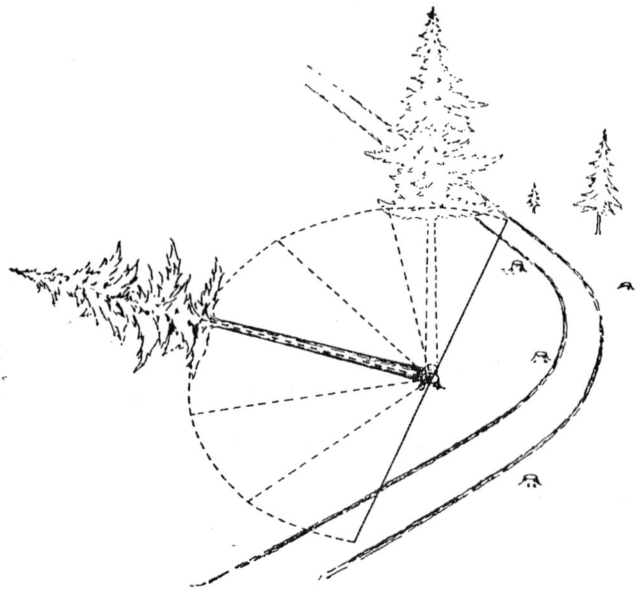

FIGURE 39. Diagram illustrating how a tree may be dropped within a quarter-circle on either side of the place where it would naturally fall. Frequently trees must be felled so as not to interfere with log-transport traffic or hung up in adjacent trees or felled on rocks, windfalls, stumps, and the like, which cause shattering.

seasons. Where the bark is to be peeled, as for pulpwood, poles, cross ties, and in the processing of hemlock and tanbark oak products, spring is the only season when work can be efficiently done. The bark

* This term is used interchangeably with falling. Also in some sections, choppers are known as fallers or vice versa.

peels easily after the starting of the spring growth, but the cost of peeling it is practically prohibitive except in spring or early summer. Logs check least in the autumn or winter, and the likelihood of sap stain and insect attack is also the least during the season of the year when trees are not growing. In the East, fuel wood cut in winter is sometimes required, because the bark is retained better on winter-cut wood.

FIGURE 40. Many small trees felled for saw logs and pulpwood are cut with hand saws. Power saws have required about 15 years for development but high wages have given them a great impetus. Much pulpwood in eastern Canada and lodgepole pine in the Rocky Mountains are felled with hand saws.

METHODS AND PROCEDURE

Felling methods in the United States may be summarized as follows:

1. Most trees, both in the smaller and larger sizes, are felled and bucked after being notched or undercut with an axe and crosscut with a two-man saw or power saw. This process permits the cutting of low stumps; the tree can be accurately directed in felling, and there is relatively little waste. The bottom of the notch is cut level, while the slope should be at an angle of 45°. The depth of the notch may represent from 40% of the diameter on small trees to 20% or less on large trees. A new type of notch, known as the Humboldt, is taken out of the stump instead of the first log, as described later and as shown in Figure 41.

2. Small trees, particularly in the spruce forests of the Northeast, and in second-growth thinning operations, are sometimes felled with the one-man crosscut saw. One-man power saws are also being rapidly introduced and used.

3. Felling trees with roots, tried unsuccessfully with redwoods, is sometimes the method used in felling walnut trees where the root and butt are valuable for manufacture into veneers. It is used also in Europe where the demand for wood is so high that the entire stump as well as the trunk is utilized (stump used for fuel or converted into charcoal).

4. Felling by means of endless chain, circular, or drag saws driven by gasoline or electric power, previously described, has been successfully practiced in all forest regions. These felling machines have not yet been universally adopted, but they may be perfected and used much more extensively.

CLIMATIC INFLUENCES

If logging is dependent on stream driving or winter snow hauling, felling must be done in the late summer or autumn as well as in the wintertime prior to the winter haul and spring drive. On the other hand, deep snows may interfere with felling operations and prevent the cutting of low stumps, as in the high Sierras in California and in parts of eastern Canada and the Northeast. Generally speaking, however, climate interferes very little with logging operations. On the Pacific Coast and in the South, the two regions where most of our lumber is produced, the climate has very little influence on the continuance of logging except for heavy rains which make truck hauling and skidding difficult and sometimes impossible.

SPECIAL FEATURES—REDWOOD

In the redwood region of northwestern California, the terrain is characterized by broken topography, sharp gulches or canyons, steep ridges, and the lack of long continuous gentle slopes. Logging is being done in steeper and rougher chances after the cutting of the more accessible timber along the bottoms and lower valleys. This condition is described by the local timber feller as "short ground." The short benches or terraces are also characterized by the frequent occurrence of holes and hummocks. Therefore, bedding is resorted to not only because of the great weight of the large-boled and tall redwoods, but equally as much because of the nature of the topography and the brittle character of the timber.

Fallen and dead timber and stumps also add to the hazards of felling the large trees which are readily shattered. Therefore, "bedding" is provided by laying brush, branches, tops, and such along the line of fall so that the impact of the falling tree will be checked, and stumps and dead and down timber avoided. Thus, the least damage by shattering will result. The position of the tree in relation to the topography, rather than its size, determines whether a bed shall be provided in advance. In general, however, all trees 5′ or more in diameter should be bedded.

For many years before redwood sap lumber was placed on the markets, because of its low comparative value the customary practice was to fell the redwood timber, then peel the thick bark, and burn the logging slash prior to log making or crosscutting. The purpose of peeling and burning was to facilitate skidding and bucking. About the year 1925 clear sap lumber attained relatively high value, and so most of the companies peeled the bark after burning the slash in order to reduce the losses resulting from fire damage to the clear sap grades of lumber.

Two and three men constitute the felling crew in redwoods. Buckers work singly or in pairs as in Douglas fir. Power saws are used in felling and bucking redwoods. The largest trees require 3 to 5 days for felling.

RULES TO BE OBSERVED IN FELLING

Great care must be exercised in felling in order to prevent undue wastage, breakage, and loss of timber. Experienced woodsmen can fell trees with great precision. Felling small timber of the East is a relatively simple operation. In the dense stands of the Pacific Coast forest * where trees attain heights of more than 200′ and diameters of 3′ to 10′ or more, the problem is greatly intensified. Felling trees in the southern pineries and in the ponderosa pine stands of the West is relatively simple because of the open character of the forest, the paucity of trees per acre, the short and open crowns of individual trees, and the generally favorable topography. Felling is made very difficult when trees are large and stand closely together, when the crowns are deep and broad, when the forest floor is interspersed with fallen timber, stumps, and the like, and the topography is rough, as in many parts of the western mountains and Southern Appalachians.

General rules for felling are as follows (local conditions or circumstances may alter these rules):

1. Trees should generally be felled along the contour or uphill on unusually steep topography. On gentle slopes, they may ordinarily be

* The largest known Douglas fir tree, almost 13′ in diameter at the base, estimated to contain 132 M.b.f., was found to be 586 years old. It required 1½ days by a 3-man crew to fell this tree with a power saw.

felled downhill if such a practice is more convenient to skidways or skidding trails. On steep slopes on the Pacific Coast, Douglas fir is felled downhill because of the danger to the men from the felled trees "jumping" and skidding down the steep hillsides. This method is very destructive to the residual stand and often to the stem.

2. Trees should be left in a position to be easily crosscut and skidded. If other conditions are favorable, they should be felled parallel to skidding trails. Much progress has been made in felling southern pulpwood for easy skidding.

3. Trees should do the least possible amount of damage to other trees and to themselves. Many trees are shattered in felling especially in the top of the bole; some are broken in falling on stumps, logs, or rocks, or across gulches and streams, resulting in waste of good lumber. Pacific Coast operators estimate their wastage in felling at 2% to 15% of the merchantable volume because of shatter, shakes, splits, and the like. Berry and Hughes in California found a breakage of 3.4% of the gross log scale in sugar and ponderosa pine.

4. Felling operations should be suspended during periods of heavy windstorms. Windfall is frequent in some regions on exposed locations, and the force of the wind is powerful in influencing the direction of the fall even when the tree is directed with wedges and a proper undercut.

5. Straight trees can generally be felled in any direction. Trees often lean or are forked or heavily branched on one side. This necessitates felling in one of three directions by the careful use of wedges to direct the fall.

STUMP HEIGHTS

There has been great loss of wood in stump heights from the felling of trees in the past. Stumps are estimated to contain 2% to 3% or more of the total volume of the tree. For many years the axe alone was used in felling, and, where deep snows occurred in the woods, stump heights of 2′ to 4′ or more were common in the northeastern softwood forests. Great loss occurs in felling Pacific Coast timber where stump heights are frequently 4′ to 10′ or more above the ground, owing to bell-shaped butts, irregular ground levels, fallen timber, and debris. Western red cedar and cypress have unusually irregular and swollen butts.

A common practice in felling American forests is to cut the stumps at a height not greater than the diameter at the cutting point on the tree. Another custom is to fell the tree just above the root swell. Even those rules involve unnecessary loss in good wood. The best lumber in the tree is usually found in the butt log. Fellers dislike to lean over sufficiently to cut stumps practically at ground level. Cutting at ground level is a common practice in Europe, except where stumps are to be removed later and utilized. Some companies on the Pacific Coast follow the rule that stumps must be cut at least lower

FIGURE 41. New method of undercutting with chain saw. This is known as a Humboldt undercut which is taken from the stump instead of the first log, thus saving some of the best lumber in the butt log.

than the surface of down logs. This generally means 2′ to 5′ or more above the forest floor.

The Humboldt Undercut. With the development of mechanized felling devices, especially with the Atkins electric-powered chain saw as applied to West Coast woods, the Humboldt undercut, which is taken out of the stump rather than from the butt log, has assisted materially in saving good wood which would otherwise be wasted. This means a saving of as much as 3% to 7% of the best lumber taken from the butt log of the large-size trees. This method, first developed with redwood and later with ponderosa pine and Douglas fir, was used in 1936 in felling large-size white pine of 26″ to 38″ d.b.h. in the Adirondack Mountains of New York. At that time, the so-called Humboldt undercut was chopped out of the stump with felling axes rather than with chain or other types of mechanical crosscut saws. Possibly other operators may antedate this. It is one further evidence of more complete utilization of the trees felled in the woods.

TABLE XVIII

STUMP HEIGHTS

Generally Used in Cutting National Forest Timber

National Forest Region	*Name of National Forest and State*	*Maximum Stump Height, inches*	*Species*
9	Superior—Minnesota	12	Norway, jack and white pine, balsam fir, aspen
2	Shoshone—Wyoming	12	Lodgepole pine, Douglas fir
3	Kaibab—Arizona	14	Ponderosa pine
4	Payette—Idaho	16	Ponderosa pine, Douglas fir
5	Sierra—California	18	Ponderosa and sugar pine
5	Stanislaus—California	16	Ponderosa and sugar pine, white and red fir, incense cedar
6	Olympic—Washington	24	Douglas fir, western hemlock, Sitka spruce
6	Rogue River—Oregon	16	Ponderosa pine
8	Ouachita—Arkansas	12	Shortleaf pine, white and red oak saw logs
	" "	8	Pulpwood
	" "	6	Posts
1	Kaniksu—Idaho	14	Western white pine
1	Kaniksu—Idaho	16	Western larch, Douglas fir, ponderosa pine, western red cedar

LOG MAKING (BUCKING)

Log making or bucking consists of cutting the felled trees into sections or logs preparatory to skidding or minor transportation. It is done largely by chain saws used in felling operations. Owing to skidding long and tree lengths, much less bucking is done in connection with felling operations than formerly. More bucking is done at landings, as described elsewhere in the text.

SWAMPING AND LIMBING, NOSING, AND ROSSING

Swamping * and limbing (also called knot bumping) are synonymous terms used on many logging jobs especially in the East, South,

* This term also refers to cutting out skid or haul roads.

and Northeast. They refer to cutting off the limbs and large conk knots preparatory to bucking and skidding. When two men are employed in felling and bucking, one man often limbs the bole while the other measures off the log lengths. The character of knots disclosed in limbing often influences the length of log to be cut. For example, if a decayed condition is disclosed in limbing, the log length may be cut back from a 16′ to a 14′ log or shorter, thus raising the grade of lumber to be produced although losing in total volume. On some Pacific Coast logging jobs no limbing is done except at the landings, because in power skidding the branches are largely broken off.

Swamping also refers to brushing or cutting out the brush or trees that may interfere with crosscutting the stem or with skid lines for horses, and less frequently for tractors.

Nosing is also done by the felling crew. It consists of rounding off the front ends of logs to be hand-logged (ball hooting) down steep slopes or for direct ground skidding by horses or tractors. Nosed logs do not "dig in" so readily where the bottom is soft or wet. Hardwood logs are sometimes nosed to facilitate skidding.

Rossing, barking, or peeling of the bark is practiced with cross ties, poles, and, to some extent, with pulpwood and saw logs. When the latter two are chuted, the logs slide much more readily. The lower side of hardwood and western white pine logs is sometimes rossed to facilitate sliding on hand-logging operations as in the Southern Appalachian Mountains and in Idaho.

PURPOSE OF CROSSCUTTING

The entire volume of the trees is seldom merchantable. It must be cut into sections that can be most economically transported and converted into salable products. Generally speaking, the purposes of reducing the bole into sections are:

1. Reduction of weight. Generally it is impossible or uneconomical to transport the entire tree to market. The weight of green lumber represents only one fourth to one third of the weight of the tree as felled. The trunk must be cut into sections that can be transported most economically.

2. Elimination of defects and unmerchantable portions of the bole. The small, knotty, and otherwise defective, tops cannot be profitably transported and manufactured into salable form. Crooks, decay, shaky butts, as in western larch, must be avoided when one is deciding on the location and length of logs to be crosscut.

3. Adaptation to method of transportation and manufacture. Power and tractor skidding and railroad haul permit the use of long lengths which can be cut and transported most economically, especially in southern pine timber and the large trees of the Pacific Coast region. Sawmills are sometimes equipped with carriages for sawing long timbers. Tractor skidding of long lengths is now generally used in nearly all forest regions.

4. Adaptation to market requirements. Logs are usually cut to suit either special orders or the general market conditions for lumber. This means that most of the logs, except on the Pacific Coast, are crosscut in the woods or at the mill to 16′ lengths or shorter. A large proportion of lumber is sold in lengths from 10′ to 20′.

FACTORS WHICH DETERMINE LOG LENGTHS

Logs are generally cut to 16′ lengths in the woods (except on the Pacific Coast) or crosscut to this length in the log pond or at the top of the jack ladder in the sawmill. There are many exceptions, however, to this general rule. Boards and dimension are customarily sold in our markets in even foot lengths from 8′ to 16′. Timbers and special orders are generally cut to longer lengths. Factors which may determine the various log lengths as cut in the woods operations are:

1. Customary demands of the market. Timbers are generally required in long lengths. For specified long lengths high prices are secured, especially export timbers which are frequently sold on a cubic average basis. Generally speaking, attempts are made to secure logs in the woods in 16′ lengths or multiples thereof, such as 32′, 48′. A large share of the demand is for boards and dimension up to 16′ lengths which make up about 70% to 75% of the total lumber shipped to our markets. As these are required in 2′ intervals generally up to 16′, it naturally follows that most of the logs should be cut to these lengths to supply this demand. Special products such as cross ties are cut to 8′ to 9′ lengths or multiples thereof. Logs marketed according to the old Adirondack standard in New York may still be cut to 13′ lengths. Logs for pulpwood are generally cut in multiples of 4′ lengths. Logs cut for poles or piling may be from 20′ to 90′ or longer.

2. Special orders. For walking beams and other specifications for oil derricks, barges, bridge timbers, and sizes to fill mine specifications, special sizes are required. Logs must be cut in the woods to conform to these specifications. Some timbers are cut in the Douglas fir region up to 90′ or longer. For a mill at Ostrander, Wash., lengths up to 100′ to 160′ have been cut. Disconnected car trucks are used to bring the logs to the sawmill from the woods operations. These require special logging and mill equipment. Logs intended for the veneer industry, known as peeler logs, are cut in special lengths for turning on veneer lathes. Export logs such as yellow poplar and oak are cut in stated lengths. For certain types of apartment-house construction in New York, Chicago, and Boston, long dimension is required, and sawmills specializing in this type of output see that their logs conform to these special orders.

3. Limitations of equipment and transport facilities. Most sawmills and logging operations are limited in the length of logs which they can saw or transport. Log cars are generally limited to 40′ log lengths. For extremely long lengths, double or even triple log cars may be used or disconnected trucks, as for Douglas fir. One company uses 70′ log cars; others 50′ cars. One large sugar pine organization in California uses small log cars on a narrow-gage railway which permits the use of only 16′ logs or shorter lengths. For the Long Bell mills at Longview, Wash., lengths up to 64′ are cut. Saw carriages and the feed and setworks in some mills are often determining factors in log lengths. The tendency within recent years in cutting virgin forests has been to enlarge the transportation or sawmill facilities so that long lengths may be brought in from the woods on account of the economy in logging of tree or long lengths (double, triple, etc., the usual log lengths), particularly in the southern pine and the Pacific Coast regions. For motor-truck logging, lengths are seldom above 25′ to 40′, and usually 10′ to 20′, especially in the South and East.

4. Method of logging. The extended use of railroads as a transportation medium from the woods to the sawmill and the development of power skidding have generally increased the length of logs brought in from the woods, notably in virgin forest operations. Long log lengths, however, cannot be loose floated or driven on streams satisfactorily, but with the decreased use of streams for transportation this factor has greatly minimized in importance. Flumes and chutes as in Idaho may provide only for stated maximum log lengths and may be a factor in determining lengths to which logs are cut in the woods. Pulpwood lengths in the Northeast and eastern Canada are frequently controlled by the stream-driving facilities.

5. Skidding costs and methods. Tractors as in Idaho are used to skid logs in double or triple lengths which are crosscut at the landing prior to the main log haul.

6. Miscellaneous. Local laws governing the lengths as in cordwood in North Carolina may have a bearing on log lengths. Logs that are to be squared for ocean transport may be required in special lengths. The butt logs of western larch are often cut short and left in the woods, as they will not float. Butt logs of sugar pine often sink in the log pond and are, therefore, cut to short lengths and sent into the mill on trucks from the log pond or storage area rather than by the usual method of floating in the log pond until hoisted into sawmill on a jack ladder.

ALLOWANCE FOR TRIMMING

Generally allowances of 2″ to 4″ are made in cutting small timber in the Southeast, the Northeast, the Lake States, and the Southern Rocky Mountains. For example, 16′ logs are actually cut to 16′ 4″ or 10′ logs may be cut to 10′ 2″. This is done to provide for uneven sawing, abrasion in logging, checks, and so on. In the large timber of the West, 4″ to 10″ are often allowed for logs up to 40′, and 12″ to 20″ in logs up to 80′ or more in length. Sometimes an allowance of an arbitrary number of inches is permitted for each 16′ in long lengths.

Because of severe abrasion, brooming, and damage, logs that are to be floated or chuted are generally given a more liberal allowance than those to be transported by rail or truck.

Pulpwood logs 8′ to 16′ long are crosscut at the pulp mills to the desired lengths, but the entire length is utilized in the pulping process. Saw logs, however, are trimmed at the sawmills to even foot lengths after being manufactured into lumber. Timbers are not customarily trimmed when cut in the mills. Cross ties are generally cut plump in length without special allowance for overrun.

In the large and long lengths cut in the South and West, allowances of 12″ to 30″ are usual. A definite allowance is made for each 16′ length.

Top diameters and conditions which determine them. The extent to which logs may be cut in the tops depends chiefly on the species and market conditions. In pulpwood operations tops are often cut to 3″ to 6″; species such as white pine are often cut to 6″. Top logs generally contain a large number of knots, and frequently the lumber produced from them is of such low grade that it does not pay to transport them to the sawmill. There has been a gradual tendency to cut smaller top diameters during recent years in order to effect the most complete utilization, especially under favorable market conditions.

On the Pacific Coast top diameters are cut down to 10″ in ponderosa pine, 10″ to 20″ in Douglas fir, 12″ to 16″ in sugar pine, and 12″ to 20″ in redwood. Hardwoods are generally cut to much larger top diameters than eastern softwoods, because of the irregular boles and large knots found in the tops. Southern yellow pine is cut to 6″ to 10″ in the tops, depending on market requirements and price conditions. Table XIX shows a list of the top utilization requirements of representative timber sales in the important National Forest regions.*

TABLE XIX

National Forest Region and State	*National Forest*	*Minimum Top Diameter, inches*	*Species*
9—Minnesota	Superior	6	White and Norway pine saw logs
9—Minnesota	Superior	4	Pulpwood-spruce, balsam, aspen
2—Wyoming	Shoshone	8	Lodgepole pine, Douglas fir
3—Arizona	Kaibab	8	Ponderosa pine
4—Idaho	Payette	7	Ponderosa pine, Douglas fir, Engelmann spruce
5—California	Sierra	10	Ponderosa and sugar pine
5—California	Stanislaus	10	Ponderosa and sugar pine, incense cedar, white and red fir
6—Washington	Olympic	12	Douglas fir, western hemlock, Sitka spruce
6—Oregon	Rogue River	7	Ponderosa pine
8—Arkansas	Ouachita	8	Shortleaf pine
8—Arkansas	Ouachita	10	Hardwoods
8—Arkansas	Ouachita	4	Pulpwood
8—Arkansas	Ouachita	3	Posts
1—Idaho	Kaniksu	6	Western white pine
1—Idaho	Kaniksu	8	Western larch, ponderosa pine, Douglas fir, red cedar

Timber-sale contracts on National Forest require that all trees shall be utilized to as low a diameter in the tops as practicable and usually to a minimum specified diameter for each species. Merchantability of tops depends on length and content as well as diameter.

FACTORS OF COST IN FELLING AND LOG MAKING

Several factors affect the costs of felling and log making. These may be briefly summarized as follows:

1. Overrun. The amount of overrun depends on the log scale used. The scale is really an expedient for arriving at the approximate log content, because log scales generally fail to record the actual content of the log. However, it is generally unavoidable, because logs may be con-

* Supplied by Ira J. Mason, Chief, Division of Forest Management, U. S. Forest Service, Washington, D. C.

verted at the sawmill into timbers, boards, mine props, cross ties, and so on, the character of which may determine the actual lumber scale. Overrun results more favorably in small-sized logs than in large ones. There is often a large variation in overrun up to 25% or more as between log scale and mill outturn scales; therefore, overrun has an important bearing on the cost of felling and log making.

2. Breakage. Serious amounts of breakage reduce the sawing output and, therefore, advance the cost of logging. The greatest percentage of breakage generally occurs in felling large trees in rough topography and wherever there are many windfalls. Breakage in species that are readily shattered in felling, such as redwood, may be as high as 25% or more.* Breakage occurs in small tops and particularly under conditions of irregular crowns. The loss from breakage is generally small in southern, western white, and ponderosa pines and most hardwoods.

3. Slope and windfall. These have a very important bearing on the cost of felling and log making. Where it is necessary to erect springboards and chopping platforms or where it may be necessary to clear away brush and windfalls prior to felling, the cost may be substantially increased.

4. Effective hours per day. This depends on the distance to work and the standard length of the working day. With the regular 8-hour day, men seldom put in more than 7½ hours of effective work and frequently still less. Figures of cost should be based on output measured on a per-hour basis.

5. Contract or day work. Contract work is always cheaper because of the absence of compensation insurance charges and willingness of the men to work more efficiently and for longer hours per day. Output per hour or per day often varies as much as 4 to 6 when day work is compared with contract work.

6. Amount of defect or cull. The amount of effort involved is practically the same in felling and bucking two trees of the same size, one being sound and the other partially defective. The net output per unit of time may depend very largely on the cull factor. This factor is very large in overmature hardwoods in both the East and South. In western softwoods, it may vary from 10% to 30%. The cull factor in

* E. Fritz of the University of California reports that breakage in redwood averaged 20% of the sound volume in a study involving more than 4 million b.f. Breakage was largely due to crossing—that is, a heavy tree falling over another and either one breaking the other.

western hemlock in northern Idaho is so large as to make costs of logging this species prohibitive.

7. Log lengths. The size factor which determines the output is the scale of the log rather than the dimensions from which that scale is derived. Cutting logs of a size that average 10 per thousand board feet into 20′ lengths results in an output considerably above that for 7 logs per thousand size when cut into 16′ lengths.

8. Softwoods and hardwoods. Many hardwoods such as oak and hickory cost more to saw per unit than white pine and other softwoods of the same size.

9. Large trees such as Douglas fir and redwood are cheaper to fell and buck, per thousand board feet, than trees of small dimensions.

The following are examples of actual output and costs in some representative regions:

A two-man felling crew in virgin Douglas fir scaling 10,055 b.f. per tree will fell 40,000 to 50,000 b.f. per day with handsaw, or about 4 to 5 trees per day. A much larger volume is produced by power saws. The felling crew works separately from the bucking crew. No limbing is done on some power cable skidding because the limbs are generally knocked off in skidding and loading.

In North Carolina on some operations a two-man crew will cut an average of 118 logs (12′, 14′, and 16′ long) per day. This second-growth pine timber averages 3 logs per tree. No trees are cut under 15″ in diameter at the butt, and logs are taken to a 10″ top diameter.

On some large operations in southern pine, especially in Louisiana, Texas, and Mississippi, the felling crew consists of 15 to 20 saws under the immediate supervision of a scaler who serves as woods or felling boss as well as scaler. The term "saw" refers to two men; thus the saw crew of 30 to 40 men works under the supervision of the scaler who lays out the work ahead of the skidding crews. On some operations, the felling crew must hire a filer and water boy. All felling is paid for by contract.

In the northeastern spruce pulpwood operations, felling and bucking are generally two separate processes, when wood is peeled. Felling is done by two-man crews: one man chops and fells the tree (by chopping entirely or with chain saw), limbs the bole to a 3″ top, and cuts into 16 and 20′ lengths, while the other peels the bark. Two good men can fell and peel 4 to 4½ cords per day in average cutting conditions of timber and topography. Felling is done from about May 1 to August 1. Bucking is done by one- and two-man crews. Two men can buck and pile 6 to 10 cords of 4′ peeled wood per day. Skidding is done with one horse between the felling and bucking operations. Therefore, bucking is done at the skidways. Pulpwood is piled along the winter haul roads ready for hauling to the stream landings by sleds or trucks. Cutting (felling) and peeling, skidding and bucking are generally paid for on a piece basis per cord. Four men can generally buck and pile as rapidly as two horses can skid in the 16′ and 20′ lengths of peeled logs to the skidway.

In cutting long-length pulpwood (12′ to 16′), unpeeled, cutters average about 3 cords per man-day for chopping (felling and bucking) in spruce stands of 11 cords per acre. During the winter the average has been increased to 3½ to 4 cords per man-day. On these operations one man and one horse will skid 5 to 8 cords per day on an average skidding distance of 300 to 400 ft. Thus, two cutters will keep one man and one horse busy skidding. Skidding follows felling and bucking, whereas with 4′ peeled wood it comes between the peeling and bucking operations. With power saws, the volume may be increased two to three times with efficient workmen and saws.

Softwoods are felled more cheaply than hardwoods and large trees less expensively than small trees, expressed in unit production as thousand board feet or in cords.

ORGANIZATION OF FELLING AND BUCKING CREWS

In general, felling and bucking are done by two-men crews except on the Pacific Coast. One man notches the tree by making an undercut while the other prepares the ground for felling. Then the two men fell the tree with a crosscut or chain saw and buck the stem, one man first measuring off the log lengths while the other limbs the bole and cuts away the brush or trees (swamping) which may interfere with the work. Sometimes three men are used. One foreman supervises the work of several felling crews and lays out the work in advance. The equipment of each felling crew generally consists of two double-bitted felling axes, one crosscut saw or chain saw, two or more wedges, saw files, a bottle of kerosene, and a measuring stick or tape.

In addition to this equipment for eastern and southern logging jobs, the fellers on the West Coast require several springboards and, in large redwood timber, a gunstick. In Douglas fir, an axe is generally used to point the direction of the fall.

The felling-crew unit on the Pacific Coast may consist of a set made up of two fellers and two buckers. Felling there is much more complicated and difficult, owing to the huge size of the trees and the density of the timber. It may require from several hours to a day or more to fell and buck one tree. A frequent procedure in Douglas fir is as follows: The cutting crew may consist of 16 men, namely, a scaler who supervises and checks on the work, 5 sets of fellers (10 men), and 5 buckers. In cutting long logs up to 64′, there are generally twice as many fellers as there are buckers. Buckers use a steel tape to measure off the logs and work singly. An allowance of 8″ of trim is made in 12′ to 24′ logs, 10″ in 26′ to 40′ logs, 12″ in 42′ to 56′ logs, and 20″ in 56′ to 80′ logs. Fellers are paid for all the tree that is sound, but buckers only for the logs sawed. No snags are paid for under 80′ or a specified length. Fellers are not paid for windfalls or "knock downs" or for shattered trees that could have been felled properly. One timber foreman and one assistant supervise the cutting and laying out of the work for 8 to 16 cutting crews.

The power-saw crew generally consists of three men and, when electric power is used, a tractor operator.

DETECTION OF METALS IN SAW LOGS

Army mine detectors from the last World War have been successfully employed to locate nails used in turpentining southern pine, spigots used for tapping sugar maple, and other forms of metals encountered in cutting timber during felling operations. In some cases, individual sawmills have encountered as many as 32 pieces of metal within a month. Considerable time is lost as a result, and also saws are badly damaged. Detection is sometimes made on the log decks at sawmills, but also on logging operations, particularly at landings.

There are two types of metal detectors. One type flashes its warning on a dial, whereas the other has an attached earphone which sounds and warns of any metal in saw logs. Many of these metal detectors are used in southern logging and sawmill operations.

PART III
MINOR LOG TRANSPORTATION

CHAPTER VI
Log Transportation

GENERAL CONSIDERATIONS

Logging is pre-eminently and fundamentally a problem of transportation, one of moving a bulky and heavy product of the forest—logs—from the stump to the mill for further conversion with the least possible financial outlay. Adequate control of transportation charges is, therefore, of utmost importance. High-cost logging operations are frequently unable to compete successfully with those of low cost. These logging costs may vary as much as 40% below to 100% or more above the average of a large number of representative logging operations in a given region. This may mean that some loggers operate at a considerable loss while others operate at a profit.

The methods to be used in bringing in the logs must, therefore, be chosen with great care, if the expense is to be kept at a minimum. As a preliminary, information must be collected about the amount of standing or available timber, the stand per acre, and the size of the average and largest trees; the topography must be thoroughly studied, and the engineering and financial policies must be reviewed. With the facts at hand, the plans for logging the tract profitably can then be drawn up.

Logging operations are generally classified for organization purposes and for cost accounting into four main divisions:

(*a*) Logging camps and supplies.

(*b*) Felling and bucking.

(*c*) Skidding (log assemblage)—minor transportation.

(*d*) Log transportation—major transportation.

The chief item of cost in these operations is for the transportation of the logs from the woods where the trees have been felled to central points known as skidways, landings, log dumps, log decks, rollways, and so on, and thence to the manufacturing operations or direct to the log market. Wood is an extremely heavy and bulky commodity in proportion to its value; therefore, transportation costs are relatively high. Transportation generally comprises 60% to 70% or more of the total logging cost, the greater share of the transportation being devoted to the major haul and the smaller part to the minor haul or the skidding operations. From 30% to 40% of the logging costs are generally distributed among felling and bucking charges not including such items as interest, insurance, taxes, overhead, and camps.

It is a cardinal principle in logging operations that logs should follow a downhill haul from the place of felling to the log storage at the sawmill or point of delivery. There are, however, notable exceptions to this general rule. High stumpage values may justify unusually expensive transportation charges such, for example, as the million-dollar railroad in northern Idaho which reaches an interior valley by a roundabout road via another valley and over a divide in order to get out valuable western white pine timber.

The outstanding factors which should be carefully reviewed in selecting the method or combinations of transportation best adapted for the particular logging operation are briefly summarized as follows:

1. Total stand of timber available with expected cut per acre and by species.
2. Size of sawmill or market to be served with daily and annual volume of log supply required.
3. Available capital for both initial investment and carrying charges as well as general financial program such as retirement of bonded indebtedness and compensation insurance.
4. Topography of the tract with natural outlets for the timber.
5. Relation to adjacent timber resources and operations.
6. Available transportation facilities such as common-carrier railroads, highways, streams, and climatic factors influencing their use.

The removal of the best and most accessible stands of timber and the demand for year-round production, together with the development of the heavy timber stands in the Far West, have served to change former methods of log transportation rapidly.

Transportation from the stump to the sawmill or market may be divided into two parts:

1. Log assemblage or minor transportation, generally known as skidding. It is also locally known as yarding, bunching, snaking, twitching, and so on. Skidding consists of bringing the logs together at a common point, a convenient skidway, landing, or deck at distances generally up to 1200 ft. and sometimes over 2000 ft. or more.

2. The main log haul or major transportation consists of hauling the logs from skidways, landings, or log decks to a destination which may

be a few miles to 100 miles or more distant. There may be two or more forms of major transportation on the same logging operation.

For more than 200 years the customary means of transportation were the streams, particularly along the Atlantic Seaboard, in eastern Canada and in New England and the Lake States. The rivers of these sections were the great highways of log transportation. Streams at the headwaters of the main rivers of the Northeast and the Lake States were used to flush the logs during the spring freshets to mills located along the watercourses or near the lumber markets. Later, rafts were commonly employed on the upper Mississippi River and its tributaries as well as on the Ohio, Allegheny, Delaware, and Susquehanna Rivers for white pine, and then on the lower Mississippi and the coastal and Gulf streams for yellow pine and cypress. The advent of large quantities of hardwoods during the later part of the nineteenth century on our lumber markets necessitated a different method of log transportation, as oak, birch, maple and most of the other hardwoods would not float satisfactorily.

Furthermore, the removal of the accessible timber stands sometimes remote from the drivable or floatable streams required different methods of transportation.

The development of the far western forests changed log-transportation methods radically. The large timber could not be successfully driven or floated down the smaller mountain streams. Disastrous results attended the attempts to use Michigan log-driving methods on the West Coast, notably the attempts to drive sugar pine on the American River in the Sierra Mountains and Douglas fir in the Wind River section of southern Washington. Where timber was located adjacent to the large streams and estuaries and bays of the Pacific, such as the Columbia River, Puget Sound, Grays Harbor, Willapa Bay, and their tributaries, economical means of transportation were accessible and are still widely used for rafting logs to tidewater mills. Much of the standing timber now lies at the higher elevations and remote locations and requires railroad transportation facilities to move logs to the mills or to tidewater and thence by raft to the mills.

Stream transportation is generally the least expensive wherever available. From the leading position in log transportation in the early days it has almost disappeared from American logging practice except for floating pulpwood in Canada and the Northeast, driving and floating in northern Idaho, and driving cross ties in Wyoming. Rafting is widely practiced in the Southeast and the Northwest, notably the latter.

Railroads and, still more recently, motor trucks, have come prominently into the industry as methods of log transportation. The motor truck has been developed to a high degree of efficiency in log transportation and has become the most important single method in all regions except in Alaska and eastern Canada.

FACTORS WHICH DETERMINE SELECTION OF METHOD OF MINOR TRANSPORTATION (LOG ASSEMBLAGE OR SKIDDING)

Cost, measured in terms of charges per thousand board feet, is the paramount consideration in determining the method of transportation. Experience has generally indicated the best practice for each logging chance. The balance of costs between minor and major transportation must be carefully considered. Unduly long skidding hauls may result in excessive charges. Experienced loggers can readily determine the extent to which truck roads or spur tracks on a logging railroad must be built to obviate the necessity of disproportionately long-distance power or tractor-skidding hauls. The factors which must be considered in determining best skidding procedure are numerous, but they may be summarized briefly as follows:

1. **Size and character of timber.** Animals have long ago proved unable to cope successfully with the enormous weight of logs on the Pacific Coast. This condition resulted in the development and use of power skidding devices and tractors. Some timber types are characterized by heavy brush and windfall which may influence the selection of the method used.

2. **Topography.** Animals are used with great difficulty on steep rocky hillsides. Sometimes the topography is too precipitous and irregular for tractors, and steam or horse skidding must be used. In swampy locations power skidding, mud boats, or tractors may be required.

3. **Silvicultural considerations or forest-management policies.** On some National Forest timber sales, tractor skidding is required. In some stands of timber, the tractor has replaced power skidding in order to protect the young growth and insure the reproduction and proper continuance of the forest as well as economy in logging.

4. **Climatic considerations.** Skidding can be done much more cheaply over snow than on the bare ground. Travoys and sleds are often used for skidding in winter where snow can be depended on.

5. Distance to major system of transportation. Frequently two different methods of skidding are employed such as both tractors and high-lead logging. Two high-lead setups are sometimes required to bring in the logs to the major system of transportation. Generally, tractors have replaced animals in all regions where the latter were formerly used and have replaced cable skidding on many operations on the West Coast.

FACTORS WHICH DETERMINE SELECTION OF METHOD OF MAJOR TRANSPORTATION (MAIN LOG HAUL)

As indicated elsewhere, major transportation on the largest operations has been largely dominated by railroad and motor-truck hauls. On small-mill operations, trucks are almost universally used. Although these are of outstanding importance, the factors which must be taken into consideration in selecting the method best adapted to the conditions are:

1. Costs. The decisive factor in logging operations is always the cost. It is often a serious and difficult question to decide which is the least expensive method to adopt. For example, there must be a large and constant flow of logs to justify the construction and operation of the logging railroad. Pulpwood in eastern Canada is generally driven or floated at low cost to the pulp mills.

2. Size, length, and weight of logs. Long lengths are often cut in Douglas fir and, to a less extent, in ponderosa pine and some other species. These generally require a rail haul or heavy and powerful trucks for a large volume flow of logs.

3. Availability of drivable or floatable streams. Water is the cheapest form of transportation, if available. The use of log rafts has practically disappeared except in Douglas fir on the Pacific Coast and occasionally in the streams of the Southeast. Stream driving, the principal method of major transportation for many years during the early days of logging, has almost disappeared except in the Northeast, eastern Canada, and northern Idaho because the timber has all been cut along the streams and has been replaced by the use of trucks. It will probably always be used, however, in Maine, New York, and eastern Canada.

4. Climatic considerations. Snow haul is relatively inexpensive. Wherever snow is available, sleds laden with logs and hauled by horses or tractors provide an economical method of log transportation. This method has practically disappeared, however, except on a very few large operations in the Northeast, eastern Canada, the Northern Rocky

Mountains, and the Lake States. Sleds are devoted largely to the haulage of pulpwood. This is due to the practical exhaustion of the large virgin stands in these regions. The snow haul is still, and probably will continue to be, the chief method of land transportation in eastern Canada on the many large pulpwood operations. Chute logging can be practiced more cheaply in the winter than in the summer.

5. Existence of transportation media such as common carriers and improved highways. Many logging operations take advantage of existing railroads or those used in connection with mining or other operations to haul logs or supplement their own railroad operations. The availability of an improved highway system may be a vital factor in using motor trucks.

6. Distance to the sawmill. If a short haul is involved, tractors or motor trucks may serve the purpose better and less expensively than other transportation media.

7. Character of the timber. Hardwoods generally do not float; they require motor-truck or rail haul.

8. Size of operation. The amount of standing timber, the daily output of the mill, and the length of the future cut may be determining factors. Installation of logging railroads must be justified by a great volume of reserve timber and the necessity of supplying a large daily flow of logs to the mill throughout the year.

9. Topography. Rough topography may require chutes in the steep ravines. On some operations, flumes may be the only economical means of transportation compared to the more costly railroad or motor-truck hauls, particularly if costs of road construction for these methods may be excessive or prohibitive.

10. Value of the stumpage. One can install relatively high-cost methods of transportation if the returns on the product justify it. Individual walnut trees are often logged and hauled for 100 miles or more to a veneer or lumber mill. The great expense involved is justified by the high value of walnut.

11. Demand for year-round manufacture. Most of the large modern mills require year-round manufacture and, therefore, a constant flow of logs to the mill. This condition generally necessitates railroad or motor-truck transportation, or both.

HISTORICAL PERIODS IN THE DEVELOPMENT OF AMERICAN LOG TRANSPORTATION

For many years during the development of the American Nation, log transportation continued on a stable basis and followed established methods. Transportation was dependent very largely on the streams as a medium, both for rafting and driving. This was especially true of the readily available white pine and spruce in the North and Northeast. Rafting of southern yellow pine in the sluggish coastal streams of the South was the common practice for many years during the colonial period and later. Then came a period during which methods began to change, and within recent years improvements have followed one another with great rapidity, chiefly as the result of the development of mechanical devices.

There has been considerable overlapping of transportation methods from a historical viewpoint, and probably this situation will continue for many years. Transportation methods which are largely dependent on the kind and size of trees, climate, topography, stands per acre, and volume of log supply must be constantly analyzed in order that the cheapest possible method of bringing in logs from the woods to the mill may be devised. A historical background often furnishes a proper perspective from which to review and determine the best methods applicable to given local conditions.

Three significant factors have been of considerable influence in these constant changes:

1. The extensive remaining volume of large-sized virgin timber on the Pacific Coast. The most important and vital changes have taken place in that region.

2. The rapid regrowth of timber in the South, notably in the southern pine, and adjacent forests in the Southeast. Rapid regrowth is found in many sections of the East and North.

3. Improved mechanical devices influencing the cost and speed of log delivery, and the condition in which the remaining forest is left.

The following are the approximate periods or eras made noteworthy by special features in the development of American log transportation:

1. The colonial and nineteenth century period characterized by stream driving and floating. Primitive and inexpensive methods of logging, using accessible and readily available stream flows, were predominant. The winter sled haul in the North and the frequent use of oxen were peculiar to the earlier stages of this development. The winter sled haul and stream driving are still the prevailing systems of log transportation in the pulpwood operations in eastern Canada, Maine, and New York, as they were 100 to 200 years ago. In the South, skidding by oxen and mules, wagons drawn by these animals and horses, and floating on rafts in both pine and hardwood forests continued for many years. The latter part of this period was marked by the coming of the railroads, the elimination of oxen to handle the big timber of the Far West, and the increased use of rafting in the Southeast and the Northwest.

2. Power skidding and railroad logging era, approximately 1890 to 1930. With the waning timber supplies in the white pine forests of the Lake States, logging was developed on a vast scale in both the South and Northwest, which rapidly became the centers of lumber production. The logging railroad dominated transportation, particularly to tap remote and formerly inaccessible virgin stands of southern pine and the big timber of the Far West. Timber adjacent to driving and floatable streams became exhausted except in the Northwest. Power skidding, which was first developed on an experimental basis in the Lake States in the latter part of the nineteenth century, was further improved, both in ground and overhead skidding methods, especially in the South and in the big timber region of the Northwest. Animals were gradually replaced. Lumber reached its peak production during this period in 1909. The motor truck first came into use on a small and experimental basis in 1913. The tractor, which had been used to a very limited extent, was gradually improved, but it remained for the World War to develop the tractor for efficient and reliable operation. During this period, practically all the large logging jobs in the country used the railroad in order to supply a large and constant flow of logs to the mill, irrespective of climatic conditions. The year-round sawmill replaced the sawmills of the previous era which were dependent largely on the logs being floated or rafted during spring freshets.

3. The present era. This was made notable by the development of the tractor and motor truck for logging purposes; improvements in felling, bucking, and loading devices; and the virtual disappearance of animals except horses used in short-distance skidding and log hauls. Small sawmills especially in the South and throughout the North and East, changed methods of logging very materially. On the larger operations both in the Far West and in the South, the railroad was still used on major log haul; stream driving still persisted in the Northeast, eastern Canada, and northern Idaho. Power skidding still continued in the heavy stands of Douglas fir, redwood, western hemlock, Sitka spruce, and western red cedar, but decreased in importance. The motor truck and tractor came prominently into wide usage throughout the country concurrently with the development of better systems of highways and the improvements of mechanical efficiency. The logging railroad continues to a limited extent only and is used only on large operations in virgin and large Douglas fir, ponderosa pine, redwood, and other trees of great size, and to a much less extent in Southern pine. Both the tractor and motor truck have been widely used as auxiliary devices to power skidding and the logging railroad. They have, moreover, gradually replaced the power skidding methods, especially in California and in many parts of Idaho, Washington, and Oregon.

MINOR TRANSPORTATION—LOG ASSEMBLAGE

PURPOSE AND OBJECTIVES

Logs are generally assembled from the felling area to central points called skidways or landings. This is usually the less expensive of the two broad methods of transportation. The bunching of logs along major routes of transport may precede or be concurrent with the main log haul. It usually requires little or inexpensive construction or preparatory work. There is a delicate balance between skidding distances and the main log haul. The decision to build additional railroad spurs, haul roads, and so on, or to extend the skidding distance is of great importance. Skidding is generally cheaper up to a given distance for each type of timber, topography, and ground conditions than the construction and operation of main-haul roads.

There are three types of skidding, classified according to the application of motive power, namely by:

1. Animals.
2. Tractors.
3. Cables attached to logs and hauled by power.

The first two are flexible in that they are widely useful for other work in connection with logging. The last is limited largely to skidding but is used to a considerable extent in loading in connection with skidding operations and for "chunking out" and grading logging railroads. With some types of loaders, skidding is done for short distances (up to 150 ft. and occasionally to 600 ft.). The advantages of using animals or tractors are largely the result of their mobility; that is, they may be moved about easily in carrying out the skidding operation; they may change locations readily, moving from one setting to another, thus adapting themselves to selective logging, avoiding impending fires, picking up logs from odd corners and meeting mill or market requirements for special orders.

HAND LOGGING

On steep slopes in small and medium-sized timber, saw logs are rolled or sent down end to end to some point below which they are transported by various devices to a skidway or landing.

Hand logging, generally called ball hooting, is occasionally practiced on the steep upper slopes in logging western white pine in the Northern Rocky Mountains, hardwoods in the Southern Appalachians, and spruce pulpwood in the Northeast and Canada. Distances may vary from a few feet to several rods or more.

Cross ties, pulpwood, extract wood, distillation wood, posts, and fuel wood are often carried up to distances of 150 ft. or more on men's shoulders. Hemlock and oak tanbark is similarly carried to skidways by hand. Sometimes a swiveled timber carrier or lug hook is used by two or four men.

Saw logs are generally nosed and barked on the underside to facilitate sliding down steep slopes or over rocky ledges, especially in the Southern Appalachian Mountains.

Hand sled logging, known as chienning, practiced in eastern Canada, is described elsewhere in the text.

CHAPTER VII

Animal Skidding

GENERAL METHODS AND PROCEDURE

Animals, chiefly horses,* are used for skidding in the East, Northeast, eastern Canada, and the South, especially with pulpwood and saw logs for small mills. For many years animals were used almost exclusively for skidding. They are still found in nearly every region except the Pacific Coast forests, and even there, on small timber operations, horses are used, especially with tie and small mills cutting second-growth stands. During recent years, they have been widely displaced by the tractor, the power skidder, and the motor truck. Animals are unable to skid advantageously the large logs of Douglas fir, redwood, ponderosa pine, and other trees of the Pacific slopes. They are still widely used in both pulpwood and hardwood logging in the Northeast and eastern Canada, and in small-mill operations in the South, the Lake States, and the Rocky Mountains, and they do auxiliary service in getting logs out of odd and remote corners in southern pine power-skidding operations.

Animals are utilized primarily for direct skidding with tongs, grabs, or chains, as well as with devices to elevate the front end of the log such as drays, bummers, mud boats (lizards), single sleds, and travoys. They work both singly and in teams and double teams, depending on the size of logs, distance, and topography.

ANIMALS: RELATIVE EFFICIENCY AND FACTORS GOVERNING USE

Horses are the most commonly used animals. Their advantages over mules and oxen are:

(*a*) They are more active, more rapid, and more intelligent, and they are easily handled.

* For details of their selection, feeding, and care, see *Horses in Woods Work*, by J. W. Sutherland, Woodlands Section, Canadian Pulp and Paper Association, Montreal. Also for details of horse skidding on pulpwood operations in spruce in eastern Canada, see *Efficiency in Skidding of Wood and Handling of Horses*, by A. Koroleff of the same organization, 1942.

(*b*) They are adapted to a wide range of climate, particularly cold weather.

(*c*) They are adapted to fairly rough and rocky ground as well as to level topography. They readily lend themselves to the problems and difficulties of snaking out individual logs as well as chaining and similar operations.

FIGURE 42. Methods of attaching chains and tongs to ends of logs for skidding with horses. The upper view shows the use of logging chains and the lower of skidding tongs. Chain dogs are often used in skidding large hardwood logs on steep slopes because tongs and chains become disengaged. A single horse is used for skidding small logs and teams for larger ones, but tractors are used for long and tree lengths and in all heavy timber stands. Chokers are generally used to attach the cables in skidding long lengths with tractors.

Horses of various sizes are used, from 1400-lb. animals for the smallest logs up to 2100-lb. ones for the heaviest work. The best weight for all-around purposes including hauling and skidding is 1500 to 1700 lb. In addition to skidding, which is their principal use, they are employed on logging jobs for:

(*a*) Hauling sleds in northern winter logging and four- and eight-wheel wagons in the South.

(*b*) Loading logs on wagons and logging railroad cars or motor trucks by the crosshaul system.

(*c*) Toting camp and logging supplies (rapidly decreasing except in eastern Canada because of cheaper haul by motor truck over improved highway systems).

(*d*) Building forest roads with scrapers and wagons (largely supplanted by bulldozers with tractors and other improved mechanical devices).

(*e*) Cable rehaul (also mules) in power ground-line skidding in the southern pines.

In pulpwood logging in Canada and the Northeast, the ratio is about one horse skidding to four to eight men cutting, with seven as an efficient average, depending on the activity and handling of the horses for skidding, size and density of timber, topography, distance, and other factors. On a 10,000-cord job 25 to 29 horses are generally used for skidding compared with 65 to 79 horses for winter hauling in spruce pulpwood operations. There is a wide variation in number of horses used, depending on distance, topography, and so on.

The amount skidded per horse-day varies generally from about 5 to 7½ cords of pulpwood. One horse will skid 2000 to 5000 b.f. of saw logs per day, or more, depending on the species, log sizes, distances, and local conditions.

Mules are used in the South but seldom elsewhere. They are used chiefly for:

(*a*) Skidding long distances.

(*b*) Hauling two-wheel carts and four- and eight-wheel wagons.

(*c*) Rehauling cable on ground power skidding.

Their advantages may be summarized as follows:

(*a*) They can work during severe and continued heat.

(*b*) They can be used for long skidding distances and on long hauls very efficiently in the heat of summer with little time lost for resting.

(*c*) They are less excitable than horses.

(*d*) They cost less to feed, require little care, and eat coarse rough food.

(e) Their initial cost is lower. Their weight varies from 1000 to 1400 lb. each. Missouri is the great center for the raising of mules.

Oxen were widely used in the early days of logging. They are adapted chiefly to primitive conditions in remote forest regions. Oxen weigh 1000 to 1200 lb. each. They have practically disappeared from American logging operations. Their advantages may be summarized as follows:

(a) Their initial cost is low.

(b) They live on rough, coarse food and require little attention and care.

(c) They operate with inexpensive homemade harness and equipment and can be maintained at little expense for food, etc.

(d) They do not mire readily in wet and swampy locations owing to their cloven hoofs.

(e) They are not excitable and are easily handled, even on very rough and rocky topography.

In early days of Pacific Coast logging they were used with two to six yoke or more, but in spite of the low initial investment required and the low maintenance charges they were largely displaced by the power skidder prior to 1900.

SKIDDING DEVICES AND METHODS

Saw logs are skidded chiefly with tongs, grabs, or log chains by dragging them directly on the ground. In order to facilitate skidding, especially in snow, in swampy locations or in rough topography, single sleds (bobs), travoys or drays, lizards (mud boats), bummers (low two-wheel carts), and high wheels were commonly used to elevate the front ends of the logs. All these devices, however, have diminished in importance and general use. The principal forms of skidding devices are briefly described as follows:

High Wheels. High-wheel carts have been used for many years past in the South, the Lake States, Idaho, Oregon, the Southwest, and California. They were primarily used for skidding long distances up to ½ mile but generally only up to ¼ mile. They were especially suitable for assembling logs on level or gently undulating topography in open stands of timber with little underbrush or windfall, such as are commonly found in the southern pineries, eastern Oregon and the Southwest.

They have been almost entirely displaced by the tractor and truck. Though primarily a skidding device, high wheels have also been used for hauling logs considerable distances.

The wheels are generally 7′ to 10′ in diameter with wooden spokes. With large heavy logs, wheels up to 11′ in diameter with steel spokes have been used. Horse and mule teams and tractors furnish motive power. The large hydraulic wheels which were an outgrowth of the two-wheel cart are briefly described in connection with tractor skidding.

Sleds. Various forms of single sleds (known locally as bobs, skidding sleds, drays, log boats, joe-hogs, go-devils, travoys, mud boats, lizards and scoots, as well as log sleds) are used, principally as skidding devices, both on bare ground in the summer and on snow in the winter in northern climates. These sleds are used to elevate the front ends of large heavy logs or a number of small logs, thus decreasing the tractive resistance. They are used for short distances, generally up to ¼ mile; on wet bottoms, under soft ground conditions; and also on steep slopes where loads of one to ten logs may be skidded more cheaply than by direct ground skidding with chains or tongs. The latter are used for small logs, either with or without nosing, for short steep slopes where the front ends of the logs do not dig into the soil unduly. On steep slopes, either with or without snow, the rear ends of the logs dragging on the ground serve to retard the forward movement of the load, thus acting as a braking device. These sleds are generally made in a local or camp blacksmith shop and are often roughhewn from the harder and more durable woods, preferably hardwoods, in order to resist the shocks and strains incident to hauling bulky and heavy logs. Travoys or lizards are sometimes fashioned from the fork of a tree, a bunk being bolted across the two parts near the crotch for supporting the front end of the log. A chain is passed through the fork and around the log and bunk and attached to the team. Cross ties, hardwood and softwood logs, poles, piling, and pulpwood are frequently skidded with these various types of sleds, especially in rugged topography. They are especially used in the Rocky Mountain region, the southern hardwood region, both in the Appalachian Mountains and the bottomland districts, as well as occasionally in the East, Northeast, and Lake States.

Motive power is supplied by single horses, teams, or tractors. In the Adirondacks, bobbing, also locally known as draying, consists of using a single rack 16′ to 20′ long and 3′ wide, the front end of which rests on the bunk of a single sled, the rear end dragging on the ground. Two to three cords or more of 4′ spruce pulpwood are hauled per load, depending on the condition of the road, the steepness of the slope, and the distance. Generally, little if any road building is required for the use of various types of these single sleds. The rear end of this pulpwood rack is shod with a 3/8" flat iron.

This method of bobbing is used for distances up to 3/4 mile, where one horse can haul single racks more easily than two sleds and a double rack can operate, especially where it is not advisable to use snubbers or a Barienger brake. On very steep slopes chaining as described elsewhere is used as an added brake (back of the bob) to slow up the speed

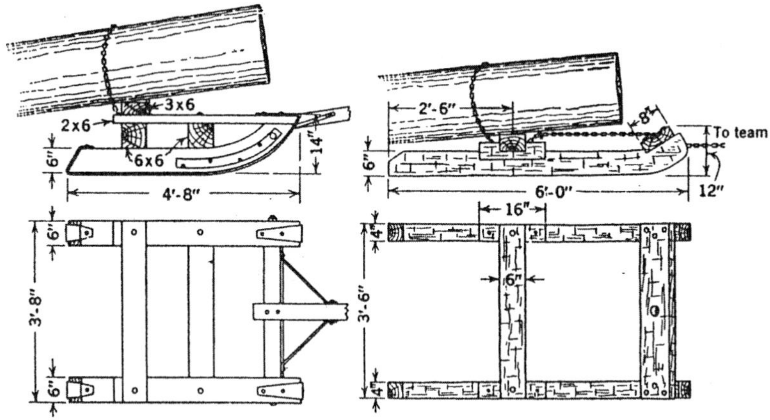

FIGURE 43. Two types of sleds known locally as drays, logboats, bobs, and so on. The front end of the log is rolled onto the bunk of the sled, the rear end dragging on the ground. Power is generally supplied by tractors and horses. Single sleds are used in skidding large logs and for longer distances than for direct ground skidding. They are generally made in the camp blacksmith shop of hewn or sawn hardwoods.

as well as to haul additional wood. When the horse holds back the load on the steepest slopes, two hinged dogs "bite" automatically into the snow roads and retard the speed of the load.

Bummers * are low two-wheel carts sometimes known as self-loading skidders used in bunching logs for wagon, truck, or railroad haul in the hardwood and pine regions of the South. Some bummers are loaded by the cross-haul method. They were formerly widely used in Idaho and other sections of the West, as well as in the Lake States and Adirondacks, but they have been replaced to a large extent by motor trucks going directly to the felling area during dry seasons, especially in the South, and by horses and tractors during the rainy season and

* Track-type bummers (crawler-wheel wagons) are described in connection with wheeled vehicles in major transportation.

in wet locations. Bummers and high-wheel carts have been largely replaced by Fairlead arches hauled by tractors in the West.

Trailing. This refers to the practice of trailing, lengthwise, one log behind the other, generally down long steep slopes. Two to seven or more logs, connected by chain dogs, are thus hauled in strings. The chain or trail dogs are released by grab skippers or mauls at the main-haul roads. Trailing is practiced usually on steep slopes in skidding 12′, 14′, 16′, and 20′ logs of small diameters, such as spruce pulpwood,

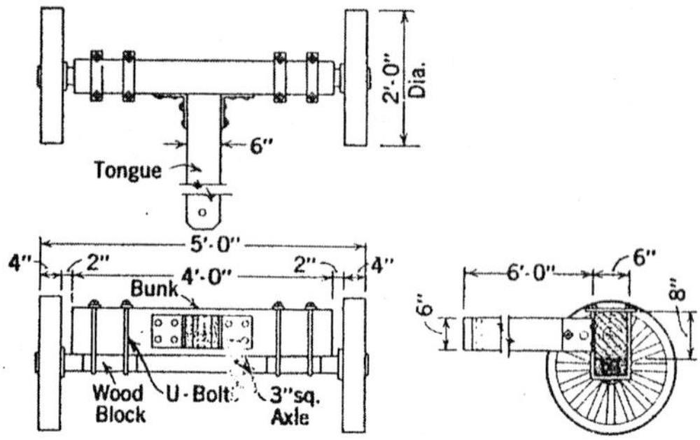

FIGURE 44. Top, front, and side views of low two-wheel bummer used in skidding logs, particularly long lengths such as poles and piling in the eastern hardwood forests. The low 2′ wheel is easily loaded. These bummers are drawn by horses or low-powered tractor units in skidding logs for medium distances. For short distances, direct ground skidding and sleds are used. No roads are built for these bummers.

with a single horse, over rough ground for distances up to ½ mile or more. It is also used on small forest tracts or odd corners where there is insufficient timber to justify building a haul road or chute.

Trailing is practiced in northern Idaho, the Southern Appalachian Mountains, the Northeast, and occasionally in other sections of the country. As it requires considerable time to dog the logs together in a trail, as well as to release them at the landing, it is feasible only for relatively long skidding distances. For shorter distances under similar conditions, direct horse skidding or draying would be used. It is not adapted to large or heavy logs.

Chaining. This is a skidding device to bring 4′ pulpwood bolts down short steep slopes. The bolts are piled in bunches of three to ten or more, depending on their size. These bunches are piled sidewise on a ⅜″ chain behind a single horse or team, loaded dray, or bobsled. Chaining is used only on steep slopes on snow where dragging the bolts at right angles to the direction of their length serves as an effective brake. It is used behind a dray or bob as an added braking device. There are generally 15 to 35 bunches (sometimes known as bobs) in a chain, totaling one-half to two cords, depending on the degree of slope and condition of the road. One chain 50′ long contained 85 spruce bolts in 17 bunches of 4′ pulpwood in the Adirondack Mountains. This

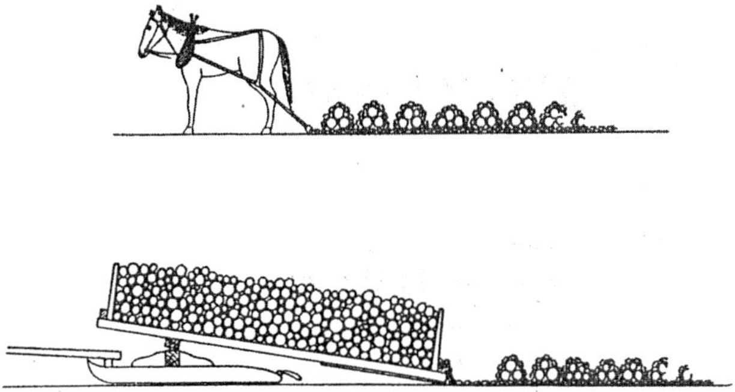

FIGURE 45. The upper sketch illustrates chaining, a skidding device to bring 4′ bolts down short steep slopes. The bolts are chained in bunches of three to eight or more and pulled sidewise. Three-fourths to two cords or more are hauled by one horse. It is cheaper than horse skidding with single logs or a trail of two to seven logs. It is seldom applicable for distances greater than ¼ to ½ mile. There are generally 15 to 35 bunches to a chain.

The lower sketch shows combined bobbing and chaining. The bob consists of a single rack 16′ to 20′ long, the rear end of which drags on the ground and acts as a brake. Two to three cords of pulpwood are generally carried per load. The bob usually has an automatic brake operated over the slip tongue. These methods are used in logging pulpwood in the Northeast generally on grades of 12% to 25% or more, where the topography is too rough or steep to operate double sleds. The cost of construction of necessary haul roads for double sleds, even in connection with the braking devices or with sanding, graveling, or haying, would be greater.

size required 19 minutes to load and 15 minutes to unload at destination.

Hand sleds.* Man power has been used for many years to a limited extent in eastern Canada, particularly in the upper Saguenay district and the north shore of the St. Lawrence River for hauling sleds. Small sleds are used on very steep grades on rough topography and for short distances where skill and experience in steering are frequently as important as strength. This method is possible where horses could not be successfully employed because of the steep rocky slopes and deep

* Although this subject is properly considered under hand logging it is generally practiced in connection with and supplementary to horse skidding.

snows. It is referred to as "chienning," a name derived from the French, meaning dogs' work. Ordinarily, one man pulls a single sled with or without logs dragging at the rear; sometimes two or three men may be used. Generally, 20 pieces of 4′ pulpwood (700 to 800 lb.) or an equivalent weight of 12′ or 16′ spruce logs are hauled down slopes having grades as steep as 60% for distances up to 250 or even 500 ft. Hand sleds are occasionally used to haul logs out of swampy places where horses would mire, during the fall before freezing.

Miscellaneous. Dog teams have been used in Newfoundland for hauling sleds loaded with pulpwood on snow. Strong sturdy dog teams have hauled ⅛ cord of pulpwood as much as ½ mile down favorable slopes to a landing along the stream.

Frequently, several of the skidding methods described are used on the same operation. For example, on one spruce pulpwood operation where there was a drop in elevation of 2000 ft. in 2.4 miles, the following were used:

(*a*) Hand logging on slopes too steep for a horse: 4′, 16′, and 20′ logs were pushed or thrown down the slopes.

(*b*) Trailing two to ten 16′ logs, one behind the other and chain-dogged together, on distances of 200 to 400 yd. on steep slopes where a single horse could operate.

(*c*) Direct horse skidding of one to six logs, 16′ and 20′ long, dogged, chained, and dragged parallel to each other—this may be used for shorter distances than trailing is.

(*d*) Chaining with three to six bolts on ⅜″ chain to 60′ to 75′ long on slopes too steep for bobbing.

(*e*) Bobbing (draying) with 22′ racks which carried two to three cords per load. This may be used for longer hauls than chaining and where slopes are too steep for double sled haul.

(*f*) Combined chaining and bobbing, where a chain load of logs was dragged behind a bobsled, thus providing an added braking device on slopes too steep for bobbing alone.

RELATIVE EFFICIENCY AND COMPARISON WITH TRACTORS

Horses are the principal animals used in skidding. Their chief competitor in logging operations has been the tractor. The rate of travel of horses is approximately 2½ miles per hour when pulling empty wagons, sleds, and similar conveyances on the level. Generally, a horse has difficulty in traveling 20 miles per day. A horse forced to cover this distance daily will probably be of little use at the end of the logging season and cannot draw such heavy loads as at the beginning of the season. Horses will make the swings or round trips indicated in Table XX on average working days on average-length hauls and with various load capacities.

TABLE XX

DISTANCE AND ROUND TRIPS MADE PER DAY WITH HORSES

Length of Haul, miles	*Number of Trips*	*Daily Mileage*
1	6–8	12–16
2	4–5	16–20
3	3	18
4	2–3	16–24
5	2	20
6	2	24
7	1	14
8	1	16
9	1	18
10	1	20

Horses are able to drag their normal load only short distances if adverse grades are encountered. The strain is too severe on grades of 3% or 4% for more than 200 ft. Therefore, it is desirable that a continuous downhill haul on favorable grades be available for horses.

In sled logging, more than two horses are frequently used to pull a train of sleds. It is difficult to get four or more horses to exert an even pull synchronously. First, the lead team is likely to pull, and then the runner team; consequently four horses are not so consistently efficient in attaining economical footage deliveries as the single-team hookups.

In northern snow logging, horses are generally used from September to March, inclusive, first in skidding logs, and, then, when cold winter weather and snow are available, for hauling. This means that there are generally 212 days for the 7 months' work, or, if Sundays are eliminated, only 180 working days.

An important consideration is the feeding of horses when idle as well as when at work and the fact that tractors can work day and night on either uphill or downhill hauls. The cost of toting in feed for the horses is more than the cost of toting in gasoline and oil for the tractors. The 30-hp. tractor burns less than 30 gal. per day. Gasoline weighs 6.2 lb. per gallon, so that 30 gal. weigh 186 lb. Not more than

5 lb. of oil and grease would be used per day, making a total weight of 191 lb. The gasoline drum weighing 34 lb. brings the total weight of fuel for a gasoline tractor to 225 lb. A horse, on the other hand, uses 25 lb. of hay and 20 lb. of oats or a total of 45 lb. per day. A team thus uses 90 lb. of feed per day. Therefore, 2½ teams will use by weight as much feed as one 30-hp. tractor. If the tractor replaces nine teams on an operation, it will use by weight one-fourth the amount of

A. Logs well placed for skidding on a well-cleared road.

B. Logs badly placed and stumps on the road are too high.

FIGURE 46. Spruce logs well placed for horse skidding as shown in *A* and those badly placed as in *B*. *Courtesy A. Koroleff.*

fuel or feed needed to keep the horses working. Furthermore, tractors are not affected by the climatic and physical conditions of the country: snow, blizzards, low temperatures, mud, rain, heat, and so on. Horses tire out as the season progresses. When horses are used on adverse grades or for long pulls, they must be rested or "spelled." A tractor is not susceptible to fatigue.

SKIDDING DISTANCES AND GOVERNING FACTORS

Horses are used primarily for small logs on short-distance skidding, on steep rocky terrain, and for small-mill operations. They are used principally for skidding distances up to ¼ mile, or with large good-quality logs for as much as 1 mile or more, particularly in odd or remote corners of the operations. In small logging operations on favorable slopes two to seven logs are trailed one behind the other. Sometimes two or three logs are bunched together and chain-skidded to the deck or landing.

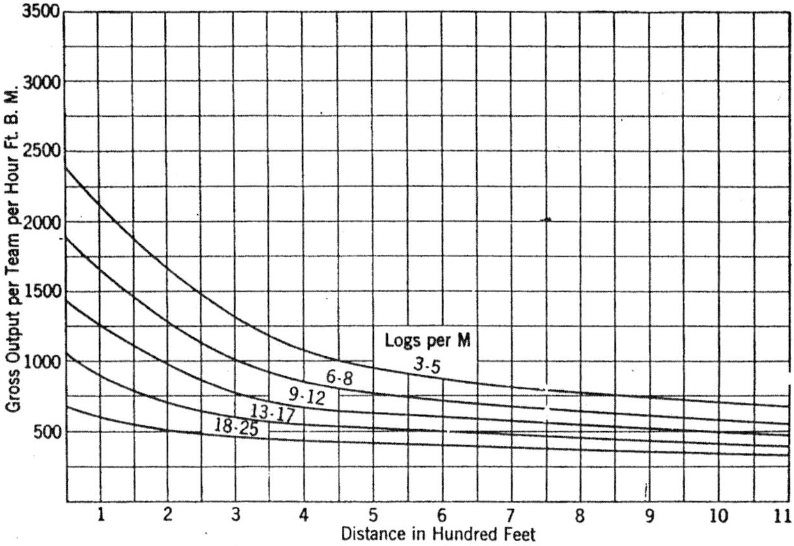

FIGURE 47. Output of winter horse-skidding team per hour for various distances and timber sizes in western white pine in northern Idaho, 10,000 to 20,000 b.f. per acre, slopes 30% to 50%, medium brush and windfall. *After M. Bradner et al.*

SUMMARY OF DISADVANTAGES

Animals are not generally used for skidding or hauling long distances because of their relative slowness compared with tractors or other devices on logging chances where either may be used, and also because they tire and are unable to maintain sustained speeds, especially when adverse grades are encountered. The great disadvantage of horses is that their maintenance cost continues whether logging operations are in progress or not. Horses are generally purchased at 5 to 7 years of age and have a work life of only 3 to 10 years.* The depreciation charges, expressed in per thousand board feet or in pulpwood cords, are often greater than for tractors or even trucks, which in many places have supplanted horses both for skidding and hauling. The timber and climatic conditions in the Lake States and northeastern and Canadian pulpwood operations are ideal for horses. Horses are used there for skidding during the summer and hauling in the winter, so that, except in the spring pulpwood river-driving season,

* Neff reports that horses are useful only 4 to 6 years on the steep slopes in western white pine skidding in northern Idaho. In eastern Canada, the average work life is generally considered to be about 5 years.

they are employed most of the year. In other eastern regions they are used the year round.

SILVICULTURAL CONSIDERATIONS

Animal skidding does the least damage to young and advanced growth when a tract is being logged. Animals are far less destructive to the residual standing timber than either the tractor or power skidder, especially the latter which generally destroys nearly all growth. In view of the fact that skidding is widespread throughout the felling area, the portion of forest floor affected in the operations is of great importance. It is likely that horse skidding will continue to be the principal method on short distances in small timber not only because it does the least damage to the young timber already growing on the tract, but also primarily because of its low cost.

In horse skidding, the trails are generally swamped out, and advance reproduction is removed. In tractor skidding, swamping is not generally done, and the advance growth is overrun with machines. Damage in skidding may be measured in its effects on (*a*) the remaining stand and (*b*) the soil. The severity and character of the damage depend on the soil type, topography, season of the year, species, volume and character of the stand to be removed and reserved, and skill employed in the logging process. No two areas of any extent present identical conditions.

In the Southwest, there is little difference between the amount of damage done by tractor and by horse skidding; occasionally horses may do more harm than tractors where conditions are comparable, because large areas are necessarily cleared for skid trails for horse skidding, whereas tractors are backed up to the logs which are skidded out with little damage. (This is in ponderosa pine averaging four logs per tree and four logs per thousand feet.)

In northern Idaho in western white pine, tractors are regarded as twice as destructive as horses. In the ponderosa pine type, tractors which are generally used do little harm to reproduction. With arch-wheel types of tractor skidding, the front ends of the logs (two or more) are elevated, thus inflicting much less damage on ground and reproduction than in direct ground skidding. Fewer trips to a given area are required with tractors than with horse skidding, thus subjecting the area and its vegetative cover to less destruction.

Power skidding is probably 10 to 50 times as destructive as horse skidding. High lines (high lead or overhead cableway systems) are far more destructive than ground lines. In northern Idaho it is estimated that horse skidding destroys and cultivates 3% to 5% of the area, and tractors (various types) 5% to 10%. Power skidding destroys most of the reproduction and 20% to 60% or more of the residual stand and has little beneficial effect, as horse and tractor skidding have, in "cultivating" the soil and making a favorable seedbed for reproduction.

CHAPTER VIII

Tractor Logging*

GENERAL AND HISTORICAL

The introduction of tractors has vitally changed methods of transporting logs in many parts of the world. They are used for performing so many different operations in the lumber industry that a tractor may be classified as one of the most flexible and valuable pieces of logging equipment.

Older methods of logging with animals, power skidders, and yarders have been replaced to a considerable extent by the tractor method.

Because of their flexibility tractors can ground-skid, pan-log, arch-and-sulky-log, bummer-log, wagon-log, winch-log, and sled-log. They also are used to build roads, load logs, and plow snow. Most lumber companies have abandoned their older equipment in favor of tractor application.

Tractors are versatile and rugged enough to log over steep hillsides and rocky terrain and in very swampy conditions. Their Diesel economy and lugging ability have made them profitable on thousands of logging shows or operations.

The first practical tractors were made in the 1880's. They employed large 8′ rear driving wheels and were powered with steam engines. These first tractors were employed primarily for agriculture but were later used in conveying heavy supplies to mines and construction projects. One of their first logging applications was made in 1893 in California. Before that time as many as 10 or 12 yoke of oxen were used in the work of skidding logs. Such a slow and expensive logging method could only be justified by the fact that lumber prices were high and labor costs were low.

The first tractors were very cumbersome and could be worked only on a relatively even-surfaced road. The wheel-type construction was top-heavy and lacked traction to pull large loads. The hauls at that

* Tractors are used principally to skid logs to landings; they are also employed in many other ways in logging operations. Because their primary purpose is to assemble logs for further transportation they are classified under minor transportation.

time were ½ to 3 miles. Logging in the early days was directly to the mill or stream. It was very evident that before much progress could be made mechanical changes must be introduced.

One of the first makes of track-type tractor was built in 1904 with a 40-hp. steam engine to power it. This tractor pulled 35% more than the previous round-wheeled 60-hp. tractor. Such performance clearly

FIGURE 48. Diesel D8 tractor with Hyster winch, arch, and Athey wheels skidding 90 M.b.f. of logs per 8-hour day at Horseshoe Basin, Ore.

demonstrated the value of track-type traction. The year 1905 witnessed the substitution of a gasoline engine for cumbersome steam power.

A tiller wheel in front of this tractor was used for turning. Such a device made the tractor so long and unwieldy on turns that log skidding or hauling was still confined to the haul roads with the oxen skidding logs to assembly points.

With timber logged out near the mill, railroad construction became necessary, but even before World War I it became evident that tractors could reduce expensive spur construction by logging scattered stands and isolated "pockets" of timber.

World War I acquainted many people with the logging possibilities of tractors, and, when steering became possible without the tiller wheel,

FIGURE 49. One of the first track-type tractors to be used in the woods, this Holt tractor was the forerunner of the present perfected machines.

FIGURE 50. Tractor of 35-drawbar-horsepower ground-skidding logs from an Alabama woods to a truck landing. Winch mounted on rear aids bunching and skidding.

the lumber industry at last had a machine that could skid and bunch as well as haul.

It was natural to find loggers substituting the tractor for animal power without changing the over-all method of logging. Short logs that had been cut short simply because animals could not skid longer

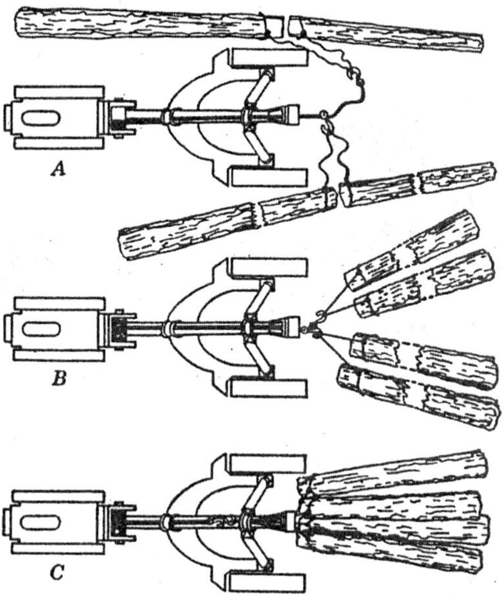

FIGURE 51. Three steps in assembling and loading logs under an arch or sulky. Logs are attached to winch line by means of chokers and are pulled into position as the tractor moves forward. *A*. Tractor and arch backed into position near logs in felling area. Chokers are attached to each log, and then each choker is secured to the winch line. *B*. As tractor is moved forward, logs bunch under the arch. *C*. Winch line then lifts one end of the logs clear of the ground. Logs are held in this position by means of locking brake on tractor winch. Tractor operator and choker setter control entire bunching and loading operation.

lengths continued to be cut short. It has only been in comparatively recent years that the industry has gone to long logs or tree lengths to utilize the full pulling power of the tractor.

Slip-tongue or stiff-tongue wheel carts were at first strengthened somewhat to adapt them to tractors. Finally one of the manufacturers of track-type tractors developed and marketed a hydraulic-wheel unit which substituted hydraulic power for leverage in lifting logs clear of the ground. With such units it was found that as much as 8000 b.f. of logs could be picked up and carried on hauls varying from ¼ mile to more than 1 mile in length. It was also learned that tractors could climb steep grades and ground-skid logs down slopes of 50% to 60%.

The fact that the logs had to be bunched by other equipment or animals for these hydraulic-wheel carts led to an extremely important design innovation. A leading lumber company on the West Coast developed the idea of mounting a winch on the rear of the tractor and placing a cable "fair-lead" on top of the arch or wheel frame. It was then possible by means of a winch line running over the fair-lead to yard logs to a spot under the wheels and then raise one end for hauling. Fair-lead logging became a definite success by 1928 as it materially reduced bunching costs by as much as 75%. These fair-lead units are now known as arches and sulkies.

In 1925 single- and double-drum winches were mounted on tractors to enable tractors to take the place of steam skidders and yarders. Tractors were then able to yard difficult places but with the added feature of extreme mobility.

Thus tractors gradually replaced animal power in the woods. They also replaced high-lead yarders and other cable skidders to a great extent.

Before the wide application of tractor logging, railroad spurs were located on tops of the hills because snaking logs downhill with steam skidders resulted in excessive log breakage. Also logs frequently ran out of their chokers. Tractor logging down grades as steep as 60% eventually changed the location of railroads to the bottoms of valleys and canyons. Railroad construction costs were reduced, and long steep grades for the locomotive were eliminated.

Some difficult and very precipitous terrain was still being logged by semistationary steam skidders, but even under these conditions tractors aided the operation by hauling logs to the main skidder line. This eliminated the need for costly changes in rigging "setups."

In northern logging, seasonal operation was the rule. Loggers waited for the winter freeze-up so that loaded sleds could be pulled by horses over frozen lakes, muskeg, and soft ground. It was natural that the first tractors displaced horses in this method of logging. Since these first tractors were powered by steam, their heavy awkwardness did little to change logging methods.

The last decade has witnessed a revolution in northern logging. As the tractors' mechanical efficiency increased, loggers began using the tractor for year-round operation. Those areas with muskeg and large lakes, however, still forced loggers to operate only in the winter, as tractor drivers experienced difficulty in staying on top of muskeg and found detours around lakes too expensive.

In 1931, the first practical Diesel tractor was developed for use in the woods. The Diesel tractor quickly became popular as it developed considerably more power owing to the inherent efficiency of Diesel operation. An example of the Diesel's economy is illustrated by the fact that a 43-drawbar-horsepower D4 tractor can ground-skid 1000 b.f. of ponderosa pine logs per trip, using only 14 gal. of low-cost fuel in 8 hours. Now, no manufacturer makes a gasoline-powered tractor in the larger sizes. Some manufacturers no longer make a gasoline-powered tractor in any size. The Diesel tractor has found universal acceptance in the woods.

There are now thousands of logging tractors in use throughout the world. Topography, size of timber, and terrain dictate the size of tractor to be used on the job. One tractor manufacturer makes five different tractor sizes to fit every job classification. Drawbar horsepowers of 25, 43, 65, 80, and 113 give a wide range of choice for tractor logging.

PURPOSES FOR WHICH TRACTORS ARE USED

The uses of tractors may be summarized as follows:

1. The chief use is for the transportation of logs or pulpwood by:

   (*a*) Arch or sulky skidding.
   (*b*) Ground skidding.
   (*c*) Log-wagon hauling.
   (*d*) Pan-bummer or sled skidding.
   (*e*) Winch; single or double-drum logging.

2. Auxiliary uses for tractors are:

   (*a*) Road construction and maintenance.
   (*b*) Building fire lines or guards.
   (*c*) Reforestation work.
   (*d*) Moving camp.
   (*e*) Loading wagons, sleds, or trucks.
   (*f*) Plowing snow or breaking out strip and branch roads.
   (*g*) Icing or tanking roads.
   (*h*) Toting supplies.
   (*i*) Powering felling and bucking saws.

If we classify uses according to region, it is apparent that the large timber of the West should find the heaviest tractors in use coupled to large track-type arches so that one end of the log or logs could be elevated. Ground skidding of all types is also common. Cable-line skidding by use of drum equipment mounted on the tractor is desirable in many hilly areas.

Southern logging is characterized by smaller trees, and so the smaller tractors, equipped with winches for ground skidding, are the rule in that region. Wheeled sulkies and track-equipped arches are also finding an increased application. Pan, bummer, and mud sled logging are not used as much as they once were. Track-type wagons are becoming more popular as the tractor has made logging in very wet and difficult places economically feasible.

Snow-country logging is finding the increasing application of tractors for year-round operation. Arches and pneumatic-tired sulkies have been successfully applied to northern operations. Ground skidding is also common. Hauling of sleds on ice or snow roads continues to be an important use for the track-type tractor.

ADVANTAGES OF TRACTORS OVER OTHER METHODS OF SKIDDING AND HAULING

Tractors are available with track tread as wide as 24″ to give extremely low pressures per square inch of bearing area. A man walking through a swamp exerts a pressure of about 10 lb. per square inch, whereas a tractor with wide tracks exerts a pressure of only about 4 lb. per square inch. Such low pressures allow the tractor to log areas so swampy that cable-line skidders can be the only alternative.

Where ground conditions are too wet for ground skidding, the use of the track-type arch or low-pressure tire sulky have proved of advantage. These devices lift one end of the logs clear of the ground and thereby reduce the sliding resistance appreciably. Track-type wagons are used when the ground conditions become extremely soft as the logs are lifted completely clear of the ground.

Tractors working in the heart of swamps during good and bad weather have reduced the need for setting up cable-line skidding equipment with their relatively high man-power requirements. Tractors leave the forest in better condition for regrowth of timber and do less damage to remaining trees than do cable skidders. With selective or partial cutting of trees becoming fixed policies of many mills, it is almost imperative that tractors be used to avoid costly destruction of young growth or seed trees caused by power-line skidding. It is inherent in skidder operation that, unless an area is clear-cut, it is difficult to secure economical operation. With tractors, on the other hand, individual trees may be selected with ease owing to their great mobility.

In the last decade the recourse to tractors has reduced the use of animals in skidding. Tractors have much greater speed and power than animals and are able to log rougher topography and steeper slopes and at the same time skid longer and heavier timber. Long distances, swampy conditions, and fallen timber are serious obstacles in animal skidding but not for tractors. It is generally recognized that a horse pulling steadily at 2½ miles per hour can exert only about 150 lb. of pull. For a very heavy pull a horse can exert a force of 450 to 600 lb. for a short period. A small D2 tractor of only 25 drawbar horsepower can exert a pull all day long equivalent to 12 horses, if it is assumed that these horses are steadily working at their maximum pull of 500 lb. Fuel consumption for this tractor is only a little over 1 gal. of fuel per hour. The fact that horses tire easily, heat up in summer, and must be fed whether they work or not makes their use a disadvantage compared to tractors. Working in soft terrain, horses, oxen, and mules have often mired themselves so completely that they must be extricated forcibly. Another disadvantage to animal operation is the high ratio of man labor to pulling ability. Current high-labor-cost trends point to the tractor as the solution to these problems since one tractor driver controls so much more drawbar pull than a team of animals.

Some loggers, especially in the South, have attempted to log directly with trucks in favorable weather, driving directly to the stump. This practice was modified considerably when loggers found that truck maintenance under these conditions rose appreciably. The trend, therefore, is toward keeping the trucks on ridges and good roads and tractor-logging to the trucks. Even favorable weather found trucks bogged down under a heavy load, necessitating a tow from the nearest tractor.

Another advantage of tractors in the woods is their versatility. After skidding logs to the landing, a tractor can load trucks in one of many ways. It can then complete the job by towing the loaded truck up steep grades or over soft spots on the haul road. Many tractors are logging with a bulldozer blade in front and an arch or sulky behind. After the arch is unpitched, the tractor can build roads or "rough-out" landings for trucks. One tractor-mounted crane is available that can be detached from the tractor in less than 1 hour. The tractor is then free to skid logs, haul wagons, or build roads with its attached bulldozer blade. Or the crane can load logs or serve as a dragline where very soft swampy ground requires improvement.

Tractors serve as an auxiliary to power-line skidding, making possible the logging of remote timber in far corners of the tract without expensive railroad construction. This practice is common in the western regions.

There are some minor disadvantages of tractors which are generally localized. Among them may be mentioned:

(*a*) A few areas may possess a very loamy soil that causes the tractor to "dig itself in." In such areas, tractor operation may be limited to dry or moderately dry weather conditions. The recent use of track-type arches and low-pressure pneumatic-tired sulkies has eliminated this trouble to a considerable extent as they lessen the pull needed at the tractor drawbar. Track-type logging wagons have also reduced this disadvantage considerably.

(*b*) The gradient or slope on which some tractors can operate is limited. Slopes of 50% to 60% have been descended by tractors while ground-skidding loads, although 30% slopes are often set as maximum by some operators. The use of arches or sulkies also tends to reduce the maximum slope that can be descended, as the logs exert less drag and tend to push the tractor to one side or the other.

(*c*) The rather large initial cost of tractor and arch or sulky combination has acted as a deterrent to some of the smaller operators. Since final costs or true costs are what the logger is chiefly concerned with, this disadvantage might be considered a minor one.

PRINCIPAL USES OF TRACTORS ON LOGGING OPERATIONS

The foregoing description has presented a skeleton outline of the wide variety of purposes for which tractors are serviceable. The following is an explanation of some of the more important phases of tractor use on logging jobs.

GROUND SKIDDING BEHIND THE TRACTOR

When logs are skidded for short distances up to about 400 ft., ground skidding is often considered the most economical method. This, of course, depends on the soil, as ground skidding in a very bad swamp might not be feasible. Where soils offer reasonably good traction, satisfactory results can be achieved.

On slopes, larger loads can be ground-skidded downhill. With greater loads, ground-skidding distances can be increased. Each particular operation determines how long ground-skidding distances should be. In some places, logs have been ground-skidded up to 1 mile, but such distances are extreme. An advantage of ground skidding over

FIGURE 52. "Caterpillar" D2 Diesel Tractor of 25-drawbar-horsepower ground-skidding 700 b.f. of hardwood logs near Michigan City, Ind.

other methods is that the tractor has no trailing vehicle to impede maneuvering in thick underbrush or cramped quarters. This makes ground skidding a good system when loads are to be bunched for large-sized tractors equipped with arches, sulkies, or wagons.

In ground skidding, tongs or chokers made of cable are placed around one end of the log and then hooked to the drawbar. Brute force is utilized to get the logs to the landing or assembly point. Cable of 6 × 19 construction made from improved plow steel and preformed is used in most places. A wire rope center is often preferred over the hemp center on choker cable because of its greater strength. Some cable flexibility is gained, however, through the use of hemp center. Di-

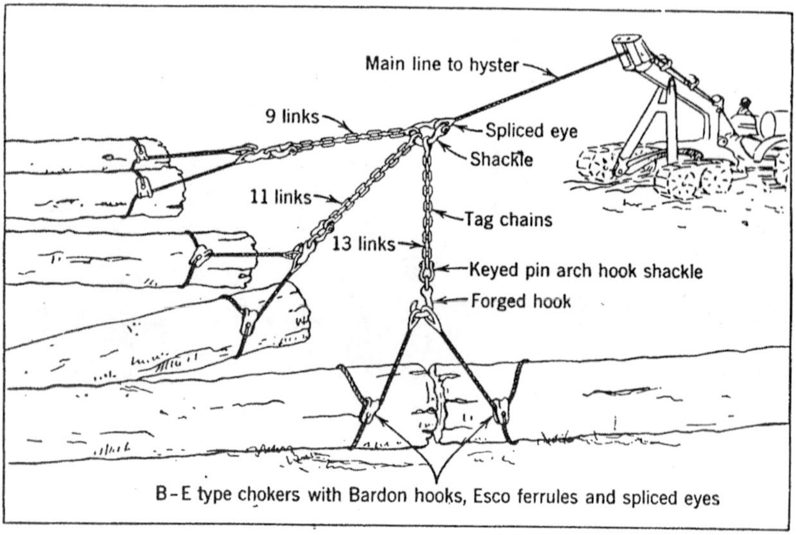

FIGURE 53. A method of attaching chokers to tractor winch line. Tag chains are of various lengths to allow hooks to roll over fairlead with greater ease.

FIGURE 54. International Tractor with arch skidding loblolly pine 600′ to a landing in South Carolina. Tree lengths are bucked into 5′ lengths for pulpwood and then loaded by means of a portable loader mill.

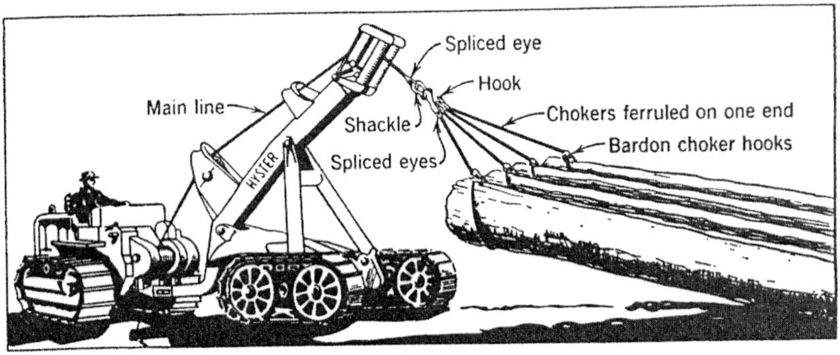

FIGURE 55. A frequently used method of attaching chokers to tractor winch line. A single butt hook is used, and chokers are long enough to allow bunching with large loads.

FIGURE 56. Logging tree length jack pine in Superior National Forest, Minn., with D4 tractor and Hyster sulky. Use of Bardon-type choker adds to efficiency.

ameters of ½″ to ⅝″ are used on smaller-sized tractors, whereas diameters of ¾″ to 1″ are used on the larger tractors.

The usual practice is to have an extra set of chokers so that they may be placed around logs while the other set is utilized in skidding. The rigger or choker men then have the logs ready for instant hitching when the tractor returns for another load. Use of chains on the drawbar instead of cable chokers is advantageous in smaller timber as chains are very flexible and do not kink easily.

A refinement of drawbar ground skidding is through the use of a single-drum towing winch mounted on the rear of the tractor. The winch line equipped with a large hook is used for bunching the logs. With drawbar skidding, the tractor must maneuver close to each log that is to be pulled out. This is often difficult, because of thick underbrush, tree tops, and irregularities in terrain. With the winch line logs can be snaked out with a force greater than the rated tractor drawbar pull without necessitating tractor movement. After the logs are winched to the rear of the tractor, all the chokers are assembled on the winch-line hook, and the tractor proceeds to the landing. Such a winch is also valuable for loading logs, pulling the tractor out of bad spots, and pulling trucks up steep grades.

As skidding distances increase or as bigger loads are needed, auxiliary equipment becomes desirable. Although a longer bunching time results from the use of arches or sulkies, it is found that bigger loads and fewer tractor trips more than compensate for lost time. The determining factor is always the final cost per thousand board feet or cord for bringing the logs in to the landing.

PAN AND BUMMER SKIDDING

Steel logging pans designed to eliminate some of the ground-skidding friction drag are of declining importance. The object of the pan is to permit the front ends of logs to ride clear of the ground. This eliminates the "digging in" action of the logs and improves the haul road with each passage. About 30% more volume per load can be skidded in this manner. In operation the pan is fastened permanently to the drawbar. The logs with chokers are secured to the drawbar, and the tractor is pulled ahead. The logs then roll onto the pan of their own accord. Disadvantages of the pan are the poor maneuverability characteristics and the fact that logs must be bunched to a certain extent.

Bummers, drays, travoys, go-devils, or bobsleds are all used for the same purpose, to elevate the nose or all of the load. Drays and sleds are characterized by runners whereas bummers have wheels or tracks.

All these devices must be loaded to a certain extent whereas pans are self-loading. Any device to elevate the "nose" of logs has the advantage of bringing cleaner logs to the mill. Many small mills in the South do not have washing facilities, and elevation of logs assumes major importance. Circular saws and band saws are quickly dulled from dirt-imbedded logs.

ARCHES

Of growing importance in all regions of the country is the application of logging arches. The logging arch is a track-mounted frame

FIGURE 57. Diesel tractor of 69-drawbar-horsepower skidding an average load of 1500 b.f. of longleaf pine near Foley, Fla. This tractor used only 2½ gal. of low-cost fuel per hour.

connected to the drawbar of the tractor. Mounted on the frame is a fair-lead for allowing the winch line to operate in a wide arc without binding. Logs are winched to the tractor and then hoisted clear of the ground on the butt end for ease in hauling. Developed on the West Coast for handling large timber, the logging arch has demonstrated its value for logging in all kinds and sizes of timber and over all types of terrain. The advantages accruing through its use are many:

(a) **Less sliding resistance.** A general rule for arches is that tractors can haul twice as many logs as they could by ground skidding.

(b) **Combines bunching, hauling, and unloading features.** The arch by means of its elevated fair-lead and winch line can bunch logs from considerable distances, although 75′ to 100′ of line is normally used. The tractor winch will accommodate lengths of 300′ to 400′ depending on the winch model. The line pull of the winch is greater than the drawbar pull, and this is important in snaking out logs that have been felled into a binding position. After the logs are bunched,

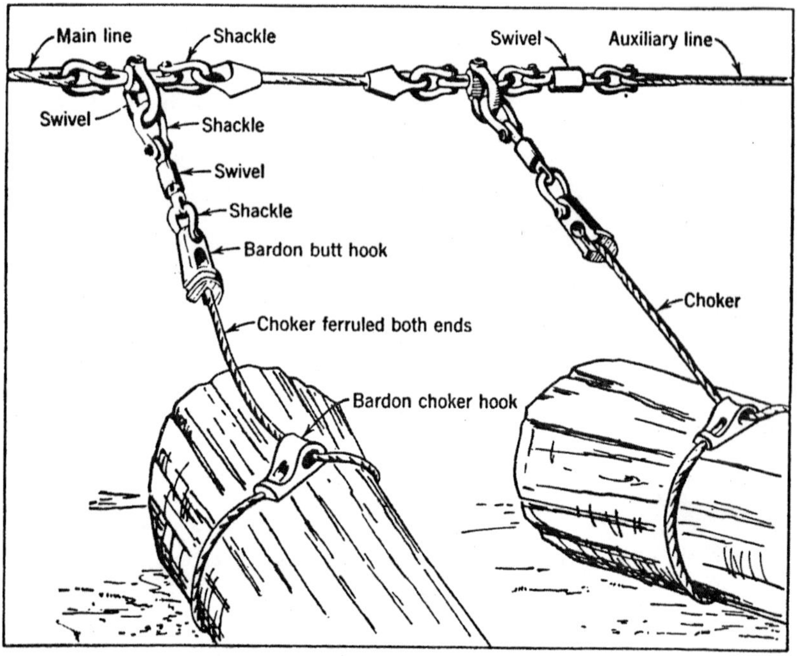

FIGURE 58. Recommended method of attaching chokers to main-line cable behind double-drum-equipped tractor.

the butts are elevated for hauling. Usually each log has its individual choker, although prebunched logs are sometimes bound with one choker.

The arch with its wide track shoes can haul over very soft underground conditions without miring. Its wide base makes hauls on hilly and rough terrain practical without tipping.

Unloading is achieved by releasing the brake on the winch and thereby allowing the logs to drop to the ground. The chokers are unhooked and the tractor pulls free of the load. If bucking of the logs is to be done at the landing, the logs can be dropped singly or in pairs. This allows the buckers room for sawing.

(c) **Higher hauling speeds.** With less draft required to pull large loads on an arch, the tractor can haul in a higher speed range. Where ground skidding requires 1st and 2d gear operation, arches can often haul in 3d gear with much greater loads.

(d) **Cleaner logs.** With all but the dragging end of the logs clear of the ground it is possible to bring logs to the mill in a far cleaner condition than through ground skidding.

(e) **Less strain on tractor and cable.** Arch logging reduces wear and strain on cable by sudden shock loads due to ground skidding. With log butts elevated on the haul, less strain is put on the tractor itself.

(f) **Allows tree-length logging.** With the ability of the arch to carry large loads has come the gradual change to logging with 32′ logs and longer. Mills have gradually realized the advantages of logging longer lengths. For example a 32′ log would have to be bunched in two operations if cut into 16′ logs. Loading would take twice as long, since two separate lifting actions are required. Long log lengths reduce the need for processing in the woods, and this in turn reduces costs.

(g) **Applicable to short and long hauls.** Arches have been used on hauls as short as 300 ft. to advantage. The arch has the disadvantage of less maneuverability compared to a ground skidding tractor but compensates for longer bunching time by greater loads in fewer trips. Hauls of 2 miles and over have been recorded with arches carrying a maximum number of logs. In the big timber of the West a 113-hp. tractor with arch has averaged 75 M.b.f. on each 8-hour shift on an 1800 ft. round-trip haul.

(h) **Arch logging proves advantageous from a silvicultural standpoint.** Destruction of young growth is less with one end of the logs elevated as compared with ground skidding.

SULKIES

A sulky is nothing more than a wheeled arch. It is used on the smaller tractors and possesses all the advantages of its "big brother." It uses low-pressure tires to get the maximum flotation in soft areas. Generally, the sulky is not able to work in such soft terrain as the arch, owing to the smaller bearing area of its tires or wheels. The sulky is commonly used with tractors up to 45 drawbar horsepower while the arch is used with tractors of greater horsepower.

SINGLE- AND DOUBLE-DRUM LOGGING

Either single- or double-drum winches may be mounted on the rear of tractors. They enable tractors to work in places that are inacces-

FIGURE 59. Double-drum Hyster winch mounted on rear of tractor for high-lead power skidding in Oregon. Decked logs were then loaded onto trucks.

sible to other equipment. Such a unit provides line pulls greater than the pull available at the tractor drawbar owing to the gear reduction present in the winches. With drum equipment, the timber can be logged off steep hillsides, out of gulches, and out of swamps. The fact that the tractor can move quickly from one set to another with considerable ease makes it an efficient skidding tool.

The usual procedure in drum logging is to use a single drum if the timber is sparse. Sheaves or "blocks" are tied to a head tree at a fair

FIGURE 60. "Caterpillar" D7 tractor of 80-drawbar-horsepower ground-skidding logs down steep grade in Montana. Logs averaged 5.5 per thousand board feet with ½-mile haul to landing. Diesel's power is unaffected by 4000 ft. of altitude.

height and the main line is run through the "block" and into the stand of timber. The line can be carried out by man power if the distance is not too great. A mule or horse is often used for longer distances if ground conditions permit.

If the timber stand is heavy, a second line is used and is known as the rehaul or haulback. Its function is to pull the empty main line back into the timber after a log is skidded in. The logs are left in a pile at the head tree. After all the timber is pulled in to this location. the tractor moves on to another tree site. An arch and tractor are then used to bring in the logs concentrated at the head trees. Important advances have been made in the logging of western timber by this method.

TRACK-EQUIPPED LOGGING TRAILERS

Track-type wagons pulled by tractors have been in use for many years in the Southeast and other areas.

When roads become impassable for rubber-tired equipment or when the timber stand is too sparse to warrant road construction, these track equipped trailers can carry full loads of logs. Some operators keep these trailers for stand-by use in case of wet weather. Also, where truck tires are often snagged by stumps and rock, the steel tracks function with complete safety.

Logs hauled completely clear of the ground have the added feature of freedom from imbedded dirt or rocks. Veneer mills are especially concerned about clean logs, as the veneer blade will be dulled or ruined when working against dirty logs. Where washing facilities are available, this feature decreases in importance.

Loading the wagon is generally accomplished by means of a tractor and tree block or through the use of a tractor crane. Truck-mounted cranes are also used.

Tractor–trailer logging is practical on hauls of 1500 to 5000 ft., depending on ground conditions. Often, the use of trailers has proved to be the only method of getting logs out of swamps unless power skidding is utilized. Greater haul distances have been recorded where conditions were favorable or essential for such hauls.

Some loggers operate two trailers in tandem so that one tractor can do the job of two. Such procedure adds to the economy in haul distance. One loaded trailer train is hauled to the mill while another train is being loaded. With such a system the tractor is not required to wait for loading.

SLED HAULING

As mentioned previously, tractors are extensively used in the north country for hauling sleds over ice and snow roads. As the tractor developed in efficiency, operators began logging timber that had previously been considered inaccessible.

The general procedure is to use small tractors for hauling single loaded sleds to a "make-up" point. The sleds are then coupled together to form a train for the long haul. Larger tractors are used to pull these sled trains. Distances of 10 miles or more are not uncommon on the long hauls.

FIGURE 61. "Caterpillar" D6 tractor of 55-drawbar-horsepower pulling train of sleds in Ontario, Canada. Each sled averages 5000 b.f. Average haul distance about 3 miles to streams.

Tractors are also used extensively in maintaining snow and ice roads. They pull plows for improving these roads and also tow icing equipment. Ruts are frequently iced to decrease sliding resistance of the sleds.

BUILDING ROADS

As timber in the West and North became less accessible, the use of tractors for road building became increasingly important. Efficient road-building tools have made truck hauls over great distances feasible. Today, more and more access roads are being opened up by private owners for the use of their huge Diesel trucks carrying tremendous loads. New low road-construction costs have resulted from the application of tractors.

The tractor tool most in favor for this road construction is the bulldozer with straight blade or angling blade. For sidehill cuts, the angled blade drifts the earth from the high side to the low side with a minimum of effort. The bulldozer with straight blade is ideal for general dozer work. Bulldozers have also been used for leveling landing sites, pioneering arch trails, and keeping trash off the landings. The earth-moving scraper is another efficient tool that is used behind the tractor for picking up loads, hauling, and then spreading.

In the South where the terrain is both low and hilly, the bulldozer and scraper have reduced logging costs. The hilly Appalachian region often requires road construction to make timber stands accessible. The improvement of old roads and water crossings is done easily with this equipment. In the low flat country, bulldozers and scrapers are

FIGURE 62. Tractor of 35-drawbar-horsepower and Hyster sulky skidding long logs near Mountain Pine, Ark. Average load was 1070 b.f. Four trips per hour were made on a 1600-ft. haul. Fuel consumption about 1 gal. per hour.

used to build roads above the wet-weather water level so that trucks can haul when the underfoot conditions deteriorate. Aggregate for clay roads can be carried for a considerable distance with tractor-drawn scrapers.

LOADING WITH TRACTORS

Tractors have proved their value as truck and car loaders. Tractors can load in a variety of ways. One of the most common and simplest is the crosshaul method. Cables are dead-ended to the front and rear of the logging trailer and then passed under a log which is in a loading position at the side of the trailer. The loose ends are then secured to a common line leading to the tractor or tractor winch. Inclined poles are set on each end of the trailer leading down to the log. The tractor driver then engages his winch or moves the tractor away from the trailer, causing the log to roll up the inclined poles and on the bunks of the trailer. This method although slow has merit in the fact that no extra loading equipment is needed as the tractor is primarily used for skidding.

Tractors can also load by means of a block and line suspended from a tree. A tree must be selected that can stand the maximum log weights. The truck is then pulled under the block so the logs can be lowered in place on the bunks. A winch line or a line dead-ended to the tractor can be used in this method.

There are many cranes manufactured today that can be mounted on tractors. One crane is a combination log loader and dragline. It is used in conjunction with the larger-sized tractors equipped with a bulldozer. The crane can be dismounted from the tractor in less than an hour if dozing or skidding work is desired. There is a definite advantage in a tractor-mounted crane compared to a fixed installation. The tractor crane's mobility allows skidding tractors to drop their loads on auxiliary landings so as not to interfere with the crane's loading action. A number of these landings may be scattered along ridges or suitable truck-loading spots. After all of the logs are loaded at one landing, the tractor crane can then proceed to the next. The bulldozer attachment allows the crane to level or drain poor landing sites and to clean off any debris that accumulates.

TRACTOR-MOUNTED FELLING AND BUCKING SAWS

Some tractors are equipped with electrically driven chain saws for felling and bucking. A tractor-mounted generator supplies current through a flexible cable to the saw itself. The saw can be used as far as 1000 ft. from the tractor that powers it. The saw is adapted to all cutting work as the complete saw weighs approximately 85 lb. Chain saws similar to this have cut through an average 54″ softwood log in 70 seconds. An added feature of the tractor-mounted generator is the use of floodlights for felling at night.

One tractor can also be equipped to handle more than one saw. Some operators use a two-man chain saw for felling and bucking plus a one-man chain saw for limbing. A weight of only 52 lb. makes the one-man saw practical on rough ground and in odd angle cuts.

PRINCIPLES OF OPERATION

Now that the various uses of tractors have been discussed, a series of operating principles are presented which should be observed for successful tractor logging. Production figures and costs are included.

BUNCHING PROCEDURE

The bunching of logs for arch or sulky skidding is one of the most important phases of tractor logging. Arches and sulkies require a relatively large number of trees or logs to utilize fully the hauling power

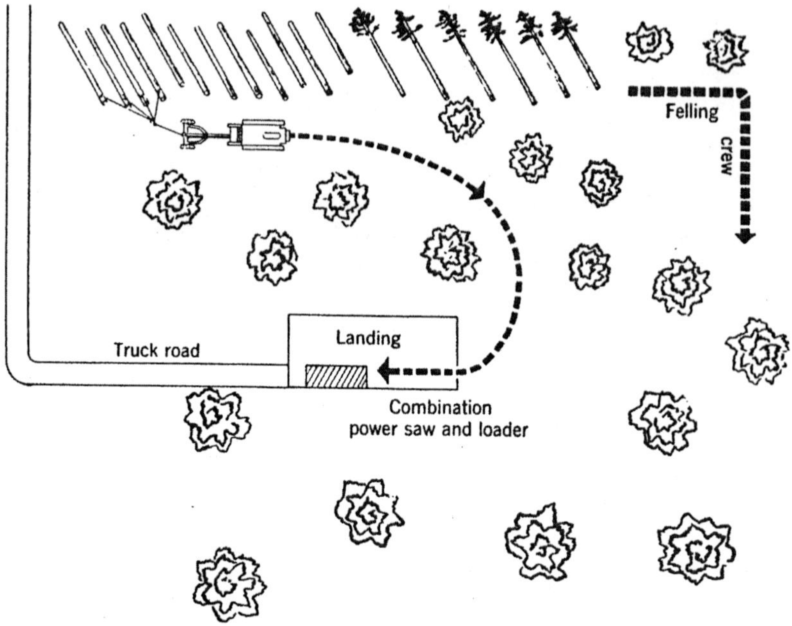

FIGURE 63. An example of how directional felling in windrows can aid bunching procedure. This operation sequence has been very successful in southern pine pulpwood operations with tractor skidding. Trees farthest from the landing are felled first in the manner illustrated, since they can then be thrown in the direction desired without hangups.

of the tractor. Bunching is usually done with the winch, and it is important that the choker man or rigger be thoroughly acquainted with all the methods for getting logs out of binds and away from obstructions. Deciding which logs to take and the proper sequence of removal is the choker man's job. Many companies feel that the choker man or rigger has one of the most important positions, as negligence or lack of thought on his part can sharply reduce production.

It is noteworthy to observe that a tree length or long log length can be taken out just about as quickly as a short log length, so that bunching operations that utilize longer logs are more efficient.

In most regions directional felling has been practiced to allow tractors to bunch more efficiently. Any tree that grows fairly straight can be felled in the direction desired to facilitate removal by tractors. Trees are either thrown with their tops away from tractor skid roads or with their tops away from the landing if such an area is close by. Felling costs are increased somewhat in this way, but operators feel that this extra cost is justified through increased tractor production.

Breakage in felling the large trees of the West has totaled as much as 20% in redwood regions, and so "fallers" generally attempt to fell trees in directions that minimize this damage. Generally, however, fallers throw the tree to facilitate tractor removal.

The use of an extra set of chokers is almost mandatory for proper operation. Some operators even have a third set of chokers on the landing to allow the skidding tractor to unhook quickly. The procedure on the landing is to remove the choker loops from the winch-line butt hook and then substitute the extra set of chokers. The man on the landing can then remove the chokers on the logs without holding up the tractor. Such a procedure is economical only if a large number of logs are carried by the arch or sulky. If the timber is large and few logs are carried, a third set of chokers is not necessary. Many operators use a smaller tractor for bunching loads for the big tractor equipped with arch or sulky. Efficient felling and bunching procedure has often minimized the need for this extra tractor.

Proper choice of chokers is advisable. Homemade chokers often fail to hold properly or are difficult to secure and release. They should be long enough to circle the log with a number of feet to spare.

HAULING PROCEDURE

In hauling, tractors should keep their skid trails as straight as possible. The use of arches and long logs results in such large loads that some binding of logs on the residual stand occurs if skid trails wind excessively. If tractive conditions are favorable, the load should be large enough to make the tractor work in second gear. It is better to make fewer trips with larger loads than to hurry to the landing with light loads, as track wear and tear is less at low speeds. The tractor should be pointed toward the landing or skid road as the last log is bunched, as tractor turns with logs behind require more time than without a load. The turn at the landing should be made after the logs are dropped; not before.

LANDING PROCEDURE

Landings should be laid out so that tractors can approach them with a minimum of turning. If logs are to be cut on the landing, it may be necessary to drop one or two logs at a time to facilitate bucking. This creates a landing in the form of a long rectangle. The loader can then be loading one group while skidding tractors can unload another. If at all possible, landings should be located at a point downhill from the felled timber. This allows extra capacity loads to be carried by the skidding tractor.

PRODUCTION

The production possibilities of tractors and related equipment are discussed in the following paragraphs. It should be remembered that

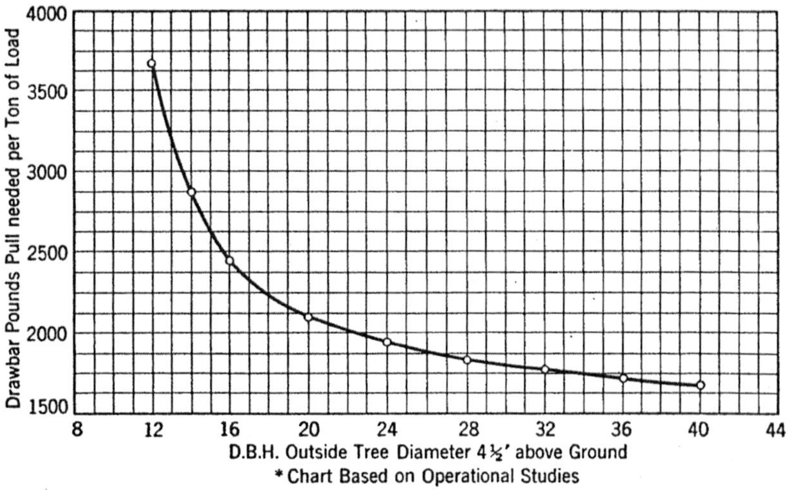

Figure 64. Drawbar pounds pull required per ton of load for various diameter trees when ground-skidding tree lengths—based on 1.90 to 2.84 logs per tree in southern pine.

the figures given are estimates only. Conditions vary so much from job to job that it is inadvisable to use the figures directly.

In ground skidding, Figure 64 illustrates the drawbar pounds of tractor pull needed for every ton of load. The pull required decreases as the diameter breast high or tree diameter increases. The chart is based on operational studies made in southern pine over a period of time and represents an average only. Actual conditions may be more or less favorable.

The following example illustrates the use of such a chart:

Timber—second-growth southern pine, forest-grown form class 79 *
Average d.b.h.—16″
Log scale—Doyle (strict)
Tree volume, average—136 b.f. (2½ log tree)
Tree weight—2350 lb. based on 36 cu. ft. at 65 lb. per cu. ft.
Terrain—level and possessing good forest tractive qualities.
Tractor—"Caterpillar" D7 (80 drawbar hp. rated at 21,351 lb. of drawbar pull in first gear at a speed of 1.4 miles per hour)

If we read up from the 16″ d.b.h. on the chart, we see that the drawbar pull needed per ton of load is 2450 lb. Dividing the available drawbar pull by the pull needed per ton gives the tons of logs that can be ground-skidded in one load.

$$\frac{21{,}351}{2450} = 8.7 \text{ tons of logs or } 17.400 \text{ lb.}$$

$$\frac{17{,}400}{2350} = 7.4 \text{ tree lengths that can be skidded}$$

Since the average tree length weighs 2350 lb., it is apparent that a D7 tractor under these conditions can ground-skid about 7 tree-length logs per trip or 950 b.f. Doyle scale.

Table XXI illustrates the method used to calculate the cost of owning and operating various sizes of track-type tractors. Final cost is shown without operator as operator costs vary so widely from region to region. The use of costs in this chapter is solely for the purpose of illustrating procedures. The use of these figures for any other purpose is not advisable as prices and costs are subject to frequent change. Since logging efficiency is judged largely by final cost per thousand board feet, a mastery of simple cost-finding technique is of considerable importance.

A study of Table XXI will show that D7 tractor of 80 drawbar horsepower under average conditions can operate at a total cost of $22.32 without an operator for a 10-hour day. An operator at $0.75 per hour will increase the tractor cost per day to $29.82.

* Form class is the percentage ratio between the diameter of the inside bark at the top of the first 16′ log and the diameter of the outside bark at breast height. It is an indicator of taper in the first 16′ log.

TABLE XXI

Typical Hourly Owning and Operating Cost of Tractors

All Based on Average Conditions

Model	D8	D7	D6	D4	D2
Working Conditions					
Fixed Cost					
Depreciation *	0.967	0.749	0.590	0.368	0.286
Int., ins., & taxes †	0.290	0.225	0.177	0.110	0.086
Total hourly fixed cost	1.257	0.974	0.767	0.478	0.372
Variable Cost					
Fuel at 9¢/gal. ‡	0.495	0.351	0.243	0.153	0.117
Gasoline (start & clean) ‡	0.030	0.030	0.030	0.030	0.030
Lub. oil at 65¢/gal.‡	0.101	0.068	0.053	0.041	0.032
Grease at 15¢/lb.‡	0.063	0.060	0.045	0.027	0.015
Repairs incl. labor	0.967	0.749	0.590	0.368	0.286
Total hourly variable cost	1.656	1.258	0.961	0.619	0.480
Total hourly own. & oper. cost exc. opr.	2.913	2.232	1.728	1.097	0.852

Data Used in Determining Some of the Above Costs:

	D8	D7	D6	D4	D2
List price §	$9,668	$7,487	$5,900	$3,676	$2,863
Approx. wt. in lb. (equipped)	35,323	25,228	17,449	10,614	7,161
Depreciation period: Years	5	5	5	5	5
Hours	10,000	10,000	10,000	10,000	10,000
Fuel cons.: Diesel fuel, gal./hr.	5.5	3.9	2.9	1.9	1.3
Filters, dollars/hr.	0.103	0.008	0.008	0.004	0.004
Lubricating oil, gal./hr.	0.10	0.07	0.05	0.04	0.03
Grease, lb./hr.	0.42	0.40	0.30	0.18	0.10

* The cost to be depreciated should be the total cost including freight and delivery.

† In estimating cost of publicly owned equipment this charge should not be made. No interest is paid on public funds. Government equipment is not (as a rule) subject to taxes. Most Government bodies make no outlay for insurance. Storage is generally carried as capital investment.

‡ Should be corrected to local rates.

§ Price is standard machine, including lighting system, front pull hook, radiator guard, and full crankcase guard. Cost of other attachments, local taxes, freight, and the like should be included.

Tables XXII and XXIII indicate the cost of owning and operating a D7 arch and winch under average conditions for a 10-hour day. These figures are then combined with the tractor cost and other labor costs to give a total cost per 10-hour day associated with the skidding of logs to a landing.

To correlate production with cost consult Table XXV for an indication of D7 Tractor performance. A 2000 ft. average haul distance will be assumed with the 32′ logs averaging 32″ in diameter. It can be seen that the D7 can average 2.1 logs per load and haul 7200 b.f. per hour under ideal conditions. If 70% of this figure is taken for a fair-to-good logging show with no adverse grades and subject to the other conditions listed, the final production figure is 5050 b.f. per hour or 50.5 M.b.f. per 10-hour day under the Scribner log scale.

TABLE XXII

COST OF OWNING AND OPERATING ARCHES AND SULKIES PER HOUR

Average Conditions

Model	D8N	D7N	D6N	D4 *	D2 *
List delivered price (incl. wheels)	$4,254	$3,566	$3,003	$1,264	$ 976
Fixed Charges					
Depreciation	0.425	0.357	0.300	0.126	0.098
Int., taxes, ins.	0.128	0.107	0.090	0.038	0.029
Total fixed charges	0.553	0.464	0.390	0.164	0.127
Variable Charges					
Repairs, incl. labor	0.325	0.267	0.220	0.089	0.067
Lubrication	0.008	0.007	0.007	0.003	0.003
Chokers					
Cable	0.372	0.242	0.161	0.101	0.076
Hooks and ferrules	0.084	0.061	0.061	0.048	0.048
Total variable charges	0.789	0.577	0.449	0.241	0.194
Total cost per hour	$1.34	$1.04	$0.839	$0.405	$0.221
Approx. weight, lb.	13,050	9,940	7,850	2,600	1,700

* With rubber tires.

TABLE XXIII

COST OF OWNING AND OPERATING TOWING WINCHES PER HOUR

Average Conditions, Logging

Model	D8N *	D7N *	D6N †	D4	D2
List delivered price	$2690	$1942	$1267	$946	$764
Fixed Charges					
Depreciation	0.269	0.194	0.127	0.095	0.076
Int., taxes, ins.	0.081	0.058	0.038	0.028	0.023
Total fixed charges	0.350	0.252	0.165	0.123	0.099
Variable Charges					
Repairs, incl. labor	0.135	0.097	0.064	0.048	0.038
Lubrication	0.023	0.013	0.009	0.010	0.008
Cable and rigging	0.323	0.218	0.179	0.143	0.111
Total variable charges	0.481	0.328	0.252	0.201	0.157
Total cost per hour	$0.83	$0.68	$0.42	$0.32	$0.26
Approx. weight, lb.	4225	2700	1740	1300	900

* Includes automatic brake and built-in drawbar.

† Includes built-in drawbar.

TABLE XXIV

D7N arch per 10-hour day	$10.40
D7N winch per 10-hour day	6.80
Tractor and operator cost per 10-hour day	29.82
Choker man or rigger @ 65¢/hour	6.50
Unhook man on landing @ ¼ of 65¢/hour	1.63
Supervision @ ⅓ of $1/hour	3.34
Total cost for 10-hour day	$58.49

TABLE XXV

ESTIMATED HOURLY PRODUCTION D7 TRACTOR, HYSTER D7N TOWING WINCH AND D7 LOGGING ARCH (80 DRAWBAR HP.) IN BOARD FEET PER HOUR—SCRIBNER SCALE, 32′ LOGS

Avg. Dia. of Logs, in.	*Avg. Logs per Load*	*Avg. b.f., per Load*	*Average Length Haul, ft.—One Way*							
			1,000	*1,500*	*2,000*	*2,500*	*3,000*	*3,500*	*4,000*	*4,500*
16	9.0	2,880	10,300	8,500	6,800	5,750	5,000	4,370	3,940	3,570
20	5.6	3,140	11,100	8,950	7,350	6,230	5,460	4,800	4,300	3,890
24	3.9	3,140	11,100	8,950	7,350	6,230	5,460	4,800	4,300	3,890
28	2.7	3,080	10,900	8,820	7,200	6,130	5,320	4,730	4,230	3,820
32	2.1	3,080	10,900	8,820	7,200	6,130	5,320	4,730	4,230	3,820
36	1.7	3,120	11,100	8,890	7,280	6,200	5,420	4,770	4,270	3,880
40	1.3	3,020	10,800	8,690	7,150	6,020	5,250	4,650	4,170	3,770
44	1.1	3,080	10,900	8,850	7,250	6,150	5,380	4,720	4,260	3,850

Logs weighing approximately 11 lb. per board foot Scribner Scale.
Average travel speeds: Loaded—second gear—194 f.p.m.
Unloaded—fourth gear—405 f.p.m.

These estimated production figures are calculated and based entirely on ideal conditions of level ground, a balanced operation, 100% operating efficiency. Fifty-minute hours are used, 6 minutes per load for hooking and unhooking, and loads of more than three trees to be bunched.

Factors to be considered when using these figures: difficulty in maintaining average board feet per load; variable conditions of terrain—i.e., adverse grades, soil conditions, etc., undergrowth, altitude; mechanical condition of equipment; weather conditions; scattered timber stands; variable weight of logs in mixed stands of timber; efficiency of over-all operation.

Seventy per cent of the above figures should be used as an estimate for a fairly good logging show with no adverse grades for the loaded haul to the landing, and lesser percentages according to the appraisal of the logging show.

Dividing total labor and equipment costs for a 10-hour day by the production per day equals a final cost in terms of production.

$$\frac{\$58.49}{50.5 \text{ M.b.f.}} = \$1.16 \text{ per M.b.f. of logs}$$

The sum $1.16 represents a total cost for skidding each thousand board feet to the landing in 32′ length logs.

Tractor production for other haul distances can be estimated in a similar manner if care is taken to give proper cognizance to the limiting factors.

If it is desired to estimate production and costs for any particular logging show, the following method may be utilized. Assume a pulpwood operation in the southern pine region where tree lengths are skidded to a landing for bucking into 5′ 3″ lengths by means of a combination power saw and loader.

Timber—southern pine forest grown form class 78
Average d.b.h.—9″
Pulpwood scale—unit (168 cu. ft. stacked wood)
Average tree volume—16 cu. ft. stacked wood *
Average tree weight—670 lb.
Terrain—5% adverse grade . . . clay, somewhat sandy, damp but not wet
Tractor—D4 and sulky (43 drawbar hp.)
  Tractor weight equipped for logging—11,400 lb.
  Sulky weight—2600 lb.
  Winch weight—1500 lb.

In order to calculate how many trees can be carried behind the D4 tractor under these conditions the rolling resistance of the sulky must be known. Rolling resistance is defined as the amount of pull in pounds needed to overcome 1 ton of weight. Rolling resistance varies according to tire size, tire pressure, and ground conditions. Table XXVI indicates a rolling resistance of 220 lb per gross ton for the D4 sulky in sandy-clay soil.

TABLE XXVI

Relation of Forest Soils to Log Sliding Resistance (One End Suspended), Rolling Resistance of Arches, and Tractive Ability of Tractors

Forest Soil Classification, Type of Soil beneath Vegetation Cover	*Log Sliding Resistance, lb. per ton of Log Weight on Ground*	*Rolling Resistance, lb. per ton, Rubber-Tired Sulky (13 × 24 Tires)*	*Traction Factor, %* *
Clay, sandy, damp or dry	1460	220	57
Clay, sandy wet	1200	270	39
Clay, no sand, damp or dry	1300	220	60

* Traction factor × tractor weight equipped for logging equals pounds pull available.

The log sliding resistance is indicated as 1460 lb. per ton of log weight on the ground. This refers to the pull required to skid each ton of log weight that is exerting a dragging force on the ground.

The tractive ability actually available for pulling logs and overcoming grade is derived from Table XXVI which shows 57% of the tractor

* Stacked wood is not solid wood.

weight in pounds as the pull available at the drawbar in this type of soil and when used in conjunction with sulkies.

$$11{,}400^{*} \times 0.57 = 6500 \text{ lb. pull available}$$

If directional felling is to be applied to facilitate tree bunching by the hauling tractor, the assumption is that all tree butts will be carried under the sulky with the tops dragging. Since the butts are heavier than the tops, 55% of the total log weight is applied directly on the sulky with 45% on the ground.

In hauling uphill the tractor must combat gravity, and the result is that, for every 1% of grade, 20 additional lb. of grade resistance per ton of tractor, sulky, and load weight must be added to the calculations. The facts just considered are given in Table XXVII.

TABLE XXVII

Grade (adverse)	5% + 3% *
Log sliding resistance	1460 lb./ton (*SR*)
Sulky rolling resistance	220 lb./ton (*RR*)
Average tree weight	670 lb.
Per cent of tree weight on sulky	55
Per cent of tree weight on ground	45
Pounds pull available for grades and load	6500 (*PPA*)
Unknown load weight designated	*X*
Tractor weight with winch and equipment	12,900 or 6.5 tons (*TW*)
Sulky weight	2600 lb. or 1.3 tons (*SW*)

* Although the tractor may be climbing a general grade of 5%, it may also be forced to climb steeper grades for short intervals owing to ground irregularity. An additional 3% grade is assumed to counteract this fact.

Since the total weight of trees and number of trees that can be carried are unknown, it becomes necessary to set up a simple equation with total weight as the unknown. The letter *X* indicates the unknown weight.

A careful study of this formula will reveal its simplicity when broken down into three sections. The first section represents the calculation for grade. Tractor weight, sulky weight and the unknown load weight all modify the grade ability of the tractor. As mentioned previously for each 1% of adverse grade 20 lb. of grade resistance per ton of weight is set up. The second section is for the consideration of the load's share of rolling resistance plus the load's sliding resistance. The

* Note that the tractor weight equipped for logging but without winch is used for this calculation.

third section deals with the sulky's share of the rolling resistance. Tractor rolling resistance is not a factor as the tractor specifications show drawbar pounds of pull after a proper allowance for rolling resistance. All three of these sections equal the pounds pull available after a traction correction so that the unknown load can be determined.

$$\overbrace{[(TW + SW + X) \text{ grade } \% \times 20]}^{1} + \overbrace{[(SR \times 0.45X) + (RR \times 0.55X)]}^{2} + \overbrace{[(RR \times SW)]}^{3} = PPA$$

Substituting knowns into the equation gives

$$[(6.5 \text{ tons} + 1.3 \text{ tons} + X)8 \times 20] + [(1460 \times 0.45X) + (220 \times 0.55X)] + [220 \times 1.3 \text{ tons}] = 6500 \text{ lb.}$$

Solving for X, we have

$$[1040 + 208 + 160X] + [656X + 121X] + [286] = 6500 \text{ lb.}$$

$$1248 + 160X + 777X + 286 = 6500$$

$$937X = 6500 - 1534$$

$$X = \tfrac{4966}{937}$$

$$X = 5.30 \text{ tons or } 10{,}600 \text{ lb.}$$

To find the number of trees that can be carried under these conditions, divide the load that can be carried by the weight of each tree.

$$\left.\frac{10{,}600 \text{ lb. total}}{670 \text{ lb. per tree}}\right\} = 15.8 \text{ trees or } 15 \text{ trees}$$

Since each tree according to Table XXVIII contains 16 cu. ft. of stacked pulpwood, and 168 cu. ft. represents one unit of wood, then the load in units or cords can be calculated.

$$\frac{15 \text{ trees} \times 16 \text{ cu. ft. of stacked wood}}{168 \text{ cu. ft. per unit}} = 1.43 \text{ units per load}$$

In terms of cords, the load can be calculated as:

$$\frac{15 \text{ trees} \times 16 \text{ cu. ft. of stacked wood}}{128 \text{ cu. ft. per cord}} = 1.87 \text{ cords per load}$$

If the average haul distance is 1000 ft. and the tractor hauls in second gear and returns in fourth at speeds of 211′ and 325′ per minute, the trip time can be calculated as 7.8 minutes.

$$\frac{1000}{211} = 4.75 \text{ minutes for haul loaded}$$

$$\frac{1000}{325} = 3.08 \text{ minutes for return empty}$$

Adding fixed time as listed below results in a total fixed time of 12.9 minutes.

Bunching time @ 0.7 min./tree	=	10.5 minutes
Unloading time @ 0.2 min./tree	=	3.0 minutes
Turns	=	3.0 minutes
Total fixed time	=	16.5 minutes

TABLE XXVIII

VOLUME AND WEIGHT CORDWOOD TABLE ALL SOUTHERN PINE

Based on 5 ft. Bolts

d.b.h., Outside Diameter of Tree at Breast Height (4½′ above Grd.), in.	*Form Class 78 Second Growth, Forest Grown Medium Height Fair Number of Limbs Medium Trunk Taper Bark—Flat Plates*		*Form Class 72 Second Growth Short Height Considerable Limbs Rapid Trunk Taper Bark, Thick Sharp Ridges, Deep Gullies*	
	Stacked Volume, cu. ft. (Not Solid)	*Weight per Usable Tree Length, lb.*	*Stacked Volume, cu. ft. (Not Solid)*	*Weight per Usable Tree Length, lb.*
6	6	260	4.5	150
7	9	360	7	220
8	12	480	10	360
9	16	670	13	500
10	20	875	16	560
11	24	1100	19	720
12	29	1295	24	1070
13	34	1630	27	1180
14	40	1890	31	1380
15	46	2200	36	1560
16	52	2560	41	1770
17	59	3210	46	2280
18	66	3560	52	2840
19	73	4080	57	3080
20	81	4540	65	3300

Fixed time plus travel time equals a total trip time of 24.3 minutes. Trips per 50-minute * hour can then be calculated as 2.06. The number of units brought to the landing per hour is found as indicated by multiplying the units per load by the number of trips per hour.

$$1.43 \text{ units} \times 2.06 \text{ trips} = 2.94 \text{ units per hour}$$

Totaling the hourly costs associated with the tractor operation results in a total cost per unit of wood for skidding on an average haul of 1000 ft. and under the conditions listed.

Tractor operator	\$0.75
Choker man	0.65
$\frac{1}{4}$ of unhooker time @ 65¢/hr.	0.17
Supervision @ $\frac{1}{3}$ of \$1.00/hr.	0.34
Hourly total cost of D4 tractor	1.10
Hourly total cost of winch	0.32
Hourly total cost of sulky	0.41
Total hourly cost	\$3.74

$$\left.\frac{\$3.74 \text{ total skidding cost per hr.}}{2.94 \text{ total units per hr.}}\right\} = \$1.27 \text{ per unit of wood deposited on the landing in tree length}$$

Such estimates can be of considerable value to the logger as every detail of his operation can be considered.

Labor efficiency is highly important, as a poorly trained crew lacking in experience can cause a marked decrease in ideal production figures. Improper rigging can also cut sharply into production. One logger in the South increased his production 40% by simply increasing the size of his winch-line hook to allow more logs to be carried.

TRACK-TYPE TRAILER PRODUCTION

Table XXIX is a type of production table that has been prepared by a manufacturer to serve as a guide to performance. It indicates the production possible in board feet with a tractor and track-type trailers in tandem.

If an operation is based on a 4000-ft. haul using a D6 tractor with two trailers in tandem, the production per hour can be found under the caption "4000" in the table. For logs whose scaling diameter averages 16″ the production per hour totals 3980 b.f. for both trailers.

* 50-minute hour used for labor and machine efficiency correction.

TABLE XXIX

ESTIMATED HOURLY PRODUCTION WITH "CATERPILLAR" D6 TRACTOR AND TWO ATHEY DT-286 (15-TON) TRAILERS IN TANDEM IN BOARD FEET PER HOUR—DOYLE SCALE

Log Dia., in., d.i.b.	*Load on Each Trailer Number of 16′ Logs*	*b.f.*	*Average Length of One-Way Haul, ft.* 1500	2000	2500	3000	3500	4000	4500	5000
10	23	828	1930	1810	1700	1600	1510	1430	1360	1300
12	19	1216	3260	3020	2800	2630	2460	2320	2200	2090
14	15	1500	4700	4310	3960	3690	3400	3200	3020	2850
16	13	1872	5870	5360	4950	4580	4250	3980	3760	3550
18	11	2156	6770	6170	5700	5270	4870	4600	4330	4090
20	10	2560	8040	7350	6780	6260	5800	5450	5150	4850
22	9	2916	9150	8380	7720	7150	6620	6240	5870	5530

Estimated production figures are calculated and based on a loading time of 1 to 1.5 minutes per log, third-gear hauling speed of 282 ft. per minute, return in 5th gear at 510 ft. per minute, unloading time of 6 minutes, and 3 minutes' delay time per trip. Fifty-minute hours are used for machine and job efficiency.

Factors to be considered when using these figures are:

1. Difficulty in maintaining average board feet per load.
2. Variable conditions of terrain such as grades and soil condition.
3. Undergrowth and altitude.
4. Weather conditions.
5. Efficiency of over-all operation.

It must be remembered that these tables are to serve as a general guide only. Conditions vary so widely from job to job that considerable care must be observed in their use. The conditions listed in the table can have a marked effect on production. Cost of production in terms of thousand board feet can be determined by dividing the total hourly cost of an operation by the production per hour, just as in previous sample problems.

ROAD BUILDING

Road construction and maintenance by tractors has become increasingly important. Timber that was previously inaccessible is now being cut through the use of access roads both public and private. Bulldozers and earth-moving scrapers not only are used for new construction but also are continually being applied to the task of improving and maintaining old roads.

In hilly terrain bulldozers are used for shelving out roadways on sidehills too steep for ordinary tractor operation. It is generally found most economical to use the bulldozer until a road 9′ to 12′ wide is made. If further cutting is desired, the tractor-drawn scraper then moves in to bring the grade down to the desired point. As mentioned previously, the tractor-drawn road-construction equipment can also be used to build up roads in areas where the groundwater is close to the surface.

Tables XXX and XXXI indicate the estimated production possible with various sizes of equipment and on different haul distances. In

TABLE XXX

ESTIMATED BULLDOZER PRODUCTION, CU. YD. PER HOUR—BANK MEASURE

Typical Excavation Production in yards per hour Return Speed 2.5 m.p.h.

Tractor Model	50-ft. Material *				100-ft. Material				150-ft. Material				200-ft. Material			
	One-Way Haul Length															
	1	2	3	4	1	2	3	4	1	2	3	4	1	2	3	4
D8	186	166	145	124	112	100	88	74	80	72	62	54	64	56	49	42
D7	169	150	132	113	102	90	79	67	73	65	56	48	56	50	44	38
D6	110	98	86	73	67	59	52	44	48	42	37	31	37	34	29	25
D4	94	83	72	62	56	50	44	37	41	36	31	26	31	28	24	20
Typical Excavation Production in yards per hour Return Speed 5.0 m.p.h.																
D7	193	172	150	128	119	107	94	80	86	77	67	58	68	61	54	46
D6	127	113	98	85	78	70	61	53	56	50	44	38	44	40	35	30

The above quantities are based on 60-minute hour efficiency, no grades, single dozer, no trench, dig and travel speed of 1.5 m.p.h.

* Material: (1) sand, (2) common earth, (3) clay, (4) broken rock.

Above quantities may be increased as much as 75% by trenching and downhill dozing.

TABLE XXXI

ESTIMATED TRACTOR SCRAPER PRODUCTION, CU. YD. PER HOUR—BANK MEASURE

Prime Mover	Pay Yards	*Average Length of Haul—One Way—ft.* 200	400	600	800	1000	1200	1500	2000	2500	3000	4000	5000
D8	17			176	152	133	118	101	82	69	60	46	38
	16			165	143	125	111	96	78	64	56	44	
	15			162	139	121	107	92	74	62	54	42	
	14			158	135	117	103	88	70	58	50	39	
	13		179	147	125	109	96	81	66	55	46	37	
D7	11		161	138	119	104	93	80	64	54	46		
	10		149	125	108	94	85	73	58	49	43		
	9		134	113	97	85	75	66	52	44	38		
	8		120	99	87	75	67	58	46	39			
D6	8	140	110	91	78	68	60	51	40	33			
	7	127	103	86	74	64	57	50	39				
	6	109	87	73	64	55	49	43	34				
D4	4	72	57	48	41	36	31						
	3	54	42	35	31	26	24						
	2	36	28	24	20	18							

Material weighing 3000 lb./yd.
No grades present.
60-minute hour efficiency.
Rolling resistance of 150 lb./ton.
Rule of thumb for loading:
15 yd. and over, 1.5 minutes.
14 yd. and under, 1.0 minute.

earth moving the cost of production is always in terms of cubic yards. A study of Table XXX will show that a D6 tractor operating as a bulldozer on a 100′ haul can move 70 cu. yd. of common earth if the 5-mile-per-hour reverse speed is utilized. The production possibilities can be increased as much as 75% by bulldozing down hill.

A study of Table XXXI will reveal that a D7 tractor working with a scraper that contains 10 pay yards of material can haul 108 cu. yd. per hour on an 800-ft. one-way haul.

In both tables the conditions listed should be observed before any attempt is made to apply the estimates of production. The tables are shown in this chapter primarily to illustrate the manner in which some manufacturers have indicated possible production in terms that are applicable to the logging and earth-moving industry. Tractors and equipment are subject to periodic changes in design, and this tends to alter production and cost data continually. Therefore, it is not advisable to use this material for actual jobs or logging operations.

SILVICULTURAL CONSIDERATIONS

The effects of various methods of skidding on the residual stand have been more carefully studied in California than in any other section of the country. This is partially due to the fact that tractor skidding was utilized there before it found favor in other sections.

As explained previously, tractors have supplanted much power skidding both for the sake of economy in skidding and for lowering the damage to seedling reproduction and the larger seed trees.

It is difficult to make direct comparisons between different methods of skidding in their effects on standing timber, owing to variations in stands, extent and character of reproduction, nature of topography, cutting practice, and the like; also the handling of the skidding devices fluctuates within a given period. Table XXXII indicates the extent of damage expressed in percentages for the different types of skidding in the California pine region.

Although horse skidding is least destructive to a growing stand, the economy of the tractor dictates against the use of animal power. The large timber in the West makes machinery of some type almost mandatory in that area.

In the South tractors have come to be regarded as extremely beneficial to successful natural seeding. The tracks expose the soil beneath the vegetation cover allowing the seeds easy access to the earth. It is generally conceded that tractor skidding is very much less destructive

TABLE XXXII

DAMAGE TO REPRODUCTION AND STANDING TIMBER BY VARIOUS LOGGING METHODS IN THE CALIFORNIA PINE REGION

Method	*Percentage of Seedlings Destroyed Less than 3.6" d.b.h.*	*Percentage of Larger Trees Destroyed over 3.6" d.b.h.*
Horse and big wheels	20–35	Negligible
Tractors and big wheels, close supervision	25–50	5–10
Donkey, ground lead, supervision	25–50	10–15
Donkey, high lead, regulated by extra ground blocks, "siwash" trees and high stumps; setting laid out in advance	25–50	10–15
Tractors with arches	25–50	10–15
Track skidders	25–50	10–15
Tractors and big wheels, unregulated	45–65	15–20
Donkey, ground lead, unregulated	50–75	15–25
Donkey, modified high lead (35′)	50–75	10–25
Donkey, high lead, high speed unregulated	50–95	25–75

than power skidding and perhaps very little more so than horse skidding.

With respect to fire control the tractor has proved of great value. In the West the general plan is to lay down a network of access roads in a forested area. These roads then serve as actual fire breaks and discourage any fire from traveling very far. These same roads can also serve as a line for setting backfires or as an access road to fire-fighting equipment.

This practice has been followed in the South to a certain extent, but here the fire-line plow is more extensively used. This plow pulled by a track-type tractor digs a wide furrow which serves both as a barrier to flames and as a place for starting backfires.

A recent development that is aiding reforestation in many areas is the tree-planting machine. Pulled by a tractor and operated by a crew of one to two men, these machines are planting seedlings with great economy and speed. With mechanical tree planters many areas that were once considered to be of little value have been put back into active use.

It is apparent that the track-type tractor is thoroughly at home in the woods. The multitude of uses that it can be adapted to has made it an essential tool indeed, a tool that can accomplish almost anything within reason.

CHAPTER IX

Power Logging—Cable-Hauling Systems*

CHARACTERISTICS OF CABLE-HAULING SYSTEMS

Cable-hauling systems for log transportation are characterized by the power machine operating in a stationary position pulling in logs from a distance by means of a wire-rope cable wound on a drum. In contrast, the direct-tractive machine, such as the tractor, moves while skidding and drags the logs along with it. The cable-hauling machine in its primary form consists of two or more drums powered through gearing by an engine. The logger's name for all machines of this type is "donkey" or "donkey engine," although they are more accurately termed by their specific function in the logging operation, such as "yarder" for hauling in logs, "loader" for hoisting logs, or "unit" where yarding and loading functions are combined in one machine. For portability the donkey may be mounted on a timber sled, a railroad car, a truck, a tractor, or a track-laying crawler. It may be powered by steam engines, electric motors, or internal-combustion engines.

The hauling pull of the donkey engine is transmitted to wire ropes (flexible steel cables termed "lines" by the logger), wound on the drums, which serve as temporary storage reels as well as power transfers. A minimum of two drums is required, a "main-line" drum for the log-hauling line, and a "haulback" drum for the line which pulls the main line, together with attachments for hooking on to the logs, out to where the logs lie. The haulback line is also called the "outhaul" or "trip" line. The lines are strung out through "blocks" in various ways to constitute the many cable-hauling systems. Usually the donkey has a third drum which carries a "straw line," a light line which can be pulled out by hand for stringing out the haulback line and rigging up the spar tree which is an essential element in most

* Most of Chapter IX was written by J. Kenneth Pearce, professor of logging engineering, University of Washington College of Forestry, Seattle, who also supplied the illustrations.

cable-hauling systems. The straw line is also colloquially called the "guinea" or "grass" line or "haywire." A "block" is what the non-logger sometimes calls a "pulley" or "pulley block" and consists of a "sheave" or wheel grooved to fit the wire rope, enclosed in a two-part steel shell joined at the top by a shackle for hanging the block and at the center by the pin around which the sheave freely rotates. The blocks are hung so as to control the "lead" or direction of movement of the lines.

Cable-hauling systems are used mainly in the Douglas fir region of Washington and Oregon, in the redwood region of California, and in southeastern Alaska and British Columbia. In these regions the initial transportation of logs from where they are bucked from felled trees, known as "skidding" in other regions, is termed "yarding." If a second movement of the logs by a cable-hauling system is required before they can be loaded on wheeled transportation, the operation is termed "swinging."

Cable-hauling systems have certain characteristic advantages and disadvantages, compared with direct-tractive systems which use animals or tractors. The advantages are:

(a) **Great power concentration.** Since power-logging machines operate from a stationary position, they may be built in any size and power within the limits of portability. They are available in a range of sizes from 2-ton gasoline donkeys powered with automobile engines to 200-ton steel-spar skidders with steam engines developing 1000 hp. Cable-hauling systems can operate at line speeds as high as 1800 f.p.m., equivalent to over 20 m.p.h. or about three times the maximum logging tractor speed.

(b) **Freedom from ground conditions.** Moving-cable systems have been developed which can operate over the most adverse topographic and soil conditions. Logs can be hauled by cable over steep, rocky, or broken ground inaccessible to tractors or through swamps or mud where direct-tractive systems would mire down.

(c) **Hauling up adverse grades.** Since the line pull is independent of direction of haul, cable-hauling systems can operate uphill as well as down. In fact some of the systems, such as the high lead, work best hauling uphill, and there is no limit to the adverse grade. Tractive systems are adapted to hauling down favorable grades, and the pulling power decreases rapidly with increase in adverse grade. There is also a maximum limiting grade up which the tractive system can return unloaded.

The characteristic disadvantages of cable-hauling systems compared with tractive systems are:

(a) **Limited hauling distance.** There are physical and economic limits to the yarding distance of moving-cable systems. These are given under the description of each system.

(b) **Logs move in straight lines.** The turn of logs in cable-hauling systems moves or tends to move straight toward the spar tree on which the lead block is hung and to damage or knock down any unfelled trees in its path. The cable-hauling system cannot follow a circuitous route to avoid residual trees as can the tractive systems. Cable systems are therefore adapted in general to logging in clear-cutting rather than tree-selection areas.

(c) **Variable control over logs being yarded.** Cable systems in general give less control over the logs than do tractive systems, and the amount of control varies with the system and at different stages of the yarding operation. Chances for damage both to the logs themselves and to residual trees are thus greater.

(d) **Fixed per acre costs.** The cost of rigging up the spar tree and of changing roads to yard a given area with cable-hauling equipment costs about the same, regardless of the amount of timber on the area. As the density of the stand decreases, the cost per thousand board feet of these and other fixed per-acre costs increases until an economic limit is reached in light scattered timber stands. Tractive systems are less affected by stand density and can thus log lighter volumes per acre economically.

HISTORICAL

Donkey logging began in 1881 in the redwood operations of the Dolbeer and Carson Lumber Company of Eureka, Calif. John Dolbeer was granted a patent on his "steam-logging machine" on April 18, 1882. The improved model built in 1883 consisted of a vertical firetube boiler with wood-burning firebox, supplying steam to a single vertical cylinder which powered a vertical spool or "gypsy" through a pinion and gear. The frictional resistance of several wraps of a Manila rope around the spool applied the power to hauling logs. The rope was pulled off the spool by hand and coiled alongside. One of these Dolbeer spool donkeys is preserved in the museum of the University of Washington College of Forestry.

The loggers of the Douglas fir region were quick to adopt the Dolbeer in preference to the horse teams which had supplanted the pioneer bull teams or oxen. In 1893 the Washington Iron Works of Seattle began building donkeys, replacing the spool with a drum powered by two horizontal steam cylinders. Wire rope soon replaced Manila, and a "line" horse pulled the line out to the logs. The addition of a haulback drum to the donkey made the line horse obsolete. The Tacoma Iron works of Tacoma pioneered compound gearing to increase pulling power, and the Willamette Iron Works of Portland, Ore., entered the field with the powerful and popular Humboldt gearing. The Washington Iron Works countered with a two-speed yarder with a low gear to start the log moving and a high gear to speed up the yarding. As donkey power and line size increased, the straw line drum was added to provide a line small enough to be pulled out readily by hand and strong enough to pull out the haulback line.

The cable-yarding system first used was the "ground lead." A "Tommy Moore" or "bull" block with a wide throat which permitted chokers and butt rigging to pass through the block was hung on a stump. The donkey was set some distance away to line up with the bull block. The main line was pulled out from the bull block along the yarding road, first by the line horse and later by the haulback which was strung out through small blocks. A "turn" of a single log was hauled in with the lead toward the bull block and parallel to the ground. The log hung up frequently behind the high stumps of those days, necessitating frequent changes of the choker hold. The names of the various choker holds, as "roll," "parbuckle," and "Sampson," are descriptive of the action of the log when the hauling power was applied to the choker hold. The butt rigging and chokers were pulled as far as possible through the bull block and then slacked off; the chokers were unhooked from the butt hooks, passed around outside the bull block, and again hooked to the butt hook on the donkey side of the block. Then the log was pulled parallel with the track if at a railroad landing, ready to be rolled on the car by a "parbuckle" or "crosshaul" as it is termed in the pine regions. If the yarding was to a skid road, the log was pulled into the trough of the skid road.

The skid road was an outgrowth of the animal-logging skid road. "Wide-face" donkeys or "roaders" with main-line capacities up to 4000 ft. were used on the skid roads which were troughlike chutes hewn from the smaller trees. A number of logs in line were trailed at a turn, and the speed at which they moved can be judged from old pictures of a boy riding the leading log and greasing the skids before him with a swab. Skid roading contributed picturesque but unprintable phrases to the logger's vocabulary which are heard today!

The first skyline or cableway-logging system is believed to have been used in Michigan in 1883. From this developed the Lidgewood skidder which was introduced in the Douglas fir region in 1904. Western loggers and machinery manufacturers teamed together to develop other less complicated skyline systems. In 1908 the slack-line system was put into operation. The need for a skyline system which could be operated with an ordinary yarder led to the invention of the North Bend system. As logging operations moved back into rougher terrain, the versatile logger invented other skyline systems to satisfy special conditions encountered, but most of these systems failed to survive. System names like Nestos, Hawthorne, Masten, Balderee, and Newby are now but a memory. Various forms of the aerial tramway also found a limited use. The double slack line developed in the redwood region around 1920. The Tyler and Dunham systems are of relatively modern origin.

The high-lead system was first used about 1905, but it did not gain general acceptance until about a decade later. Since World War I the high lead has remained the most popular cable-hauling system. The principle of hanging the main-line lead block high on a spar tree to give a lift on the log to overcome obstructions obtained double the output of the ground lead. Simplicity of road changing and type of donkey remained, and the high lead involved less capital investment and overhead than skyline systems.

Rising stumpage and log values and wages of wood buckers and splitters made wood fuel uneconomic, and about 1917 the industry changed to oil fuel for the steam donkey. In a few operations coal was used, and the few favorably situated with respect to cheap electricity replaced steam power with electric motors. In 1922 the Skagit Steel & Iron Works of Sedro-Wooley, Wash., began making gasoline donkeys, and in 1924 the first Diesel donkey entered the woods. As internal-combustion engines improved and logging camp mechanics learned how to service them, these found a wider acceptance in logging. The first general use of gasoline and Diesel donkeys was on "cold decking" yarders operating at a distance from the railroad where fuel and water supplies for steam power presented a problem. They also eliminated the fireman and the water-pump man from the crew and reduced the forest-fire hazard.

Until the depression years of the 1930's the logger had demanded and the manufacturers had supplied bigger, faster, and more powerful yarders and skidders. The ultimate was reached with the 200-ton interlocking skidder carrying a 110′ steel tower, requiring a crew of 21 or more men to operate, and an investment of $100,000. The recovery years following the depression witnessed a complete reversal from the previous trend toward heavier equipment. A number of factors contributed to the new trend toward lighter donkeys and rigging. The rapid development of motor-truck transportation demanded light donkeys which could be moved on trucks. Rapidly rising wage rates favored the use of lighter chokers and rigging which could be handled by crews of fewer men. More extensive use of tractors changed the operational organization. Development of high-speed gasoline and Diesel engines of low weight-to-horsepower ratios enabled the manufacturers of logging equipment to meet this demand. To approach the flexibility in power and speed of steam multispeed transmissions, hydraulic couplings, torque converters, "torq-masters" and automatic transmissions have been added to the modern donkey. The steam donkey is now virtually obsolete.

The West Coast logger had tentatively experimented with the crawler tractor as early as 1912, and the gasoline tractor was accepted to a limited degree during the 1920's while the tractor was rapidly supplanting cable-hauling methods in the western pine regions. The development of the powerful Diesel tractor, the arch, and the bulldozer provided a combination which could satisfactorily handle the big logs of the redwood and Douglas fir regions. The "arch cat" with its greater flexibility and mobility displaced the yarding donkey wherever topographic and soil conditions permitted the tractor to operate. Tractor roading on bulldozed "cat" roads competed with skyline swings. The tractor even moved into the donkey's own field by mounting two or three drum "tractor donkeys" and operating high leads.

The latest development in moving-cable machines is in experiments in "prelogging" small understory timber which would be broken in logging the larger trees, and in "relogging" the slash in clear-cut areas to salvage pulpwood, which formerly was economically unfeasible. Steel towers have been mounted on tractor donkeys and tubular steel spars on wheels towed by motor trucks carrying light gasoline donkeys; A-frames mounted on light donkey sleds are being used for relogging and prelogging. All these use the high-lead principle with the main-line lead block elevated 40′, 50′, or 60′ above the ground. In the pulpwood operations in the Northeast and Canada, experiments in mechanizing skidding with machines similar to the relogging equipment of the West are being carried on. The principle of mounting the donkey on crawlers carrying a steel tower, first applied by the Clyde Iron Works of Duluth in 1925, is now being revived for use on the larger donkeys to reduce the high labor cost of lost time in moving and rigging up.

A moving-cable machine entirely new in principle has recently been invented by a logger. The hauling machine rides on two tight skylines, hoisting and carrying the turn of logs suspended under the machine at speeds up to 35 m.p.h. Open-sided jacks permit the skylines to be supported at any number of points and to be extended to distances far beyond the span of the conventional skyline supported at two points. This system has been named the "Skyhook."

The development of loading systems kept pace with the increasing production of yarding systems. The parbuckle (crosshaul) which simply rolled the logs onto the car gave way to the crotch line which was faster and controlled the logs better. With the rigging up of spar trees, guy-line loading systems were invented. The single tong was succeeded by the Duplex with two tongs and a special donkey with reversing steam engines. Motor-truck logging increased the popularity of the swinging-boom systems. To pick up logs from anywhere on the large landings required on tractor-yarding operations, the old crotch line with spreader bar and haulback was revived. The latest are mobile loaders, adapted from the self-propelled power shovel on crawlers, and crane-type loaders, mounted on rubber tires.

CABLE-HAULING EQUIPMENT

Cable-hauling equipment may be divided into three major classifications:

(*a*) Donkey engines, variously termed logging engines, logging hoists, or yarders and loaders, which supply the hauling power. The name "donkey," however is not applied to the specialized interlocking skidder.

(*b*) Wire rope and attachments, including lines, straps, chokers and the various devices for connecting them. These comprise the moving elements which handle the logs.

(*c*) Blocks and carriages which are devices with sheaves through which the lines are threaded to determine the "lead" or direction of movement.

DONKEY ENGINES

The essential elements in a donkey engine are a power plant, a transmission connecting power plant and drums, drums for actuating and storing the lines, and the operator's controls. They must be built on a strong frame to withstand the stresses and shocks of logging service and mounted so the donkey can be readily moved. The relationship of the various elements and their operation will be clarified by refer-

FIGURE 65. Bird's-eye view of light gasoline yarder with gear and chain guards removed.

ence to Figure 65. This is a bird's-eye view of a typical modern light gasoline yarder with gear and chain guards removed. The power plant mounted crosswise of the frame is a 6-cylinder gasoline engine developing a maximum of 106 hp. and a maximum torque of 312 ft.-lb. Through a single plate clutch the engine drives a 4-speed helical gear transmission enclosed in the case to the right of the engine. A multiple-strand roller drive chain connects a 15-tooth sprocket inside the transmission case with a 68-tooth sprocket on the rear shaft. This shaft carries a haulback drum between the roller-shaft bearings and a straw-line drum on the outboard end. A gear on the rear shaft drives a pinion on the jackshaft below, which meshes with a main-drum gear on the forward shaft. The jackshaft pinion serves the dual purpose of reducing the speed ratio, thus increasing the power on the main drum, and spooling the line on the drum from the top.

FIGURE 66. Side view of Skagit gasoline yarder.

As all gears are rotating, in one constant direction, when the yarder is running, the drums float on bronze bushings on their shafts, in order to wind in on one drum and permit the other to unwind. To apply pulling power to a drum, the friction-control lever, seen at the left end of the shaft, is pulled toward the operator who stands at the side where the checkered foot pedals are. The position of the various control levers may be understood better by reference to Figure 66. The friction lever turns a screw-type friction against a thrust collar and pushes the drum against oak or composition friction blocks bolted to the gear. The drum flange has a tapered slot into which the ring of friction blocks fits. As long as the friction is engaged, the drum rotates with the gear. When the friction lever is released, by being pushed away from the operator, coil springs on the gear push the drum away from the friction blocks, and the drum then idles on the shaft. The straw-line friction lever on the operator's side moves the chain on the far right to apply the straw-line drum friction. Band-type brakes encircle drum flanges on the operator's side and are applied by foot pedal. The short engine-throttle lever and the heavier and slightly longer gear-shift lever are seen between the friction levers. A foot pedal operates the clutch. The frame is welded from structural steel I-beams. On the larger gasoline and Diesel yarders the frictions and brakes and some types of clutches and transmissions are operated by compressed air, controlled through valves by small hand levers.

The yarder illustrated in Figure 65 weighs 9700 lb., is 9′ long by 7′ 6″ wide, and has drum capacities of 1050 ft. of 1″ main line, 2250 ft. of 5/8″ haulback line, and 3000 ft. of 3/8″ straw line. Average main-line speed varies from 62 f.p.m. in first gear to 285 in fourth gear. Average haulback-line speeds range from 159 to 735 f.p.m. Bigger yarders in the 200-hp. class have top average line speeds of around 476 f.p.m. on the main line and 1770 on the haulback. Additional drums for operating skyline systems are usually mounted in tandem on a lengthened frame. A number of drum combinations and a wide range of sizes of yarders and loaders are manufactured to meet every condition. One manufacturer offers machines ranging in weight from 4200 to 58,000 lb., powered with gasoline or Diesel engines rated at 34 to 301 hp., with twin Diesels used in the largest yarders to develop 500 hp. or more. Standard automotive and industrial gas engines, Hesselman cycle engines, high-speed automotive and industrial Diesels are used. The logger has a choice of a variety of modern transmissions to obtain the desired performance under the conditions and with the cable-hauling system used. He may use 2-, 3-, or 4-speed gear transmissions, "automatic" transmissions such as the 3-speed air-controlled Carcomatic and Gearmatic, or the 2-speed Torq-Master. With high-speed engines operating above 1500 r.p.m., fluid couplings, familiar as the "fluid drive" on automobiles, and torque converters may be used. The torque converter is a hydraulic turbine, which offers, in addition to the shock-cushioning and nonstalling advantages of the fluid coupling, an infinite number of speeds and torque ratios within its working range and an increase in output torque up to approximately 2½ times engine torque at one quarter of the engine speed.

The steam donkey in use today has a vertical fire-tube boiler with injector water feed from a tank mounted on the rear. Oil is the preferred fuel, and the oil feed is usually controlled automatically by steam pressure on diaphragms operating the fuel valves. Fuel-oil feed may be by vacuum or pressure from an oil tank. In railroad operations an oil-tank car is spotted on a spur track behind the donkey. Steam cylinders range in size up to 13″ bore by 18″ stroke and steam pressures up to 225 lb. Piston- or ball-balanced valves are driven by outside valve gear distinctive with each manufacturer, by which the make of donkey may be identified at a glance. A pinion on the crankshaft drives the drum gears through various arrangements of gear trains. Instantaneous-change two-speed gears are actuated by frictions similar to drum frictions. The remaining parts of the machine are similar to the gasoline or Diesel donkey. Steam frictions may be used as well as air or hand frictions.

Donkey sleds are made of two logs strongly braced by cross timbers and steel tie rods. "Fair-leaders" are mounted on the front end for running the lines through when moving, and the sled should be long enough so the lines running from the fair-leaders will spool evenly on the drums. Steel car mountings have 8 to 24 wheels on full swiveling trucks and usually hydraulic jacks to transfer the weight from wheels and track to timber blocks when yarding. Full crawler track-laying mountings, such as are used on power shovels, and several types of rubber-tired wheel mountings are available for truck-logging operations.

WIRE ROPE

The development of efficient power-logging systems has hinged on the development of wire rope. The Manila rope adequate on the early days of power logging could have served only a small fraction of the cable needs that have developed with the various power systems. Modern wire rope now serves a multiplicity of uses under conditions that are more severe and abusive than those that any other logging equipment is required to bear. Though often considered a static material, wire rope is in reality a rather complicated piece of machinery with a large number of moving parts and with problems of lubrication and co-ordination of its separate parts.

Construction and types. Reduced to its simplest elements, wire rope consists of numerous individual wires combined into strands which are in turn grouped around a central core to form a flexible metal rope. The functions and arrangements of the various parts are as follows:

Wires. The base of wire rope is the individual wire. The ultimate strength of the rope is dependent on the material and size of these wires.

The flexibility and toughness of the whole rope also depend on the individual wires. Individual wires are drawn from the following materials: (1) improved plow steel, (2) plow steel, (3) cast steel, and (4) iron. The strength of these is in the order listed, improved or special plow steel being highest and iron lowest in value. Wire rope for logging purposes is generally made from plow steel, though in recent years the number of special or improved plow-steel ropes for conditions of severe usage has increased.

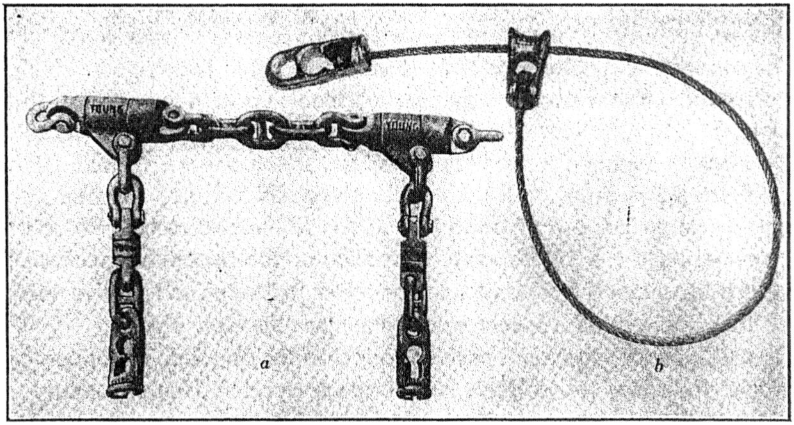

FIGURE 67. (*a*) Typical high-lead butt rigging made up of clevises, swivels, and links. The main line is attached to one end and the haulback line to the opposite end. The suspended portions terminate in butt hooks to which chokers will be attached. (*b*) Typical choker assembly. Ferrule on one end of the choker wire rope engages sliding choker hook; ferrule on other end engages butt hook. *Courtesy Young Iron Works, Seattle, Wash.*

Sizes of individual wires vary not only in ropes of different sizes but also in ropes of the same size constructed for different uses. Also, it is common to find wires of two sizes in the same strand. Fewer wires of large diameter give greater strength, whereas more wires of smaller diameter make a more flexible rope of lower strength. In the preformed type of rope, the individual wires are preformed to the shape they will take in the finished strand; thus, some of the stress a straight wire is subjected to when forced into its place in the strand is eliminated, and a more even distribution of stress among the individual wires is secured. The average strength efficiency of wire rope is about 82% of that of the metallic area of its cross section.

Strands. The number of individual wires in a strand is usually designated by a number; the relative size and position of these wires are designated as its construction, which has certain type names. For logging purposes, a 19-wire strand is most common. In arrangement, this 19-wire strand might be built up from one large central wire surrounded by 9 smaller wires and these in turn by 9 large wires forming the so-called Seale construction, whose large outer wires are well adapted to withstanding abrasion. An alternative, known as Warrington construction, consists of 1 central wire, surrounded by 6 similar-sized wires which are surrounded in turn by 12 alternating large and small wires to make up a 19-wire strand. Wire rope is also built with fewer or more wires to the strand; 7-wire strands, as well as 37-wire ones, are useful, though not generally for logging purposes.

The number of strands to a rope is also designated by number; thus 6 × 19 wire is the common logging rope made up of 6 strands of 19 wires each. Another more flexible rope known as 8 × 19 with 8 strands of 19 wires is commonly used for chokers. The manner in which the strands are arranged around the core is known as the lay. Right lay, which is standard, has the strand laid in a right-hand direction around the core while the wires in each strand lie in the opposite direction. Lang lay has both the strands and the wires in the strand laid in the same direction, giving a longer wearing surface on each wire.

Core. Cores of wire rope perform the twofold function of supporting the strands and providing a lubrication reservoir. Hemp cores are common and furnish an excellent means of storing a large amount of lubricant. Hemp is not entirely satisfactory as a support for the strands where the rope is subject to lateral pressures of high tensity. For uses such as skylines. or for long lines where the drum pressure is great, independent wire-rope cores are often employed. This core has the further advantage of adding some 10% to the strength of the rope.

Some rope sizes and types in common logging use are given in Table XXXIII.

TABLE XXXIII

Use	*Diameter*	*Material*	*Construction*	*Core*	*Breaking Strength, tons*
Skyline	1¾″ to 2″	Plow or improved plow steel	6 × 19	Independent wire rope	115 to 175
Main line or skidding line	1⅛″ to 1⅜″	Plow steel or improved plow steel	6 × 19	Hemp or wire rope	45 to 85
Haulback	¾″ or ⅞″	Plow steel	6 × 19	Hemp	20 to 28
Straw line	⅜″	Plow steel	6 × 19	Hemp	5
Chokers	⅞″ to 1¼″	Plow steel	8 × 19	Hemp	24 to 48

Connections and miscellaneous wire-rope eauipment. The most common connection used in attaching rigging to wire ropes in logging is the "eye" splice. This can be made up in the field in short time with a minimum amount of equipment and when complete does not add greatly to the weight or bulk of the line. It does not develop the full strength of the rope, however, and wears out rather rapidly unless the "eye" is supported by a steel thimble. Socket connections in which the rope is held by separating the individual wires and filling the socket with molten zinc develop the full efficiency of the rope; the use of these is increasing.

The following miscellaneous wire-rope equipment is essential in connecting the various lines or attaching logs to the hauling lines:

Butt rigging. A combination of swivels, shackles, chain or straps, and butt hooks which performs the following functions: (1) permits connection between main and haulback lines with swivels to prevent kinking or tangling; (2) provides place for attachment of one or more chokers.

Choker. Short length of flexible wire rope with connection for attachment to butt hook on one end and choker hook on the other. Chokers are used to form the noose around the log and hold it during skidding operation. Choker hooks may be flat open hooks or one of a number of patent types developed to prevent accidental unhooking of chokers while the log is in motion. Of the patent types, the Bardon and Peters hooks are commonly used.

Strap. Short piece of wire rope with "eye" splice in each end, for attaching blocks to stumps and similar purposes.

BLOCKS AND CARRIAGES

Blocks serve two main purposes in power-logging systems: They change the direction or angle of line pull, or they furnish a means of increasing the mechanical advantage of a system. The most common use of blocks in a logging system is the first. Normally a power-logging engine is designed to have sufficient power for operating purposes. However, for moving occasional heavy logs with light equipment, for moving donkeys, for tightening skylines, and similar operations, it is necessary to increase the line pull available. Here the mechanical advantage gained by adding more moving lines to a system is utilized. A one-block hold gives a mechanical advantage of 2; a block and half, of 3, and so on.

Blocks consist of two part-steel shells, inside which hardened-steel sheaves turn on bronze roller or ball bearings and have shackles or swivel yokes for attachment to the stump or spar. Most blocks are provided with a lubrication reservoir so that they may run some time without attention. Blocks range in size from the small riggers or pass-line block, weighing 15 to 25 lb. and built for ⅜″ or ½″ line, to high-lead blocks with sheave diameter of 30″ to 36″, weighing 1500 to 2000 lb. and adapted for line sizes of 1⅜″ to 1¾″. Between these limits, blocks are built in a wide variety of sizes for different uses.

Skyline carriage. Used in skyline systems, carriages or "bicycles" as they are sometimes called are the carriers which travel along the skyline or track cable. They may have one or more track sheaves and vary from simple inverted blocks to the modern slack-pulling canyon-type carriage used in the West with the Lidgerwood skyline skidders.

Tree shoes. For suspending skylines in spar trees, tree shoes are frequently used instead of blocks. Tree shoes are steel shells with shackles to permit attaching to spar trees by means of straps and they have round wood blocks or metal-troughs to support the skyline. These are, of course, adapted only to stationary skylines, large lead blocks being used to suspend the skyline of a slack skyline system from the head spar. Tree jacks support the skyline on three to six rollers or sheaves forming an arc.

CABLE-HAULING SYSTEMS

Although cable-hauling systems are rigged in many ways to meet the requirements of varying logging conditions, certain elements are common to all systems. A "main line" or skidding line hauls in the logs. A "haulback line" pulls the main line out to where the logs to be hauled can be hooked on. The log is grasped by a wire-rope "choker" placed around the log near one end. The choker is attached to a "butt hook" which, through various forms of "butt rigging," is pulled by the main line. The haulback line is strung out through haulback blocks hung on stumps to lead the lines in the desired direction. The haulback blocks may be designated as "side," "corner," or "tail" blocks, depending on their location. The lines run out from the donkey drums through "lead" blocks hung on a "spar tree," "spar pole," or tower."

SPAR TREES

The spars are usually rigged from tall trees, although portable steel spars or towers are used on light mobile yarders for small logs and on huge interlocking skidders. The tallest stoutest tree available at or near the place where the spar is to be rigged is selected in advance of felling of the timber to be logged and is marked with a big "X" chopped into the bark. If no suitable tree is growing in the required place, a spar tree may be "jumped" distances up to 100′ by rigging guy lines on the desired tree, sawing it off at the butt, and then pulling the butt into position while holding the spar tree vertical by alternately slacking off and tightening the guy lines. Or a tree may be hauled in and raised by a multiple-block purchase from a "dummy" tree rigged for this purpose. Before the spar tree is rigged, all the surrounding trees in a circle with a radius equal to the height of the spar tree are felled for guy-line clearance.

FIGURE 68. High climber topping a spar tree along a spur line. Insert shows the spar tree guyed and rigged for power skidding. The diameters of the trees where topped may be from 24″ to 30″, and this point may be from 150′ to 200′ or more above ground. *Photo by John D. Cress.*

The spar tree is rigged by a "high climber," or "head rigger," assisted by a rig-up crew and a donkey. Standing trees are first limbed and "topped" by the high climber. Wearing long-spurred climbing irons and a wide heavy leather belt to which is fastened a Manila rope with a wire-rope safety center, he climbs by leaning against the tree while standing in the embedded spurs, flipping the slack rope up, and then leaning away from the tree against the taut rope and taking several steps upward. The high climber carries a short-handled double-bitted axe and a short felling saw hung on ropes from his climbing belt so as to dangle well below his feet. As he ascends, he chops or saws off the limbs until he reaches the desired height or top diameter at which the tree is to be "topped." For use with the larger yarders in old-growth Douglas fir this point is usually about 24″ to 30″

in diameter and 150′ to 200′ in height. "Topping" is a spectacular and hazardous operation. The top is felled in the same manner as trees are felled on the stump by hand. The undercut is sawn and chopped out and the back cut sawn until the top falls. The spar snaps back from the thrust of the falling top and sways in a wide arc while the high climber hangs on grimly.

A light "pass line" or rigging block is pulled up by the high climber hand over hand and hung on a wire-rope "strap" or sling near the top of the spar. A ⅜″ pass line is threaded through the block, and thereafter the high climber ascends and descends the spar on the pass line, sitting in a "bos'n's chair." The pass line is also used to hoist the other straps, the blocks, and the guy lines. The top guy lines, usually six in number, are shackled to the spar tree just below the pass-line block. The lower ends of the guy lines are pulled around notched stumps spaced equidistant from each other and at a distance from the base of the spar approximately equal to its height. The guy line is pulled tight by the main line of the rig-up donkey and spiked with railroad spikes. Several more wraps are taken around the stump, tightened, and spiked in turn. The main-line lead block, also termed the "bull" block, is hung on a heavy wire-rope strap a few feet below the top guy lines, and an additional "safety strap" is fastened to a loose shackle on one of the top guy lines. In the event of failure of the main strap the bull block will slide down the guy line instead of falling on the men working around the base of the spar tree. In rigging a head spar for a skyline system the skyline "shoe" or "jack" is hung first below the guy lines, and the bull block is hung below the skyline shoe. The haulback lead block is hung below the bull block. Other blocks required for various skyline systems are hung as shown in the illustrations. On spar trees used with the more powerful donkeys an extra set of "buckle" guys are hung at about two-thirds the height of the spar to reduce the length of the unsupported column and prevent the spar from buckling under a hard pull on the main line. If the spar tree is to be used for loading as well as yarding, the loading guys are hung with or below the buckle guys, and the loading lead blocks and other rigging of the loading system below the guys. The straw-line lead block is generally hung near the base of the spar.

Steel spars usually have the top guy lines and the lead blocks permanently attached. The steel spar or tower is moved into position, usually with the spar lowered and resting on a wheeled support, and raised to the vertical with special mechanisms. The guy lines are pulled out and anchored to stumps.

When the spar is rigged, the yarding donkey moves in and is set facing the spar tree at a safe distance from it on leveled ground or a cribwork of logs forming a level platform. The rear end of the yarder sled is anchored to stumps to prevent its sliding forward while yarding. Usually the weight of yarder and sled is enough to prevent it from lifting up. The lighter tractor donkey, however, must have the drum end tied down, usually by a hole being dug under a large stump, between the roots, and a holddown cable run under the stump.

TECHNICAL TERMS

Certain technical terms which are common to all systems are used for the various elements of moving-cable logging. The following terms used on the Pacific Coast are repeated frequently in the descriptions of the various cable-hauling systems:

Landing. The place to which logs are pulled by a cable-hauling system for loading on a wheeled transportation system. The landing is immediately adjacent to a road or railroad.

Yarding. The operation of pulling logs from where they are felled and bucked to the first concentration point. Yarding on the Pacific Coast is synonomous with skidding in other regions.

Deck—cold deck or hot deck. A pile of logs yarded together at a place distant from the landing to which they must be pulled in a second handling. If all the logs coming to a given deck are yarded before the operation of moving them to the landing begins, the pile of logs is termed a "cold deck." If the two operations are carried on simultaneously, the logs being "swung" away as fast as they are yarded, the place to which they are yarded is a "hot deck."

Swinging. The operation of moving logs from a cold deck or hot deck to a landing by a moving-cable system. If this operation is performed by a tractor, it is termed "roading." Roading was originally used for the operation of donkey hauling over pole-skid roads.

Setting. The area yarded to a given spar tree. Sometimes where swinging is practiced, the entire area from which logs come to one head spar tree at the landing is called the setting.

Turn. The one, two, or more logs yarded or swung at one time.

Road. The path which logs follow while being yarded or loaded. Also applied to the alignment of the main line when being pulled toward the tail block. In yarding, when all the logs that can be reached along one "road" have been yarded, "road changing" is performed to route the lines along a new "road."

FIGURE 69. Aerial view of clear-cut settings in the Douglas fir region yarded with cable-hauling systems. Note yarding roads radiating from landings along truck roads. Compartments of standing timber were left to provide seed source for natural reforestation of logged areas. *Courtesy St. Paul and Tacoma Lumber Company, Tacoma, Wash.*, on whose operations this photograph was taken.

Side. All the men and equipment engaged in logging a given setting. "Sides" are designated by number or by system, as the "skidder side," the "slack-line side."

JOB NAMES

With the exception of the interlocking skidder crew, the job names of the crew members making up a cable-hauling side are the same on the various systems. The usual yarding crew contains one each of the following with the exception of choker setters, whose numbers vary with the power of the donkey and the size of the chokers used in consequence:

Hook tender. Working foreman in charge of the side.

Rigging slinger. In charge of the choker setters; selects logs to make up a turn; shouts signals to the "whistle punk."

Choker setter or chokerman. Places choker around log and hooks choker to butt hook.

Whistle punk or signalman. Sounds horn or whistle on donkey through electric system or jerk wire to inform yarder engineer when to wind in the various lines or to stop. The signal is loud enough to give warning to all the members of the crew when a line is about to move.

Gopher or powder monkey. Digs or shoots (with dynamite) choker holes under logs partially buried in soft ground. On small crews one of the choker setters will do this work as required.

Chaser. Unhooks the choker from the log at the landing.

Yarding engineer. Operates the yarding donkey.

Steam donkeys require the addition of a fireman to attend the fuel-oil valves and the water injector and steam gages. During extreme fire-danger periods a watchman is added to the crew to watch for fires set by sparks from steam boilers, around haulback blocks, and from smokers around the landing. The following men usually divide their time among two or more sides: camp foreman, in charge of all the sides operating out of one camp; donkey doctor or mechanic who makes repairs on donkeys. High climber, who, in addition to rigging up the spar tree and taking down the rigging when the setting is finished, goes up the spar to oil the blocks at intervals.

The loading crew which also belongs to the side is generally composed of one head loader, who selects the logs to be loaded in turn and by hand signals directs the loading engineer operating the loading donkey when to hoist and where to place the logs on truck or car; one, two, or more second loaders, depending on the loading system and the output, who hook on and unhook the tongs or straps used to grasp the log; and one third loader who brands the loaded logs and assists the other members of the loading crew in spotting cars or truck trailers, cutting wood chunks to block loads, and so on. The third loader is omitted from the smaller crews. A scaler may be employed to scale the loaded logs to determine daily production or to make out the scale bill for the load when contract truck haulers are employed.

GROUND-LEAD SYSTEM

The ground lead, originally used for yarding clear-cut settings, as described under the heading, "Historical," is obsolete in the West. In the southern pine and ponderosa pine regions, the two-line and four-line skidders which were practically ground leads have been supplanted by the tractor. The pull-boat slack-rope system of the cypress region is occasionally used in swamp logging. A donkey mounted on a barge skids logs by ground lead from swamps into water channels where the logs are rafted. Hollow steel cones placed over the front end of the log prevent it from digging into the mud and help it to nose past stumps. Cones are sometimes used in yarding poles out of cedar swamps. In southeastern Alaska donkeys are mounted on log rafts for yarding along the shore line. The lead blocks are hung on A-frames tall enough to give a high-lead lift on the turns for the lower part of the yarding range.

A "low high lead" which is virtually a ground lead is used on tree-selection settings under conditions where tractors cannot be used. The main-line block is hung only high enough to enable the logs to be handled and piled up at the landing. Careful planning of the layout of the yarding roads and supervision of operations can leave the residual stand with but little more damage than if logged with arches and tractors. However, output is low since slow line speeds must be used to minimize damage, and hangups are frequent in the absence of a lift on the log. Yarding with A-frames mounted on sleds is used in relogging in the West and as experimented with in Canada and the Northeast constitutes a "low high lead," as there is no lift on the turn until it nears the A-frame except when it is tight lined by braking the haulback line.

HIGH-LEAD SYSTEM

The high lead is the most commonly used and, therefore, the most important of the cable-hauling systems. Despite the great advances made by tractor-arch yarding, and the opposition to the high lead of certain conservationist groups, more logs are yarded on the Pacific Coast, from the redwoods through the Douglas fir region to Alaska, with the high lead than with any other system. There are sound reasons for the popularity of the high-lead system. Much of the timberland is too steep, broken, or rocky for tractors. On clay or clay–loam soils tractor operations are hampered during the long rainy season. The high lead is the simplest of the moving-cable systems used today, requiring less time to rig up and change roads, a smaller crew, and lower capital investment. If the appropriate-size donkey and rigging are chosen, the high lead is adaptable to a wide range of timber size and density. If the main transportation roads are located to the best advantage, the high lead is adaptable to varying conditions of topography. Finally, in logging shade-intolerant species, such as Douglas fir, demanding clear cutting for regeneration, high-lead logging is not incompatible with good silviculture. The state laws of Washington and Oregon require burning of the slash on clear-cut settings, and so the unmerchantable trees knocked down by the high lead would otherwise become fire-killed snags. As logging operations move farther back in the mountains to rougher and steeper ground, the high lead promises to continue an important logging method.

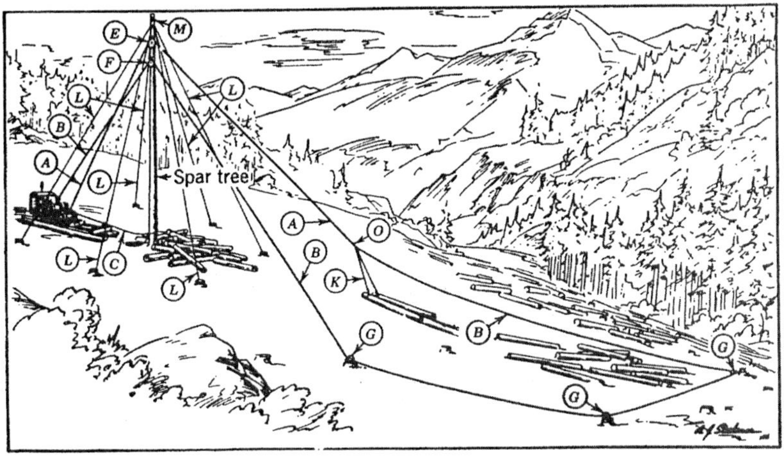

FIGURE 70. High-lead yarding.

A. Main line.	*G*. Haulback corner or tail block.
B. Haulback line.	*K*. Chokers.
C. Straw line.	*L*. Guy line.
E. Main-line block.	*M*. Pass-line block.
F. Haulback block.	*O*. Butt rigging.

Prepared for Skagit Steel & Iron Works under direction of J. Kenneth Pearce, professor of logging engineering, College of Forestry, University of Washington. Copyright 1947, Skagit Steel & Iron Works, Sedro-Woolley, Wash.

The successive steps in the operation of the high lead subsequent to the rigging of the spar tree and moving into position of the yarder previously described are as follows: The logging of the first quarter of the setting begins on the semicircle opposite the yarder, and the quarter on the same side of the spar as the lead blocks are hung. A haulback tail block is hung on a stump or tree at the edge of the setting and at the back end of the first road to be yarded. This road may be on the "fair" lead on an extension of the line of sight from the donkey to the spar, or on the "square" lead perpendicular to the fair-lead. One or more haulback corner blocks are hung at the ends of subsequent roads in the direction which yarding will progress as roads are changed. The straw line is pulled out by hand through the tail block, then through the corner blocks, and back to the foot of the spar tree. To lighten the labor of pulling by hand the straw line is usually handled in sections of 500 ft. or so which are fastened together with a "Hindu" hook or a "Milly Hogan," a single strand of wire rope twisted around itself in a loop joining the "eye" splices in the ends of two wire ropes. The end of the straw line at the spar tree is hooked or "Mollied" to the end of the haulback line. Reeling in on the straw-line drum pulls the haulback line out through the blocks and back to the spar tree. Then the end of the haulback is clevised to one end of the butt rigging, the other end being attached to the main line. One or two chokers are hooked on to the butt hooks on the butt rigging. Reeling in on the haulback line pulls butt rigging and main line out along the yarding road to where the logs lie. The chokers are placed around the logs by the choker setters, directed and helped by the rigging slinger, who, when he sees that everyone is "in the clear," shouts the signal to pull on the main line. The signal is transmitted by the "whistle punk" to the whistle or horn on the donkey. The yarder engineer throws in the main-line drum friction, starts in low gear, and the turn of logs moves toward the spar tree. As the resistance to the turn decreases, the shift is made into higher gears to increase the line speed. The logs drag their length along the ground until a stump or other obstacle is encountered. Then the vertical component of the force in the main line leading in the direction of the elevated block lifts the front end of the log up until it noses over or around the obstacle. Obviously the magnitude of this vertical component of force varies with the angle of the line between the turn and the high-lead block, which is determined by the difference in elevation and the distance between the turn and the block. The ability of the high lead to lift the turn over obstructions decreases as the distance from the spar tree increases and is greater yarding uphill than down. Above the altitude of the block there is no lift when logging is downhill.

While yarding in a turn the yarding engineer shifts gears as required to provide the power necessary to overcome resistance. When the log reaches the landing, the chaser indicates by hand signals where to drop it. The lines are slacked off, and the chaser unhooks the chokers from the logs.

The operation of pulling out the main line with the haulback and yarding in the turns of logs with the main line is repeated until the back end of the first road is reached. Then the lines are slacked off, the bight of the haulback line removed from the tail block, and the haulback reeled in until it leads toward the block set at the back of the second road to be yarded. The operation of yarding the logs which can be conveniently done along a road and then, changing roads, proceeds until an arc of 90° has been covered and the yarding of the first quarter completed. Yarding does not always proceed so smoothly as the foregoing might indicate. Occasionally a turn "hangs up" when the lead is not high enough to lift it over an obstruction. Then the lines have to be slacked and the choker "hold" changed to roll or turn the log to free it. If the hangup occurs at some distance from the chokermen the hook tender will change the choker hold.

On completion of the yarding of the first quarter, the lead blocks on the spar are "swung" or shifted to the opposite side of the tree, and yarding of the second quarter proceeds. When the yarding of the second quarter is completed, the donkey is "turned" or moved to the opposite side of the spar tree so that it faces the unlogged semicircle of the setting. The third quarter is yarded with the lead blocks on the same side of the spar as the second quarter. To yard the fourth quarter the blocks must be swung to the opposite side of the tree to the position occupied when the first quarter was logged.

When the yarding of the setting is completed, the spar tree is unrigged, and the yarder is moved to a new setting, either by stringing out its lines and pulling itself along on its sled runners, or by loading on a truck or car. Sometimes the spar tree is felled and bucked into logs, but often it is left standing, since no means of loading it remains when the rigging has been taken down.

The feasible yarding distance depends on the topography, the height of the spar tree, and the density of the timber stand. The usual economic external yarding distance for large donkeys is 700 or 800 ft. and for small yarders 500 ft. or less. On the "long corners" on a square setting the yarding distance may be 1000 ft. or more. Full settings vary from 20 to 40 acres, depending on the size of the donkey. Partial settings are often necessitated by topography or to avoid yarding across truck roads or railroads. Since the high-lead system operates more efficiently uphill, where the bull block is always above the logs, than it does downhill, where there is no lift on the logs near to and above the altitude of the bull block, sidehill settings are often rectangular in shape with the longer haul on the lower side.

The output of a high-lead side varies with the yarding distance, topography, density of timber stand, log size, and relative power and speed of the yarder. Daily average production varies from 25,000 or 30,000 b.f. with small donkeys and low volume per acre to 200,000 b.f. with the most powerful yarders in the best "shows." Probably 75,000 or 80,000 b.f. constitutes a general average daily output of the common Pacific Coast high-lead side today.

It is evident that the conditions to which the high-lead system is adapted are yarding clear-cut areas, where topographic or ground conditions are unsuitable for tractors, where a preponderance of the yarding is uphill, and where the yarding distance is within the economic yarding distance of the high lead. The high lead may also be used for swinging from cold decks if the swing is uphill over fairly smooth ground or if the decks are too small to justify rigging up a skyline swinging system.

To summarize the advantages of the high-lead system: It can log terrain unsuitable for tractors with an ordinary two- or three-drum donkey and lower capital investment than other cable-hauling systems. Rigging up and road changing are simpler and quicker, and, if roads and landings are properly located, the high lead is adaptable to a wide range of conditions of topography and timber stand.

Among the disadvantages of the high-lead system compared to the tractor are the relatively short economic yarding distance, requiring closer spacing of truck roads or railroads; its destructiveness to the residual trees and, hence, its unsuitability for selective logging; and its larger crew, resulting in lower production per man under comparable conditions. Disadvantages compared with other cable-hauling systems are (1) lack of control over the log on steep downhill or sidehill hauls, (2) decrease in the lift on the log to overcome obstacles with increase in the distance from the spar tree, and (3) its restriction to shorter economic yarding distances.

SKYLINE SYSTEMS

All skyline systems have the following elements in common: A heavy wire-rope skyline is suspended by "shoes" or "jacks" or blocks hung on straps on the spars between a head spar or tower at the landing and a tail-spar tree at the back end of the yarding or swing road. The lift on the turn of logs is provided by a carriage riding on the skyline. Skylines may be classified as (*a*) tight-line systems and (*b*) slack-line systems. The tight lines have both ends of the skyline anchored to stumps. The North Bend, the modified or downhill North Bend, the Tyler, and the interlocking skidder are all tight-line systems. The slack-line systems have only the tail end of the skyline anchored, and the skyline passes through a big block on the head-spar tree from which it leads to a large drum on the donkey by means of which the skyline can be raised and lowered or "slacked." Slack-line systems include the MacFarlane Wirkalla or single slack-line, the Lawson, Bourne, Duplex or double slack line, and the Dunham.

Skylines may also be classified by use as (*a*) swinging systems, and (*b*) yarding systems. The yarding systems may also be used for swinging, but the swinging systems are not adapted to yarding. Systems used principally for swinging from cold decks or hot decks are the North Bend, the modified North Bend, and the Tyler. The other systems previously named are used for yarding.

The yarding or swinging distance over which any skyline system may operate is limited by the deflection or sag in the skyline. Deflection is determined by the height of the tree shoes above the ground, the "span" or horizontal distance between head and tail spar, the profile of the ground between them, and the allowable tension in the skyline. Deflection is expressed in percentage of the span. Decreasing the deflection increases the tension in the skyline and lowers its load-carrying capacity at a given factor of safety. The logger's rule of thumb limits deflection to 5% of the span. Thus a span of 1400′ would require a deflection of at least 70′. If this deflection would not enable the carriage to clear the ground, the span would have to be shortened. Permissible deflections for given loads are better calculated from tables and formulas given in the wire-rope manufacturers' handbooks. The tables are compiled for horizontal spans. Tensions for steep-inclined spans should be calculated, as at a given deflection tension increases rapidly with increase in the slope. For example, in a 2″ cable with a 5% deflection the tension per 100′ of cable, owing to the weight of the cable alone, increases from 1800 lb. on the horizontal to 2410 lb. on a 50% slope and 3900 lb. on a 100% slope.

Skyline swinging systems. *North Bend system.* The most frequently used skyline system is the North Bend. It is rigged by threading the main line through a "fall block," and shackling the end of the main line to the carriage. The fall block is a heavy block designed to work upside down compared with the other blocks. The butt hooks, chokers, and haulback line are all attached to the fall block. The haulback tail block is hung so as to pull the fall block down to where the chokers can be attached to the logs to be hauled. The fall block rides in the bight of the main line and up and forward when the main line is reeled in. The carriage rides freely on the skyline, moving in accordance with the forward component of the force in the main line between the fall block and the carriage. The load of the turn on the fall block is divided between the main line and the skyline. The turn of logs drags along the ground until an obstacle is encountered. Then, as the

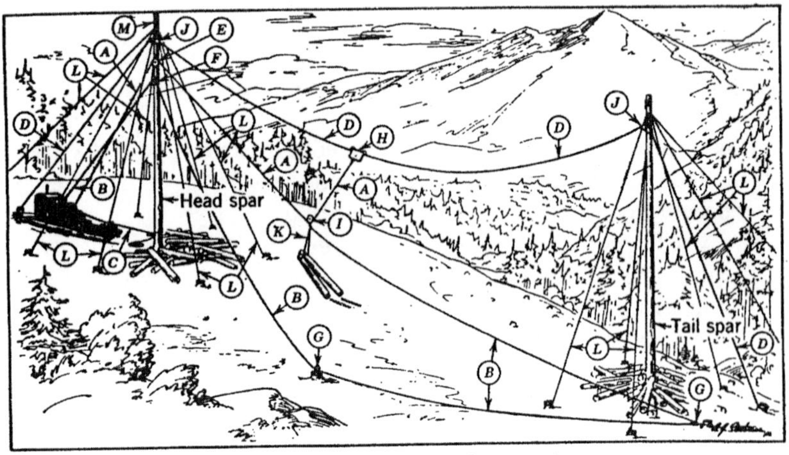

FIGURE 71. North Bend system.

A. Main line.	*G.* Haulback corner or tail block.
B. Haulback line.	*H.* Carriage.
C. Straw line.	*I.* Fall block.
D. Skyline.	*J.* Tree shoe.
E. Main-line block.	*K.* Chokers.
F. Haulback block.	*L.* Guy line.
	M. Pass-line block.

Prepared for Skagit Steel & Iron Works under direction of J. Kenneth Pearce, professor of logging engineering, College of Forestry, University of Washington. Copyright 1947, Skagit Steel & Iron Works, Sedro-Woolley, Wash.

main line continues to wind in and the forward movement of the turn of logs is stopped, the fall block rides up the bight of the main line, carrying the turn with it, until the turn rides over the obstruction. The turn may also be raised by holding the brake on the haulback line and going ahead on the main line. The turn also tends to follow a straighter road under the skyline than it does with the high-lead system.

The North Bend is adapted to swinging uphill or on horizontal spans or on moderate downhill slopes. The lifting effect previously described is not realized on steep downhill slopes. The customary swinging distance is twice high-lead yarding distance or 1400 to 1600 ft. A North Bend will swing more logs a day than a high lead can yard, operated with a comparable donkey. For this reason swinging from cold decks, which permits the North Bend to work to maximum capacity, is favored over swinging from hot decks, which limits the output of the swing to that of the yarder. The principal advantage of the North Bend over other skyline systems is that it can be operated with any

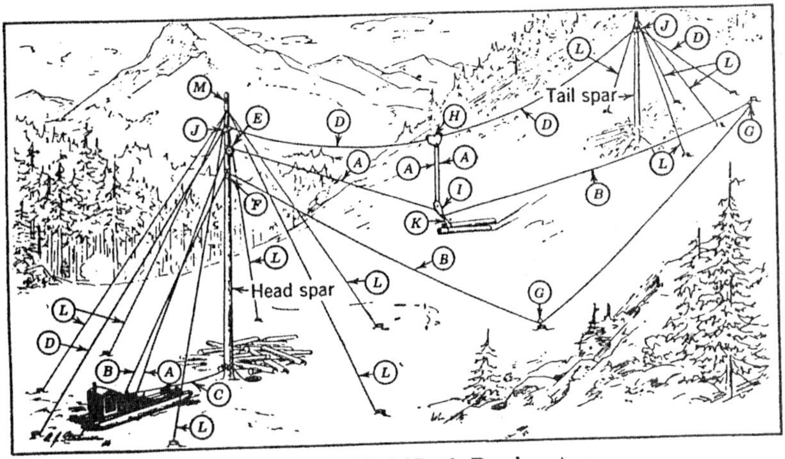

FIGURE 72. Modified North Bend system.

A. Main line.
B. Haulback line.
C. Straw line.
D. Skyline.
E. Main-line block.
F. Haulback block.
G. Haulback corner or tail block.
H. Carriage.
I. Fall block.
J. Tree shoe.
K. Chokers.
L. Guy line.
M. Pass-line block.

Prepared for Skagit Steel & Iron Works under direction of J. Kenneth Pearce, professor of logging engineering, College of Forestry, University of Washington.

two- or three-drum yarder with sufficient line capacity for the desired swinging distance. Logging operators with limited capital and equipment can use a donkey to build a cold deck and then move to the landing and swing the decked logs with the same donkey. The North Bend gives better control of the log than the high lead. However, it is not adapted to yarding because the time required to rig up the skyline would waste too much time in changing roads for yarding.

Modified or "downhill" North Bend. For swinging down steep slopes the North Bend is rigged to give a greater lift on the turn of logs. The end of the main line is fastened to the top of the fall block, then passes through a sheave in the bottom of the carriage, down through the sheave of the fall block, and forward to the main-line lead block. This gives more lift on the turn to overcome obstructions. However, it puts more stress on the skyline and thus requires more deflection than in the regular North Bend for the same tension under load. Owing to the wear on the portion of the main line passing through the carriage and fall-block sheaves, a "tag line" of more flexible construction than the main-line wire rope is usually fastened to the end of the main line with a "Molly Hogan."

The downhill North Bend has also been rigged by threading the main line first through the carriage sheave, then down through the fall block, and back up to the carriage where the end of the main line is attached. This gives the maximum upward lift on the turn, but has the disadvantage of putting all the stress of the turn load on the skyline. Both modifications of the North Bend are operated in the same way as the regular North Bend, the haulback line pulling the fall block with attached chokers out and down to the logs. The brake on the haulback drum is used to prevent the turn from running out of control down steep slopes.

Tyler system. The Tyler system requires a special donkey with a second main-line drum with good brakes to handle a "lifting" or "hoist" line. This line runs from the drum through a lead block on the spar tree hung just below the skyline shoe, then back through the rear sheave of a set of two offset sheaves in a special Tyler carriage, down through a fall block, up through the second carriage sheave, back through a shoe of the tail tree, and down to a stump where the end of the lifting line is anchored. The tight-line skyline is suspended on shoes hung just under the top guy lines on head and tail spar, and both ends are anchored to stumps, in the same manner as the North Bend skyline. The main or hauling line is fastened to the fall block as are the haulback line and chokers. The main-line block is hung lower on the head tree below the haulback lead block.

In operation the fall block is pulled out and down to the decked logs by the haulback line as in the North Bend. When the chokers are set, the lifting line is reeled in, lifting the fall block until the turn is elevated to the desired height. The brakes are set on the lifting line, and the main line hauls in the turn. As the turn comes in, the elevation of the turn may be varied by slacking off or reeling in the lifting line. The Tyler system is specially adapted to swinging over swing roads with cliffs or sharp drops where, if the North Bend system were used, the turn would drop and break the chokers. The Tyler permits the turn to be lifted entirely clear of the ground before such a drop is reached. Where there is no condition necessitating lifting the turn clear of the ground, the front ends of the logs are elevated just enough to clear obstruction, and the rear ends drag along the ground.

The Tyler gives much better control of the log than the North Bend and, consequently, less breakage of brittle species such as cedar. Dis-

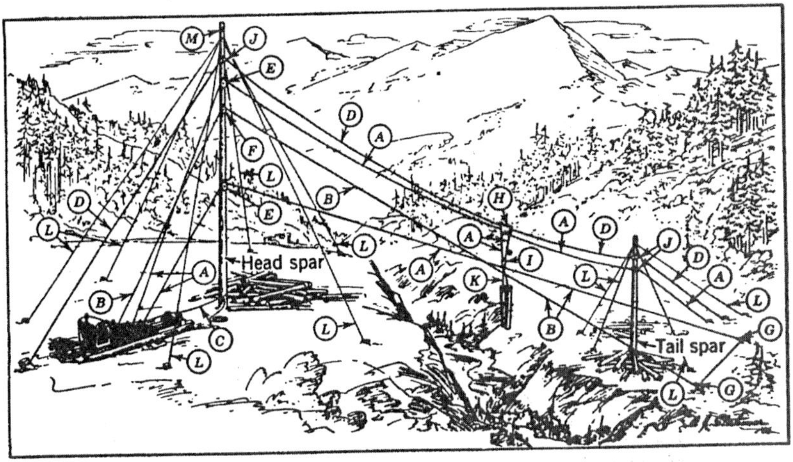

FIGURE 73. Tyler system (uphill).

A. Main line.
B. Haulback line.
C. Straw line.
D. Skyline.
E. Main-line block.
F. Haulback block.
G. Haulback corner or tail block.
H. Carriage.
I. Fall block.
J. Tree shoe.
K. Chokers.
L. Guy line.
M. Pass-line block.

Prepared for Skagit Steel & Iron Works under direction of J. Kenneth Pearce, professor of logging engineering, College of Forestry, University of Washington.

advantages of the Tyler are the added capital investment in donkey and rigging, greater stress on the skyline, severe wear on the lifting line, and longer time needed to rig up. However, the Tyler will operate satisfactorily uphill or down, over any topography, as a swinging system. The length of time required to rig up precludes its use as a yarding system.

Tyler gravity system. On downhill slopes steep enough for the carriage and turn of logs to run by gravity a modification of the Tyler system can be operated with an ordinary high-lead donkey by using the main line as the lifting line and omitting the main hauling line.

The speed of descent is regulated by brake on the haulback drum. Under the very limited conditions to which it is adapted it offers the advantages of the Tyler with no special equipment required except the carriage.

Skyline yarding systems. *Slack-line system.* The usual slack-line system uses a special donkey with a skyline drum large enough to

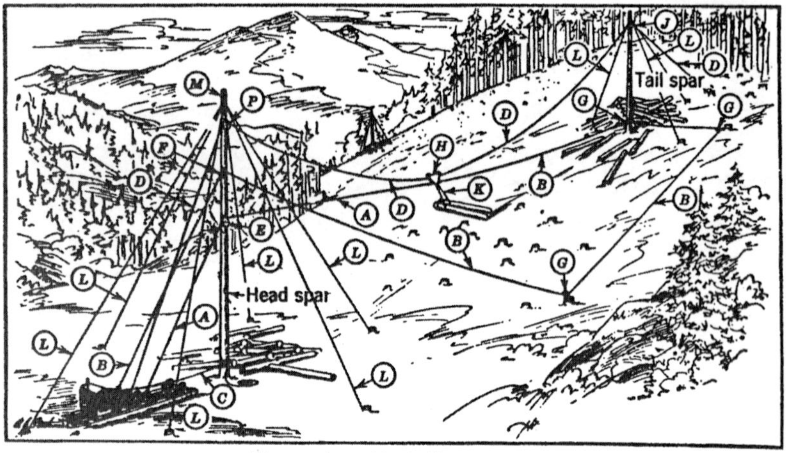

FIGURE 74. Slack-line system.

A. Main line.
B. Haulback line.
C. Straw line.
D. Skyline.
E. Main-line block.
F. Haulback block.
G. Haulback corner or tail block.
H. Carriage.
J. Tree shoe.
K. Chokers.
L. Guy line.
M. Pass-line block.
P. Skyline block.

Prepared for Skagit Steel & Iron Works under direction of J. Kenneth Pearce, professor of logging engineering, College of Forestry, University of Washington. Copyright 1947, Skagit Steel & Iron Works, Sedro-Woolley, Wash.

carry the desired length of skyline, and powerful brakes to hold the skyline up in the air. In rigging up this system the skyline is pulled through a lead block on the head-spar tree, hung just below the top guy lines, threaded through a one- or two-sheave carriage, then pulled out to the tail tree, and threaded through a shoe and anchored to a stump. Sometimes in yarding across canyons a tail tree is not required, and the skyline runs directly to an anchorage on one or more stumps. The end of the main line is clevised to the front edge of the carriage and the end of the haulback line to the rear edge of the carriage. The butt hooks and chokers are attached to the bottom of the carriage.

The skyline is raised off the ground by reeling in on the skyline drum, the brakes are set to hold the skyline up, and the haulback is reeled in to pull the carriage out to a position over the logs to be yarded. Then the skyline-drum brake is eased, and the skyline slacked down until the chokers can be set on the turn of logs. Then the skyline is raised until the front ends of the logs are elevated enough to ride over obstructions, and the carriage and attached turn are hauled in by the main line. If desired, the turn can be lifted completely off the ground. As the turn is yarded in, the skyline is raised or lowered to adjust the position of the turn to the profile of the yarding road. When the turn arrives at the landing, the skyline is slacked off until the chokers can be unhooked. In road changing the skyline tail hold is loosened and the skyline reeled in and then hauled out along the new yarding road. Or the end of the skyline may be pulled sideways directly to the new tail hold at the back end of the new road.

The slack line is better adapted to yarding than any of the tight skyline systems excepting the interlocking skidder. The slack line is specially suited to yarding downhill or across steep-sided canyons. It works equally well uphill or under almost any conditions where deflection can be obtained. The only exception is across narrow canyons where the skyline would rest on the ridges on each side, preventing the carriage from dropping down to logs in the bottom of the canyon. Advantages of the slack line include better control of the turn than with the high lead or North Bend and less lost time changing roads than with any of the tight-line systems excepting the skidder. Being a much lighter-weight yarder than the skidder the slack line can be used in truck-logging operations and for cold decking. Slack lines commonly operate over yarding distances of 1200 to 1600 ft. and have yarded as far as 3500 ft. across canyons where ample deflection was obtainable. The slack line is used to a greater extent in the redwood than other regions. Disadvantages of the slack line are the higher capital investment and greater weight involved in the special skyline donkey and the larger crew required on the side as tail trees must be continually rigged ahead.

Double slack line. The double slack line, also called the Lawson, Bourne, or Duplex Flyer, is similar to the slack line in principle and operation. The special donkey required for this system is equipped with two skyline drums, and the carriage has two sets of cable sheaves. A smaller-diameter skyline is used, usually the same size as the main line, and both ends of the skyline are wound on drums. At the tail tree or tail stump the skyline runs through a large equalizing block, and two skyline lead blocks are hung on the head-spar tree. Developed for yarding across deep canyons to speed up the raising and lowering of the skyline, this system is now believed to be confined to the redwood region. Because of the huge size and weight of the double slack-line donkey it is limited to railroad operations. The double slack line holds the cable-hauling yarding distance record in the Douglas fir region of 5480 ft.

Wirkalla system. Named for Oscar Wirkalla who patented it, this system is a slack line operated by two donkeys. One donkey raises and lowers the skyline, and the other operates the main line and haul-back line. The main drum of the common high-lead donkey is not large enough to carry the entire length of skyline but can hold enough skyline to raise and lower it while yarding. If the main-drum brakes are not powerful enough to hold up the skyline, a block purchase may be applied to the skyline by running the main line of the skyline-operating donkey through a block attached to the end of the skyline and back to an anchorage on a stump near the donkey. This doubles the pulling and braking power of the donkey on the skyline but halves the speed of raising the skyline. As it ties up two yarders on one side, this is a makeshift and is used only where a slack line is required so infrequently that the purchase of a regular slack-line donkey is unjustified.

Gravity slack line. The slack-line principle can be applied on steep downhill slopes where the turn will run by gravity by using the main line of a high-lead donkey as the skyline. The skyline lifts the turn until it runs down by gravity, the descent being controlled by the haul-back line which also pulls the carriage and attached chokers back up the yarding road. This is a makeshift modification of the slack-line system as only a few roads on a sidehill setting would provide the necessary conditions for gravity operation. The weight of the turn and the yarding distance are also limited by the smaller size of the main line as compared with the regular skyline cable.

Dunham system. The Dunham is the newest of the cable-hauling systems and has so far been used in only a few operations. The main line serves both as a skyline and a hauling line. The end of the main line is attached to the front edge of the special three-sheave Dunham carriage. Then the main line is run forward through a lead block on the head-spar tree, back through the upper carriage sheave, on back through two tail blocks hung in the tail tree and separated so as to spread the two parts of the line, forward through one of the two lower carriage sheaves, down through a fall block, up through the other lower sheave, forward through a lead block on the head spar, and down to the main drum on the yarder. The chokers and haulback are attached to the fall block. The haulback line runs through a tail block set and changed as required to lead the fall block down to the logs, then back to a "bearing" block hung in the tail tree above the main-line tail blocks, then forward through a bearing block attached to the upper edge of the carriage, on to the haulback lead block on the head spar hung just below the top guy lines, and down to the haulback drum. Reeling in on the main line both lifts the fall block and the turn of logs and pulls the carriage forward.

This system may be operated with the common three-drum donkey, providing the main-line drum is of sufficiently large capacity to carry the long line required, of three times the yarding distance. The Dunham system can operate under almost any topographic conditions and gives good control over the turn of logs. The rigging is complicated, and line wear on the main line is rather severe. It is limited to shorter yarding distances than other skyline systems.

Interlocking skidder. Originally developed by the Lidgerwood Manufacturing Company and used in the past in many regions of the United States, the skidder is now used principally in the large-scale railroad operations in the Douglas fir region. With the shift from railroad to truck logging and the trend even in railroad operations toward lighter yarders requiring smaller crews, the importance of the skidder system has declined in recent years. The interlocking skidder unit is still the champion "log-getter" and will yard and load more logs a day than any other cable-hauling system.

The skidder uses a tight skyline, the tail end anchored to a stump. The head end is tightened by multiple-sheave heel tackle which permits rapid changing of roads. The main hauling line, termed the "skidding" or "tong" line, runs from the lead block out through a sheave in the special skidder carriage and terminates in attachment to butt rigging and chokers. A "slack-pulling" line runs from a lead block to one or more sheaves in the carriage and forward to fasten to the skidding line at a distance from the carriage equal to the height of the logs to be hauled. The "receding" line, as the haulback is called by skidder crews, is attached to the rear edge of the carriage and runs through a block on the tail tree, then forward to the receding-line lead block on the head spar through such corner blocks as are required to keep the receding line clear of the skidding road. In operation the receding line pulls the carriage back until it is directly above the logs which will make up the turn to be skidded. Reeling in on the slack-pulling line pulls slack in the skidding line, allowing butt rigging and chokers to lower to the ground. With the brake set on the receding-line drum, the skidding line is reeled in until the turn of logs is raised the required height to clear obstructions. Then the receding-line brake is released and the interlocking gear engaged so that the receding-line drum unwinds at the same rate as the skidding drum winds in. This holds the

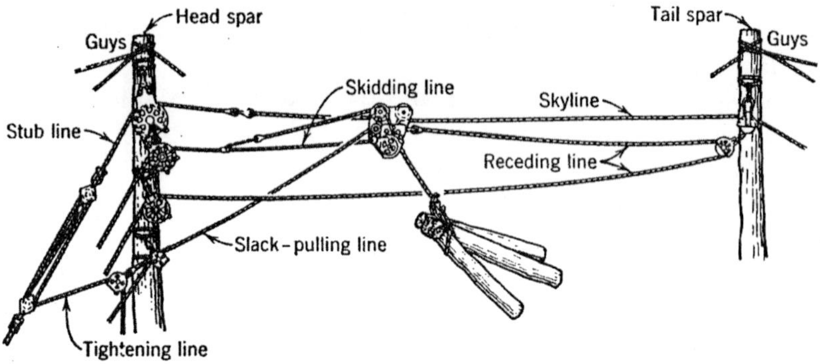

FIGURE 75. Interlocking skidder system with Berger slack-pulling skidder carriage. *Courtesy Berger Engineering Works, Inc.*

front ends of the logs off the ground at the height from the carriage to which they were raised at the time the interlock was engaged. As the relative diameter of the lines on the two drums changes, the turn gradually lowers, and the interlock is disengaged while the receding line is held and the skidding line continues to wind in to raise the turn again. At the landing, slacking off the skidding line lowers the turn to the ground for unhooking the chokers.

When logging is across deep valleys, where the weight of the long length of skidding line required between the carriage and the point of attachment of the slack-pulling line to the skidding line is greater than the weight of the butt rigging and chokers, special carriages are required to give a positive downward feed to the skidding line. In the Berger carriage, the slack-pulling line is threaded around an extra sheave keyed to the skidding-line sheave axle. As the slack-pulling line is reeled in on the skidder drum, it rotates the skidding-line sheave and also raises it on a pivot to bring the skidding line into contact with idler sheaves, thus feeding the skidding line through and down to the ground in an action similar to laundry feeding through a wringer.

Two types of skidders are the "tree rig," using a spar tree, and the "steel spar" or "tower," mounted on the skidder car and hinged so that it can be lowered for moving. The tree-rig skidder is made up of three units, each with its own steam engines, mounted on a single frame and taking steam from a common boiler. The skidding unit has skidding-, receding-, and slack-pulling-line drums. The rigging unit has transfer line, for pulling the skyline in changing roads, straw line, and heel tackle and utility drums. The loading unit has loading and car-spotting drums. The tower skidder has additional drums for raising the tower and tightening guy lines and a swinging heeling boom mounted on the front end for loading.

Two skylines are used with each skidder. While one is in use on the road being skidded, the other is transferred from the last completed road to the next road to be logged. The next tail tree is rigged and the tail end of the skyline anchored in advance of road changing. To change roads the heel tackle is slacked off, lowering the skyline on the road just completed to the ground, and the skyline is disconnected and removed from the carriage. The skyline on the new road is threaded through the carriage and fastened to the heel tackle, and the receding line is pulled in along the new road and shackled to the carriage. Tightening the heel tackle raises and tightens the skyline, and skidding proceeds on the new road. The skyline is reversed end for end when it is transferred to a new road, thus equalizing the wear of the carriage on it.

The combined skidding, rigging, and loading crew generally consists of 21 men. In addition to the job names of the crew members common to all cable-hauling systems, the skidder crew includes a "head rigger" who is in charge of the crew, a "second rigger" who climbs and rigs the tail trees, sometimes a "second second" assisting, and several "third riggers" who are the ground men of the rigging-up crew. As the skidder in medium-sized timber hauls five or six logs to a turn, more chokermen are required than in other systems.

The interlocking skidder is adapted to any topography where deflection can be obtained. Steel-spar skidders with towers 60′ to 110′ high usually work on spans of 1000′ to 1600′. The taller tree rigs can reach out 1800 or 2000 ft. or more Tail trees are generally spaced 150 to 200 ft. apart around the perimeter of the setting. Advantages of the skidder include excellent control over the turn of logs, eliminating hangups and minimizing breakage; long yarding range permitting wider spacing of railroad spurs; and high daily production due to fast line speed, big turns, and quick road changing. Disadvantages are high capital cost and overhead; railroads must be constructed and maintained to a high standard to carry the weight of the skidder car, particularly of steel-spar type; and, in times of labor shortage, large and highly skilled crews are needed.

"SKYHOOK" SYSTEM

The newest logging system, involving both cable-hauling and direct-tractive principles, is the "Skyhook," invented by Phil Grabinski, one-

FIGURE 76. Skyhook operating in southern hardwood swamplands in North Carolina. *Courtesy Pointer-Willamette Co.*

time world's champion high climber.* The Skyhook is an automotive vehicle which rides suspended on four sets of sheaves from two skylines. It pulls itself along with two stationary traction lines each lead-

* The first patents on the Skyhook were issued to Mr. Grabinski in May 1944.

ing through two blocks and around two traction wheels on the Skyhook. The tight skylines are hung on open jacks supported by trees, poles, or towers at intervals of 500 to 2000 ft. The skylines are clamped on the jacks so that the deflection is unchanged when the Skyhook travels from one span to another. The Skyhook travels through the open jacks, and so the length of the "Skyroad" is physically unlimited. Moderate horizontal bends in the "Skyroad" are feasible. Powered by an industrial gasoline engine, the Skyhook hoists loads up to 20 tons and travels at speeds up to 35 m.p.h. With rubber tires mounted on the traction wheels, it moves on the ground to the job, and then, with tires removed and the Skyhook on blocks, it runs out the lines from wire-rope reels and rigs its own Skyroad. The operator of the machine (who might be termed the "Skyrider"!) rides seated in the Skyhook. The Skyhook is a swinging system, and, as it carries the turn of logs suspended in a horizontal position, it can load them on truck or barge at the terminus of the Skyroad. The first applications of the Skyhook were in the Douglas fir region and in the hardwood swamplands of the South. The Skyhook is manufactured by the Pointer-Willamette Company of Portland, Ore.

OPERATING COMBINATIONS OF CABLE-HAULING SYSTEMS

The various power-logging systems are used singly or in a variety of combinations to complete the power transportation of logs from tree to car or truck. The most common combinations are mentioned here.

DIRECT YARDING TO RAIL, TRUCK ROAD OR WATER

This means using one of the systems adapted to yarding as the only step in moving a log from where its parent tree was felled to where it is loaded on some long-distance transportation system, such as rail or truck. This is the simplest type of operation possible, but may require extreme skill in locating the rail or truck spurs. In rough ground it may be impossible to place all the area within single yarding reach of the main transportation system without involving prohibitive construction expense. Likewise, there is difficulty in arranging operations in direct yarding so as to secure a uniform output from each unit, since there is no concentration of logs from scattered stands or rough ground. Hence, at present most operations that were at one time based on direct yarding have now been forced into other combinations.

DIRECT YARDING WITH COLD DECKING AND SWINGING OR YARDING AND SWINGING FROM INACCESSIBLE OR DIFFICULT CORNERS

This is merely an extension of the direct yarding by placing additional machines to pile on to cold decks or yard together the logs from those areas which cannot be reached by the main yarding machine. In this combination, spurs are located to give direct yarding access to the major part of the area, and auxiliary methods are adapted to reach areas impossible to yard directly to the track.

COLD DECKING AND SWINGING

With this combination, the major portion of the timber logged is first yarded on to cold-deck piles, usually by light high-lead yarders. These cold decks are so located as to furnish readily accessible reservoirs of logs which will permit the final swinging and loading operations to be maintained at a uniformly high rate of production. This assures a more nearly balanced output than is possible with direct yarding and also permits the concentration of logs in light scattered stands. The method extends the reach of any yarding system and is quite commonly employed in scattered stands or small timber.

Chief objections to this combination are the increased fire danger and increased breakage inherent in the cold-deck pile. Felled timber may be burned over with only a 25% to 50% loss, but logs in cold decks are completely consumed. Breakage of some species such as cedar is excessive in building and removing large cold-deck piles.

YARDING AND SWINGING

With this combination two machines are used as in the cold-deck system, but they work together. In other words, the logs are yarded in by one machine and then relayed directly to the landing by the swing machine. With this system the fire danger and breakage of cold decks are eliminated, but the output at the landing is determined by the output of the yarder and may vary within wide limits from day to day. There is no reservoir of logs for maintaining production on days when the yarder is working in rough ground or light stands.

COMBINATIONS OF TRACTOR AND CABLE-HAULING SYSTEMS

Various combinations of arch tractors and cable-hauling systems are common on the Pacific Coast, each being used to log under the conditions to which it is best adapted. In country mainly suitable for tractors small areas of sharp canyons, steep sidehills, or rock outcrops are often found in which tractor road building is costly or steep grades would be required. Such areas are high-lead-yarded to a "hot deck" at a point which can be reached by a favorable tractor road. Light gas donkeys or tractor donkeys operate the high lead. From the "hot deck" the logs are "roaded" to the landing on truck or railroad. Roading is often combined with direct-tractor yarding of logs on better ground to maintain uniform production. Logs are not cold-decked by the high lead when they are to be roaded with tractors because time is wasted in pulling logs out of a cold deck to make up a tractor turn.

In selective logging on steep sidehills or where uphill hauling is involved, tractors are used to "feed" skyline swings. The narrow skyline road is clear felled to avoid knocking down trees during swinging and to avoid damaging the residual trees adjacent to the skyline road. Contour tractor roads are bulldozed out from the skyline road. Arch tractors yard the logs adjacent to the tractor roads and drop them under the skyline. The logs are swung "hot" to keep the tractor roads clear.

SILVICULTURAL CONSIDERATIONS OF CABLE-HAULING SYSTEMS

Cable-hauling systems are inseparably connected with clear-cutting methods. For those forests best handled by clear-cutting methods, power-logging methods are fully as effective from a silvicultural standpoint as any other logging method. The radiating roads expose mineral soil, offering seedbeds well suited to establishment of new stands of such species as Douglas fir. However, when the desirable silvicultural treatment is some form of tree selection, based on removing part of the stand and leaving a residual stand, power logging is not so desirable as one of the tractor systems. This applies particularly to their use in yarding where the straight-line load movements of cable-hauling systems necessarily result in greater injury to the remaining stand. For use in swinging or roading alone, the difference between cable-hauling and tractor systems is negligible. Special applications of power yarding have given satisfactory silvicultural results in selection stands. In certain operations in California sugar pine forests, carefully applied ground-lead methods have compared favorably with tractor logging as far as damage to the remaining stand is concerned. In some mixed stands of Port Orford cedar and Douglas fir, the high-lead system operated with light gasoline donkeys has proved quite satisfactory where the remaining stand is largely mature fir, not readily damaged by the light equipment. In general, cable-hauling yarding systems are satisfactory in connection with silvicultural methods depending on regenerating a new even-aged stand on removal of the present stand but are not satisfactory where a considerable portion of the stand is left to form a major part of the volume for a future cutting cycle as in selective logging.

LOADING SYSTEMS ADAPTED FOR USE WITH POWER LOGGING

Brief mention of the loading methods commonly used with cable-hauling systems is necessary to a complete understanding of the power-logging cycle. Obviously these loading systems are possible with any yarding system, and some of them are common in tractor or horse-logging operations. Others have advantages which adapt them to power-logging conditions, so that they are largely confined to that type of operation.

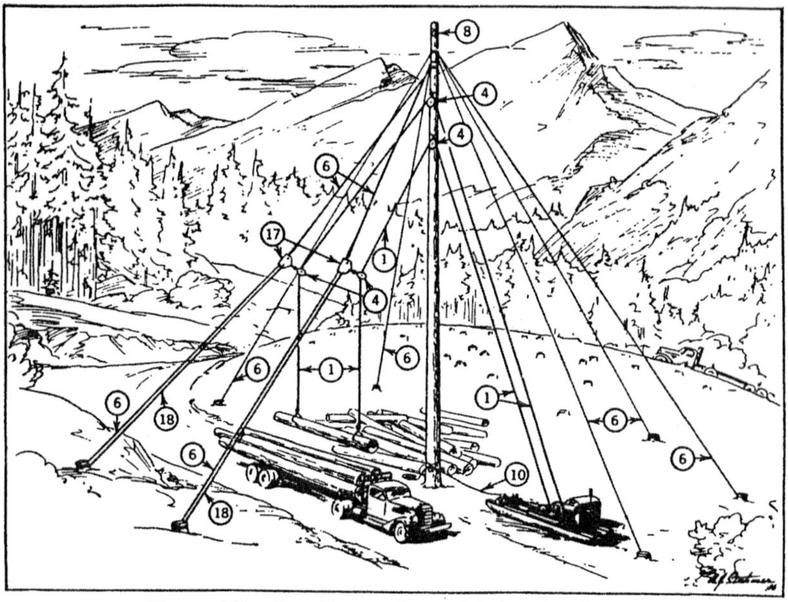

FIGURE 77. Guy-line loading with Duplex drums.

1. Loading line.
4. Loading-line block.
6. Guy line.
8. Pass-line block.
10. Straw line.
17. Loading jack.
18. Jack guy.

Prepared for Skagit Steel & Iron Works under direction of J. Kenneth Pearce, professor of logging engineering, College of Forestry, University of Washington.

GUY-LINE SYSTEMS

Guy-line loading systems are arranged to utilize the guy lines of a spar tree for supporting the loading blocks in position over the rail or truck road. They are possible only with power-logging systems.

Duplex system. This system is perhaps the most common in connection with high-lead yarding. It makes use of two loading lines operating through loading blocks suspended from two guy lines. A duplex loading engine with reversing drums controls the two loading lines. Tongs hold the logs while they are being loaded, although with exceptionally large logs it is necessary to supplement the tongs with straps.

Single-tong system. The single-tong loading system was used with the early skidders. It employed a single loading line with the loading block suspended from a loading guy line. In the hands of a skillful loading lineman the system gave satisfactory results with small logs. It has now been largely replaced by other systems.

FIGURE 78. McLean boom loading with power swing in both directions.

1. Loading line.	6. Guy line.
3. Swing line.	8. Pass-line block.
4. Loading-line block.	12. Traveling block.
5. Swing-line block.	13. Boom-suspension guy.

Prepared for Skagit Steel & Iron Works under direction of J. Kenneth Pearce, professor of logging engineering, College of Forestry, University of Washington. Copyright 1947, Skagit Steel & Iron Works, Sedro-Woolley, Wash.

BOOM SYSTEMS

Boom-loading systems are also commonly found in connection with power logging. They have the advantage of being able to place the tongs in position over the log to be loaded, thus requiring less effort on the part of the loaders in hooking on.

Double-tong or McLean boom. In this system the boom is made of two logs arranged to swing around the base of a spar tree and having the outer end supported by a guy line. Two loading lines operate through blocks on the boom which is swung into position for hooking on or loading the log by means of boom swinging lines. Frequently a counterweight suspended from a line passing through a block on one of the spar-tree guy lines serves to swing the boom in one direction.

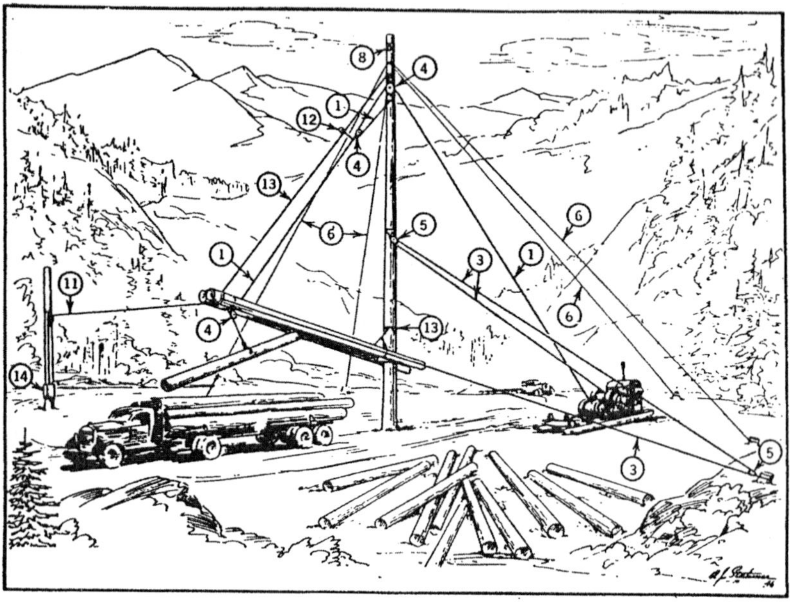

FIGURE 79. Heel-boom loading counterweight swing.

1. Loading line.
3. Swing line.
4. Loading-line block.
5. Swing-line block.
6. Guy line.
8. Pass-line block.
11. Counterweight line.
12. Traveling block.
13. Boom-suspension guy.
14. Counterweight.

Prepared for Skagit Steel & Iron Works under direction of J. Kenneth Pearce, professor of logging engineering, College of Forestry, University of Washington.

Single-tong or heeling boom. This system, particularly as it is adapted for use on the steel-tower skidder, is the fastest loading system for large logs. It is employed either with a double-log boom which pivots around the base of a spar tree or with the steel loading boom of the tower skidder. In this system a single loading line and tong are required. The log is hooked at a point somewhat behind its center of gravity, so that the back end of the log rises up and bears against the boom itself. The log, suspended from the loading tong and bearing against the heel of the boom, is swung into position over the load. The system has the advantage of having only one tong to handle and at the same time of providing for complete control of the logs being loaded.

The Duplex system and the two-boom systems load by far the major portion of logs yarded by power methods. The following systems, how-

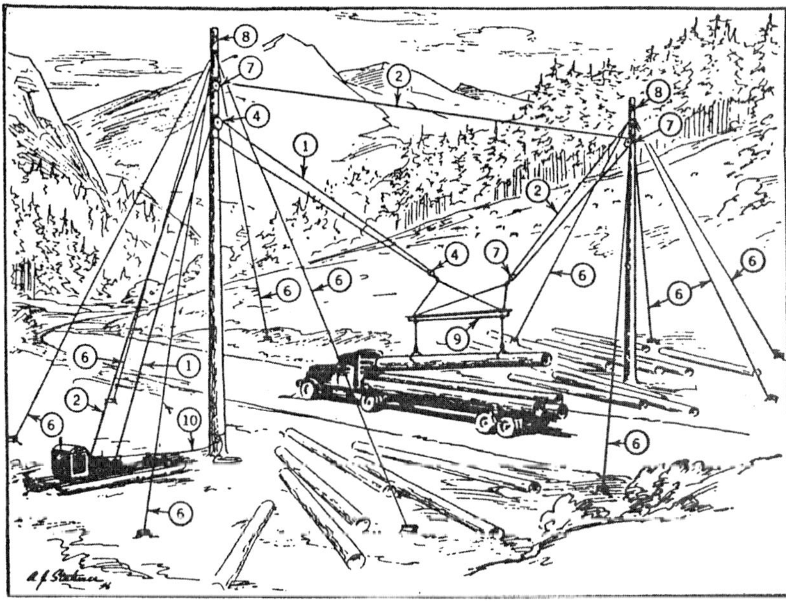

FIGURE 80. Tight-line loading with block purchase.

1. Loading line.
2. Haulback.
4. Loading-line block.
6. Guy line.
7. Haulback block.
8. Pass-line block.
9. Spreader bar.
10. Straw line.

Prepared for Skagit Steel & Iron Works under direction of J. Kenneth Pearce, professor of logging engineering, College of Forestry, University of Washington.
Copyright 1947, Skagit Steel & Iron Works, Sedro-Woolley, Wash.

ever, are used to a sufficient extent to be mentioned in connection with power-logging systems.

CROTCH-LINE OR TIGHT-LINE SYSTEM

The version of the crotch-line loading system used today is the tight line. Loading tongs or straps are hung from a steel spreader bar made from a short section of rail. The ends of the spreader bar are connected by short wire-rope "bridles" to a tail fall block and either a head fall block or a shackle. The haulback line is threaded through the tail fall block and the end anchored to a tail loading spar, giving a block purchase. Depending on the donkey power in relation to the size of the logs the main line is similarly rigged on the head fall block and head spar, or a straight-line pull to the bridle shackle is used. In operation the haulback spots the spreader bar over the desired log. The tongs are set so that the log will lift in a horizontal position, and the main line is reeled in while the brakes are held on the haulback, thus hoisting the log. It is then "tight lined" to the truck by easing the haulback brakes while pulling in the main line. The system is adapted to loading tree-length logs or poles and piling on moving-cable landings and to wide tractor landings. A two-drum yarder with double brakes on the drums is generally used as the loading donkey.

SELF-PROPELLING LOADING SYSTEMS

Locomotive crane. Locomotive cranes most frequently serve as auxiliaries to power-logging systems. They are extremely useful in picking up logs along the track where they may be difficult to reach during the regular logging operation, as well as in handling wrecks and innumerable other jobs. They are not customarily employed for loading in conjunction with regular power-logging systems since it is necessary to arrange for switching each car as it is loaded.

Shovel loaders. Mobile shovel loaders are converted from crawler-mounted power shovels by substitution of a longer boom. Usually an A-shaped heeling boom is used, with a single tong on the loading line. Shovels rated 1½ yd. and larger are required for loading Douglas fir. Shovel loaders serve the same purposes in truck-logging operations as locomotive cranes do in railroad operations. They are also used for yarding short distances.

CHAPTER X

Combined Minor and Major Transportation

CONDITIONS WHERE SINGLE HAUL IS USED

With the cutting of the larger timber tracts in the North, South, and East, the smaller remaining areas and especially the regrowth of timber have invited the introduction and operation of many small mills.

FIGURE 81. Sled logging to one of the many small sawmills in Michigan. The sled loads are generally 400 to 800 b.f. per trip. Note elevated ramp from which logs are rolled to log deck and carriage of sawmill shown in right rear of picture. Most of the 60,000 logging operations in the United States supply logs for the smaller mills.

These have resulted in the literal moving of the sawmills to the timber in place of the formerly prevailing practice of moving the timber to the mills.

This fact has generally reduced the distance of the log haul and in many instances has eliminated the usual practice of minor and major transportation. Logs are being transported directly to the mills in one operation rather than being assembled at convenient skidways or landings and then transported by some other method to the mill. Thus minor and major transportation are combined in one operation.

Improved transportation devices, notably the motor truck and the tractor, have contributed to this situation. Moreover, in open stands of timber on level topography, particularly in some sections of the

FIGURE 82. Good example of tractor hauling logs directly from felling area in woods to sawmill. Thus landings and loading are eliminated, and there is one single haul. This method is especially adapted for small mills. The maximum was 1¼-mile haul—the tractor operating 9 hours per day on less than 1 gal. of fuel oil per hour. Small mill saws up to 10 M.b.f. per day. The South contains more than 15,000 of these small mills.

southern pineries and the ponderosa pine timber of the West, trucks have been used to go directly to the felling areas, load the logs, and haul them to the mills. This system has also been successful in small and scattered stands of hardwoods and softwoods in the East, the Lake States, the Southern Appalachians and the South, and is employed to a limited extent in Douglas fir in the Inland Empire and California.

Improved mechanical devices in connection with the tractor, especially in the matter of speed and loading facilities, and its use as an auxiliary to the motor truck on uphill hauls have also contributed to the present situation.

Direct skidding to small and portable mills without any other minor or major form of transportation has been the common practice in the East, South, and Rocky Mountains for several years. But the rapid regrowth of cutover lands and the cutting of second-growth stands for the better logs have given rise to a remarkable increase in the single haul, especially with the motor truck.

FIGURE 83. Diesel tractor hauling average loads of about 1000 b.f. of yellow poplar, oak, and other species about ¼ mile average haul to small sawmill in Maryland. These homemade sleds elevate the front ends of the logs and reduce tractive resistance.

METHODS

The single haul may be summarized in its application as follows:

1. Horse skidding directly to the mill is found chiefly in logging with small and portable mills. Distances up to ¼ mile are the customary practice, but large logs of high quality and value may be skidded on drays or bummers for distances of 1 mile or more.

2. Tractor skidding exists especially in rough topography, on uphill hauls, in swampy or wet locations, and for distances up to 1½ miles. Tractors often serve as auxiliaries to motor trucks on the single haul especially on adverse grades, hauling over low or wet places or where the woods roads are not in a satisfactory condition. Tractors are also

used with horse skidding, the horses assembling the logs and the tractors hauling them on drays (sleds, log boats, or mud boats) to the sawmill.

3. Motor-truck haul has proved to be the most useful and extensively adopted method of single log haul, especially since trucks are powered with winches, cables, and similar devices to do their own loading and sometimes skidding. Trucks have been found valuable not only in the open parklike stands of southern and ponderosa pine, but also in

FIGURE 84. Diesel tractor used for ground-skidding logs from woods directly to small sawmill in Oregon. The same tractor is used for driving the sawmill and for general farm work. Thus tractors are used for multiple purposes—logging, powering sawmills, farm work, and many others.

northern white pine, western white pine of Idaho, spruce and hardwoods in the Adirondacks and New England, hardwoods in the Southern Appalachians, and heavy Douglas fir and associated species in the Northwest. Little if any expense is incurred in building roads for the short distances to improved highways. Even in Canada and the Northeast during the winter, trucks have been found to operate successfully when driven directly to the felling area, loading logs and hauling them to the mill. Plows or tractors are sometimes used to break or pack the snow, and small trees or brush are removed with the axe. Furthermore, when the ground is frozen in the winter, trucks can be driven in swamps and low places where logging is impossible during the spring and summer. On short hauls, trucks sometimes can drag logs directly to nearby sawmills.

PART IV
LOADING FOR TRANSPORT

CHAPTER XI

Landings*

GENERAL

Landings are locations to which logs are skidded, chuted, or flumed, preparatory to being loaded and transported to mill or destination. They are generally at some point between minor and major transportation.

FIGURE 85. Lodgepole pine poles peeled and decked for motor-truck hauling. Photo taken in the Medicine Bow National Forest, Wyo.

They are the places where logs are unloaded from one transportation medium to await loading on another or major transportation device. For example, logs may be horse-skidded to a landing along a main-haul road where sleds may transport them to another landing or banking ground along a stream. In the Northwest, logs are generally skidded by tractor or power skidder to a landing along a railroad track where they are loaded on railroad cars and transported to the mills.

* This subject is described in connection with various methods of transportation elsewhere in the text. Landings are also known as skidways, log dumps, rollways, log decks, yards, and banking grounds.

Landings are generally provided by clearing the ground of standing and fallen timber so that the logs may be unhooked and then loaded with the greatest facility. When the logs are to be rolled from one point to another, skids or long round logs are provided to facilitate the movement of logs. For example, the terminus of a chute may be at a railroad landing. A skidway is provided so that the logs may be rolled from the chute terminus to within reach of the log loader.

FIGURE 86. Landing along pond near Pierce, Idaho, where Diesel tractor equipped with bulldozer and cable control is pushing logs from landing into water. This illustrates one of the many auxiliary uses of the tractor.

Landings vary greatly in size from a few square yards to a mile or more in length. In Wisconsin, in winter-sled logging, skidways are generally about 20′ wide and 50′ long.

Landings may be from a few rods to a mile or more in length along a driving stream. On the Pacific Coast and in the South, where tractor skidding and railroad haul are customary, little primary work except tree felling may be necessary, as the logs are dropped at the landings in miscellaneous and disorderly fashion. Where motor-truck haul is used in small timber, elevated skidways or high banks are preferred so that the logs may be rolled onto the trucks by hand with the least expenditure of time and effort. Elevated ramps may constitute the landings on pulpwood operations where loading is directly onto railroad cars as in the Northeast. Some chutes and flumes may land logs in a stream; they then float away or pile up in a log dump, depending on the depth of water. In eastern Canada, landings are constructed as high as 700 and 800 ft. above a driving stream on minimum slopes of 50% and maximum slopes of 80%. These landings are cleared for widths of 200′ to 300′, and the furthest logs may be 100 to 160 ft. from the edge of the slope. The logs on large pulpwood landings are sent down the steep high landings during the spring.

The frequency of landings depends on the type and density of timber and the areas that are topographically tributary to them. Thus, only a few logs may be assembled on one landing, whereas several million feet may be skidded to another type. The illustrations show the widely varying nature of these landings. Their location is generally determined by the forest engineer or the camp foreman.

FIGURE 87. A distant view of two high landings 700 ft. above the river where approximately 1250 M.b.f. of spruce pulpwood piled along the edge of these landings is rolled down these steep grades to the river to be driven to pulp mills. An area 200′ to 300′ wide is cleared of timber, stumps, and so on, and the logs are rolled down before the frost leaves the ground. Maximum grade is 80%, and minimum is 50%.

Where hot logging is practiced, landings may be much smaller in area than where logs are cold-decked or left for some time awaiting further transportation.

Within recent years the prevailing practice has been to skid tree lengths to landings where they are loaded directly on trucks or other means of transportation or are bucked up into shorter log lengths by power saws as required by the sawmills. This has meant the designing and clearing of much larger landing areas and provision for sawing long or tree lengths into shorter lengths. Generally, the trees are bucked up by chain power saws. Bucking can be done much more cheaply and effectively and therefore at a lower cost per thousand board feet at landings than in the woods. Tractor skidding with its large available power has made possible the skidding of more long and tree lengths to landings in all forest regions.

FIGURE 88. Landing in selectively cut southern pine forest where long logs are skidded from near-by woods and then crosscut and loaded on trucks for delivery to pulpmill. Ford tandem-drive truck equipped with loading boom powered by the truck engine to load pulpwood. In the foreground is a portable-type circular power saw used to cut tree-length logs, skidded by tractor, into pulpwood lengths.

FIGURE 89. Logs averaging 500 b.f. each cut from stand of about 5000 b.f. per acre being loaded on 10-ton Athey wagons by means of crosshaul. Loads average about 2000 b.f. each. Taken near Vicksburg, Miss.

DURATION OF STORAGE—COLD DECKING AND HOT LOGGING

On single log hauls, there is obviously no decking. Hot logging refers to operations where the flow of logs from the woods to the mill is continuous, no expense being involved in clearing land or building up log decks or landings for storage purposes.

FIGURE 90. Landing in eastern Canada where mechanized operations are increasing. To this landing, tree lengths of spruce are skidded for distances of up to 800 ft. with a winch and slack line, as shown in right background. Then tree lengths are cut into 8′ logs by the slasher and loaded mechanically on trucks, as shown in the left foreground. This slasher and loader has a capacity of 6 cords of spruce per hour with a 5-man crew. A line roller moves the logs to the slasher saw. *Courtesy I. F. Fogh, logging engineer, Canadian International Paper Company.*

Cold decking is practiced where logs are left for some time because the major transportation facilities are unable to keep up with the minor method of skidding or because the operation can thereby be more efficiently organized or because, as in the Northwest, railroad construction may thereby be avoided. Cold decking is frequently practiced in the Douglas fir region of the Northwest where a pile of logs is often yarded to a point between the stump and the logging railroad and then later swung to the landing for loading on cars with a separate power-skidding machine.

Because of climatic conditions, cold decking may be necessary. Logs decked along haul roads during the summer and fall awaiting the winter sled haul furnish a good example of long-continued cold decking. Another is the pulpwood banked along driving streams in eastern Canada and the Northeast from early fall to March or April, awaiting the spring drive.

CHAPTER XII

Loading*

After the logs have been assembled by one of the three major methods of skidding, they are generally loaded on vehicles: railway log cars, motor trucks, animal- or tractor-drawn wagons, double sleds, and so on; they are then hauled directly to the sawmill or to a landing along a stream or to further methods of transportation. In case of railroad hauls, logs are loaded directly for transport to the mill. In small-mill logging, power skidding direct to tidewater as in Alaska, or motor-truck logging direct from woods to mill, logs are not usually decked, and no separate loading at assemblage points is necessary.

The following are the chief methods of loading:

ROLLWAY DIRECT TO TRANSPORT FACILITIES—HAND LOADING

Elevated ramps or skidways are used to roll small short logs directly onto log cars, sleds, motor trucks, and similar vehicles. These ramps are often constructed so that the logs skidded or hauled by tractors or horses may be rolled onto the major transportation device with the least expenditure of labor, time, and effort.

Cross ties, fuel wood, short-length pulpwood, small poles, posts, and bolts used for various products have been generally loaded by hand. Small saw logs are also rolled by manual labor down skids (and occasionally up short low grades) directly to the railway cars, motor trucks, and so on. This is particularly true in sections of the Central Rocky Mountains and in the second-growth forests of the Southeast and East. Hand loading has been largely superseded by portable mechanized loading devices, as described in the last part of this chapter.

* Other data on loading are included in minor and major transportation where necessary to clarify the description.

FIGURE 91. Crosshaul method of loading piling near Snow Hill, Md. One tractor loads tree lengths while another tractor pan-skids logs to landing from which the piles are loaded onto heavy truck and trailer.

FIGURE 92. Hyster winch attached to Diesel tractor loading logs on truck in southern pine forest in Georgia. This device can also be used in skidding logs out of swampy or wet locations. A Diesel engine also drives the small sawmill used for sawing these logs.

ANIMAL POWER AND TRACTOR LOADERS

For small and medium-sized logs and timber products too large for hand loading, the following devices are used, power being generally supplied by horses or tractors.

CROSSHAUL

This system is usually employed on small and short-length logs. A chain or cable attached to a truck, wagon, or log car is placed around the log, and then over the vehicle to a horse, mule, or tractor, which pulls at right angles to the direction of the main haul. Logs are generally elevated on skids assisted by men with cant hooks or peavies. A 3/8″ or 1/2″ crosshaul chain 40′ to 80′ long is generally used. A crew of five men and a team or tractor can load 30,000 to 50,000 b.f. per day. A crew of three men can load 10,000 to 20,000 b.f. or more per day. This method is entirely too slow and inadequate for large-capacity mills, especially on railroad operations. It is used primarily for loading wagons and trucks.

MISCELLANEOUS

The principal methods used on small operations and particularly with small logs are the A-frame and gin pole, generally with animal or tractor power, but occasionally with engine power.

ENGINE-POWERED LOADERS *

Engine-powered log loaders, including locomotive cranes and donkey engines, are used principally for large and long-length logs, particularly for Douglas fir, redwood, ponderosa pine, also for the heavier timber of the Northern Rocky Mountains and for hardwood and the long length timber in southern pine. Some form of engine-powered loader is practically always required for railroad logging. Also several kinds of hoists are attached to tractors and trucks for loading, principally in connection with motor-truck and tractor haul. Motor-truck loading is often done by means of a cable run through a block secured to a guy rope extending from a tree or the head spar. This is the method where logs are yarded by means of a high lead or other systems of power skidding. The power is usually furnished by a drum on the gasoline or steam yarder.

* Loading large long heavy logs on the Pacific Coast is also described in connection with the chapter on power logging.

Logging operations are generally located in remote places where repairs are made with difficulty and considerable loss of time. Delays are expensive, and they affect not only the loading machines but also other machines, transport facilities, and even the mill operations. Therefore, it is highly desirable that log loaders as well as most power machinery used in logging possess the following qualifications:

1. Strength, flexibility, and resilience to withstand sudden and violent strains incident to lifting great weights for short distances.

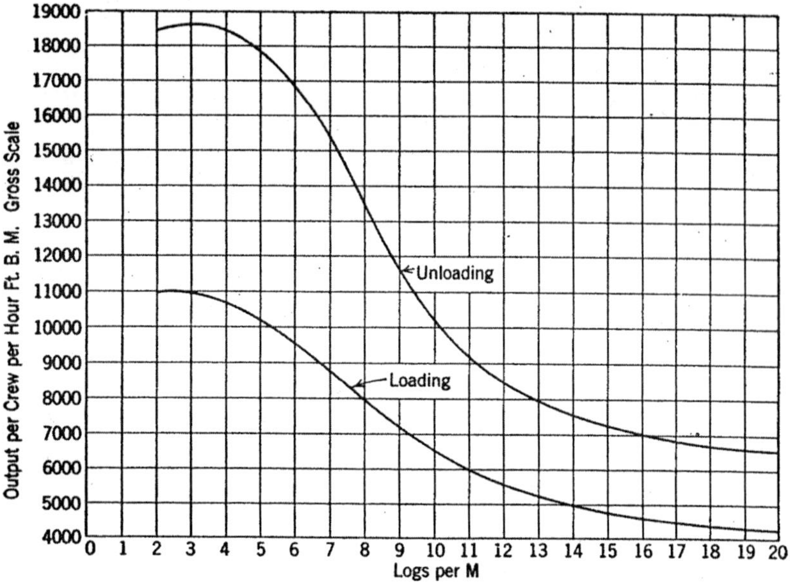

FIGURE 93. Output per crew per hour in loading and unloading logs of various sizes. Average load 2620 b.f. Loading with end horse jammer on motor trucks, day-labor basis. Crew consisted of team and teamster, top loader, two hookers and a tail-down man. Inland Empire.

2. Durability of construction, so that few repairs and little attention will be required. Loaders are often operated by unskilled labor far from repair shops.

3. Fireproof qualities, both in themselves and in relation to brush and debris incident to logging jobs.

4. Simplicity and ease of operation. Mechanical parts should be few in number and simple in operation since the loaders may have to be handled by several men of varied experience and knowledge.

Loaders with both rigid and swing booms can be used to skid logs without guy ropes up to 150 ft. For distances beyond this and up to 600 ft., guy ropes are generally required.

Where the volume of logs to be loaded exceeds 50 M.b.f. per day, railroad cars are usually loaded by means of steam or gasoline loaders.

In the Douglas fir and redwood forests, much of the loading is done by means of donkey engines mounted on sleds. In an exceedingly heavy stand of timber with a comparatively large quantity of logs to be yarded to one point, the necessity of having a quickly movable loading apparatus such as steam loaders mounted on railway trucks,

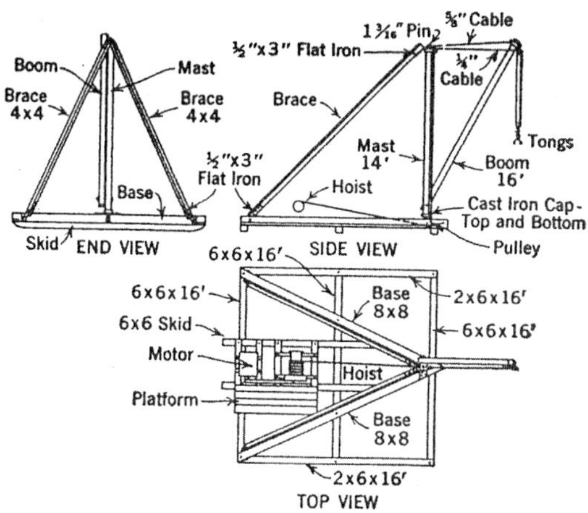

FIGURE 94. Details of log jammer used with motor hoist on small-mill operations in the East. This has been used to load up to 30 hardwood logs 10′ to 16′ long on two trucks in 20 minutes under favorable circumstances. Two truck drivers and operator were used in loading. One man hitches tongs in the middle of the log, the operator elevates it and swings it over the truck and lowers it, and the top loader breaks tongs and adjusts the log on the truck.

or on a track as in the McGiffert loader, is reduced to a minimum. Ponderosa pine in California, Oregon, and other western states, as well as southern pine in the Southeast, particularly in long lengths, and hardwoods in the Appalachians, Mississippi Delta, and Lake States are generally loaded on railway cars by mobile or readily portable loaders. Steam log loaders generally have a loading capacity of 30 million b.f. of logs or more per year. On the West Coast, much loading is done by a loading engine mounted on a railway car with the yarding engine which together forms a unit.

More than a thousand machines of one type are used in power log loading. The actual amount of logs loaded probably does not exceed 50% to 60% of the capacity amount in various forms of engine-powered loaders. The number of men generally required to operate engine-powered loaders varies from three to seven. Where coal is the fuel as in the Appalachian Mountains, or where oil is burned under boilers as in Oregon and California, no fireman is required, so that one operator on the machine is sufficient. Where pine knots or other wood supply the fuel, as in the southern pine territory, a fireman is required on the machine as well as a log loader. If end hooks are used in loading, two men are required on the ground, and, if the logs are large, a top loader is necessary on the car. If no fireman or top loader is required,

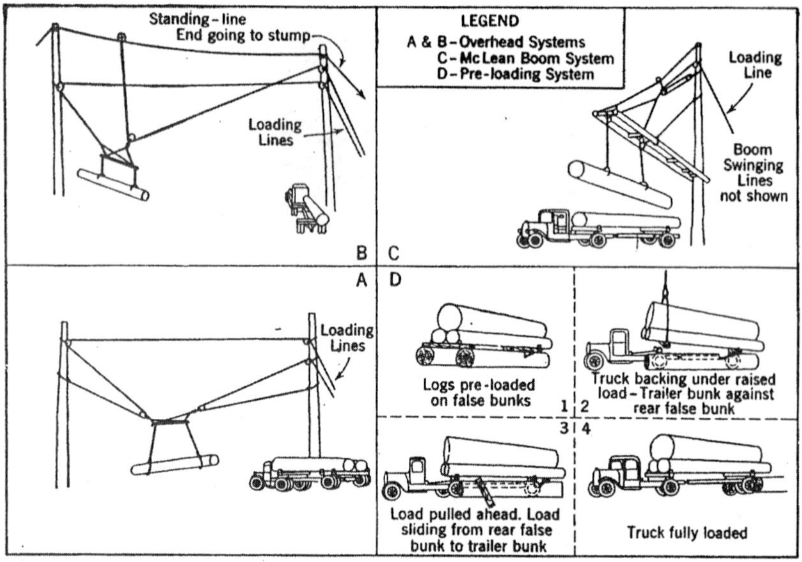

FIGURE 95. Three methods of loading large and long-length Douglas fir logs in the Northwest. *After E. F. Rapraeger.*

a total crew of three men can operate the machine. This is also true when loading is done by means of a center tong, in which case a tong man on the ground is necessary as well as a top loader to disengage the tongs on the car. Under favorable conditions, one top loader, two hookers, and one extra man to handle the binder chains and "spell off" with the end hookers should be able to load upwards of 100 M.b.f. per day. Where logs and cars are plentiful and the machine is pressed nearly to capacity, one or two extra men are usually provided to spell the end hookers and to assist in other ways. Small logs are sometimes handled in bunches, two men usually being employed to run the chains around the bunches of logs.

Log hoists or conveyors (endless chains) are used to haul pulpwood, cross ties, and mine timbers either laterally or longitudinally from log dumps or from a stream directly to railway cars—a method similar in principle to the log hoist at sawmills. In Wyoming cross ties are elevated from driving streams directly to railway cars. In the Adirondacks and in eastern Canada endless chains take pulpwood out of the water and load it both in 4′ and longer lengths directly onto railroad cars. In the East cars are loaded with mine timbers by log hoists which

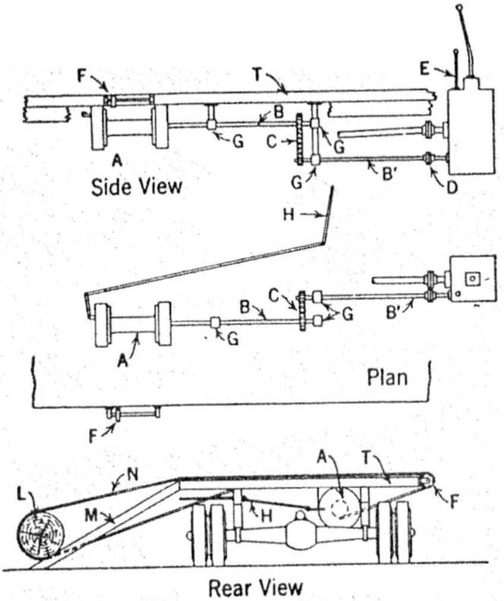

FIGURE 96. Diagrammatic sketch of truck motor-driven loading and skidding device as seen in the Northeast. This is being installed on motor trucks to provide facilities for loading and short-distance skidding as well as hauling.

A. Drum and cable.
B. Drive shaft.
B′. Drive shaft.
C. Chain drive.
D. Double universal joint.
E. Transmission gear-shift lever which controls the operation of drive shafts.
F. Idler pulley.
G. Bearings.
H. Clutch lever to engage the drum with the shaft.
L. Log.
M. Skid pole.
N. Cable.
T. Truck platform.

elevate the logs from a log yard or rollway and then convey them to the cars. A top loader or two assist in arranging the logs as they are delivered.

Aside from donkey-engine hoists and locomotive cranes, which are used primarily in the heaviest timber, two types of loaders are common on railroad operations.

LOG LOADERS OPERATING FROM LOG CARS

The Marion or Barnhart loader, as well as the American and Rapid loader, rest at all times on one of the cars of the log train to be loaded. Resting on each car in succession, and loading from this position, these machines require strongly built cars and have a tendency to strain or even wreck the log cars. They must be moved with each car loaded and ordinarily are incapable of spotting the train when moving themselves along the track, except on the top of the logging cars, which require a locomotive constantly in attendance. Furthermore, cars on the track must be closely coupled, and, if long timbers are handled, long drawbars are necessary between adjacent cars of the train. This type has the advantage that, when the logs are to be loaded on the same day

FIGURE 97. Loading large truck at the landing of power-skidding operation in Douglas fir stands near Mt. Baker, Wash. *Courtesy U. S. Forest Service.*

at widely separated points, a machine mounted on a railroad car can be run from point to point at comparatively high speed by a locomotive.

Other types, such as the McGiffert, are limited in their movement to about 7 or 8 m.p.h. In rough topography where it is impossible to deliver logs conveniently along the side of the railroad track, a full-circle machine operating from log cars similar to the Marion or American types has an advantage.

LOG LOADERS RESTING ON RAILROAD TRACK

The McGiffert loader is an outstanding example of this type. It is self-propelling and entirely independent of the cars to be loaded. It does not rest on or touch the logging cars and is capable of loading several cars at one place without moving. It spots its own cars, and no independent car is required for transportation. This type can be moved from one place to another without the assistance of a locomotive and can be used as well with long as with short cars, where the cars are entirely disengaged, or where they are connected by means of long drawbars. It has much greater stability and weight than full-circle machines and can readily handle long and heavy logs which cannot be loaded by full-circle machines resting on the car deck.

The Decker loader, also of this type, is applicable principally to the comparatively small logs of the Lake States and the short-length logs of the South, and to skeleton logging cars which are relatively low in height. In handling long and heavy logs it has less stability than the McGiffert and similar loaders because of its narrower base, and, where flat cars are used, difficulty is encountered in passing the car through the machine unless special construction is provided. With standard flat cars, long and heavy rail aproaches are required for the Decker loader.

CAPACITIES

The capacity of the log loader depends on the size and weight of the logs to be loaded, the type of machine in use, transportation facilities, and the number of logs available. Many log loaders operated by steam can handle 50 M.b.f. or more per day. The McGiffert loader can handle 100 M.b.f. or more per day. The amount to be loaded is generally fixed by the requirements of the mill or by the amount of logs it is convenient to handle at a single location. The McGiffert loader has handled 73 M.b.f. of logs scattered along the right of way at seven different landings in 1 hour and 20 minutes. The logs were cut in long lengths and averaged about 2000 b.f. per log. A stiff boom machine operating in western Montana averaged 145 M.b.f., log scale, in 9 hours, handling short logs averaging 8 to 9 per thousand board feet. In the Sierra Mountains in California, a McGiffert loader with a crew consisting of a loaderman, a fireman, and six hookers loaded 56 cars with 1229 logs scaling 521,690 b.f. in 7 hours and 43 minutes and had to stop at this point because no other cars were available for loading. In southern pine, one loaderman, one fireman, one man to supply wood, together with six tong hookers—a total crew of nine men—loaded approximately 500 M.b.f. of logs in one day. These are extreme rather than average illustrations of performance.

In loading cypress in Florida, 240,684 b.f. of logs were loaded on 50 cars in 9 hours by a converted Clyde-Rapid-Decker machine with a 36′ boom.

DEPRECIATION

Many logging companies charge off depreciation on log loaders in 5 years and others in 10 years, but this is rather conservative practice. The 5% depreciation charge, estimating a life of 20 years, is probably more accurate. Some engine-powered loaders have been in operation for 25 years or more. The annual cost for repairs is generally about 2½% of the purchase price per unit. In some cases, this may run from 5% to 9%.

POWER REQUIRED

The power required for loading varies with the weight of the logs. For light work in the Lake States and the Southeast, the double-cylinder steam engine with 7″ × 10″ cylinders and a boiler carrying 100 lb. of steam furnishes ample power. For medium-sized logs and longer lengths such as are found east of the Pacific Coast States, a 9″ × 10″ engine is ordinarily used with a boiler capable of carrying 150 lb. For the larger western timber a 9″ × 10″ engine with a gear ratio of about 7½ to 1 from the crankshaft to the drum is sufficient. Steam for the engine should be furnished by a boiler of 200-lb. capacity.

PORTABLE MECHANIZED LOG LOADERS

Within recent years, many types of portable log loaders have been devised, especially for the conditions existing in southern pulpwood logging and in eastern Canada and the Lake States. Ingenious logging superintendents and woodsmen have been responsible for the introduction and use of many laborsaving devices which have proved very economical and successful. Among the portable log loaders recently developed are the Camp, Brunswick, Crossett, A-Frame loader mounted on truck, the Dierks as developed in Arkansas, the Rite as developed in Alabama, the Bouffard, Drott, Logger's Dream, and the Witherell types. There are also many other loading devices commonly used in the woods.

The great advantage of these log loaders is that they supplant the slower and more expensive manual effort, provide portability, and reduce the time required for loading and, therefore, the cost of loading. With the disappearance of many logging railroads and the accompanying large log loaders operated in connection with them, and the concurrent introduction and use of the motor truck for log hauling, there has been a considerable increase in the use of these mechanized port-

FIGURE 98. One of the several types of mechanized bucking and loading devices developed to reduce costs. This is the Allen portable pulpwood mill developed by the Florida Pulp and Paper Company for processing tree-length pine logs into pulpwood. Logs are rolled into endless-chain conveyor shown in right foreground by the tractor-powered crosshaul winch mounted on the conveyor. Logs are fed through the cutoff saw, up the two-chain elevating conveyor onto truck in background. Production time for truck load of two standard cords is 15 minutes.

FIGURE 99. Drott portable skid loader mounted on International tractor. Used for assembling and loading small logs and pulpwood in southern pine. The sharp points of the loader pick up logs or pulpwood and drop them on a truck located in the rear of the loader. It is widely used in the Lake States, Canada, and the East, as well as in the South.

able log loaders. The initial expense for manufacturing these types is relatively low. Overhead charges such as maintenance and operational costs are reduced. At the same time, the loaders are equipped to do the work efficiently and very economically, and they reduce the need for much high-priced manual labor.

BOUFFARD LOADER

One of the types widely used in Canada, as well as in the South and in the Northeast, is the Bouffard truck-mounted log loader used for

FIGURE 100. Cord-a-minute pulpwood mill in operation. In the foreground is the landing to which full-length logs are skidded from the woods. The circular saw shown at left cuts the logs into desired pulpwood lengths which are dropped on the conveyor and elevated to the truck shown in right background. This is one of several mechanized devices developed to reduce the cost of manual labor and speed up the operations of crosscutting and loading. *Courtesy B. L. Montague Company, Sumter, S. C.*

hardwood and softwood logs, as well as for pulpwood, piling, and other forms of primary forest products. This loader consists of a pair of loading arms hinged to bunks at the side of the truck and raised by a cable. The arms are curved at the ends in order to hold the logs in place when they are elevated from the ground onto the truck. A standard power take-off from the truck engine to the loader winch is operated by the truck driver. This device can be attached to any standard make of truck. It has a maximum pull of 4000 lb., a drum speed of 150 f.p.m. at 1800 r.p.m. truck engine speed. The drum generally carries about 100′ of ½″ cable.

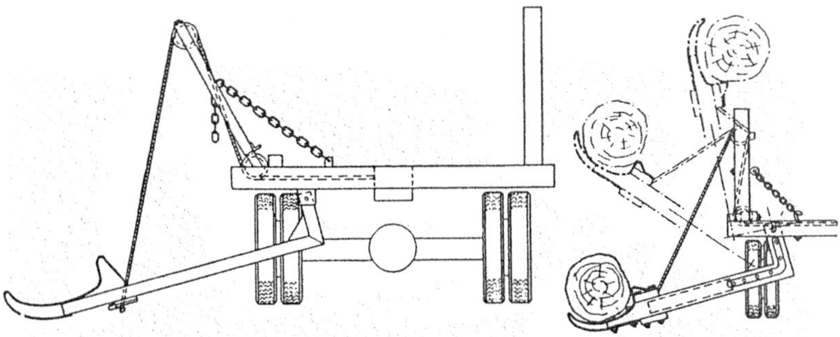

FIGURE 101. Diagrams illustrating the Bouffard truck-mounted log loader and how it operates. This is extensively used in Canada and the eastern and southern forests. Power is transmitted from the truck engine through a standard power take-off to the loader winch. The truck driver and helper normally form the loading crew.

The advantages operators have found in the use of the Bouffard log loader are:

1. It reduces the loading time, requiring about one-half the time needed for jammer loading and about one-sixth the time needed for crosshaul loading.

FIGURE 102. The Dixie pallet system as developed and used in pulpwood operations in Georgia. Time and labor are saved by loading each pallet separately and then moving it with its load directly onto the truck-body as shown.

2. It reduces labor costs because a two-man crew consisting of the truck driver and one helper can efficiently operate the loader.

3. A truck equipped with this loader is ready to load logs at any time, eliminating the need for jammers, jammer crews, gin poles, horses, tractors, and so on.

FIGURE 103. Two views of the Montague trailer pulpwood mill in operation near Georgetown, S. C. Long-length pine pulpwood is hauled to the lower part of the conveyor and cut off to desired lengths with a circular swing saw. Then they are mechanically elevated to the truck where the top loader places the bolts in position. This type of pulpwood mill is widely used.

It is used only for logs up to 16′ long, hardwoods up to 20″ in top diameter and softwoods up to 22″ diamater. It is not adapted for use in the big timber of the Pacific Coast.

THE LOGGER'S DREAM

The Logger's Dream is a combined skidder and loader mounted on a truck chasis and is widely used in parts of the Rocky Mountains, as well as in the East and South. There are 130 machines in operation in Mississippi alone. There are two different models, one with and the other without cable rehaul. It has a loading device on the rear of a truck, giving it quick mobility. Thus it is driven directly into the

FIGURE 104. The Logger's Dream which is a combination skidder and loader mounted on truck chassis and quickly and inexpensively moved and set up is widely used in the South and in some other sections of the United States.

woods for skidding logs short distances and loading them on trucks. It has definitely cut down the cost of skidding and loading on many operations. It skids logs for distances up to 300 ft. With the rehaul cable, it will take logs out of or across canyons, ravines, swampy areas, and rough terrain up to distances of 800 ft. The machine is powered with a 100-hp. V-8 Ford industrial unit. It is used with both center tongs and end tongs, but the latter are recommended for greater efficiency and safety. The skidding boom is approximately 24 to 26 ft. above the ground. This elevated point has a tendency to lift the front ends of the logs as they are skidded, usually for distances of 300 to 500 ft.

An important feature of this machine is the ⅝″ guy lines generally 50′ long, used to support the boom and prevent its bending forward when in action. The machine complete with truck weighs between 10,000 and 12,000 lb. It can be mounted on any 1½-ton or 2-ton long-wheel-base truck. The great advantage of this mobile skidding and loading unit is that it will skid and load more logs with less labor and gasoline. Production figures indicate that costs are 40% to 60% lower than those of other methods formerly used for similar purposes under the same conditions.

It has the following characteristics (Fig. 104):

1. It is mounted on truck chassis and therefore is quickly movable and may be driven into the woods for skidding and loading purposes.

FIGURE 105. A portable-type swing-boom log loader used by the Camp Manufacturing Company to supply its sawmill and pulp mill at Franklin, Va. Note the end tongs being placed by men in foreground with the swing boom used to elevate the logs and drop them on truck. At right is a wheel-mounted circular bucking saw cutting long lengths into shorter lengths. The same saw is used for felling as well as for crosscutting at the landings.

2. It does not require any road construction in the open level stands of southern pine and very little swamping out of brush and stumps in other types.

3. It is not adapted to the big heavy ponderous timber of the West Coast.

4. It is operated by three men.

5. It may be used to lift overturned trucks or to tow trucks in unusual circumstances.

6. A rear guy line to stump or tree gives support to the loading boom, thus adding stability to the machine and preventing undue wear and tear on the truck body.

It also has the following advantages:

1. It skids and loads logs up to distances of 800 ft. but operates to best advantage up to 300 ft. It will handle logs that weigh up to 12,000 lb.

2. There are four models which are designed to handle almost any logging problem in the East, South, or Rocky Mountains, except in very rough terrain and swampy locations.

3. It has quick and inexpensive mobility and travels at the usual truck speeds on the highways. It is readily and easily set up and taken down. It will load a truck in 10 to 17 minutes with three men.

4. It is exceptionally well adapted for selective logging.

5. If used with a rehaul cable, it saves time, effort, and man power, although the initial overhead for investment is larger.

6. It is adapted not only to saw logs but also to pulpwood, poles, piling, and other timber products.

PART V
MAJOR LAND TRANSPORTATION

CHAPTER XIII
Chutes

Chutes, known also as slides and sluices, are generally pole channels or troughs used for transporting saw logs, poles, cross ties, pulpwood, bolts, and fuel wood down relatively steep grades, usually for short distances. A few are made of earthen channels, lumber, and metal, but probably 98% of all the chutes in the United States are of logs. They are primarily adapted for use in small or medium-size timber in rough topography where logs must be transported down steep gradients. Under these conditions the cost of construction of roads for sleds and motor trucks or of flumes or railroads may be prohibitive. Chutes have been used in mountainous regions in limited sections of the Northeast, the Southern Appalachians, and the Rocky Mountains. They are found chiefly, however, in Idaho, western Montana, the Northeast, and eastern Canada. They have been used in the mountain sections of northern and central Europe and India for many centuries.

TYPES

On the basis of gradient, chutes may be classified as (1) gravity or running chutes and (2) trailing chutes. If the gradient is sufficiently steep and the logs are of such weight that they slide down unassisted, the chute is known as a gravity chute. If the slope is insufficient to permit the logs to slide of their own weight, it is known as a trail chute. Various parts of the same chute may be either trail or gravity chutes. Trailing chutes in the summer often become gravity chutes in the winter. When the frictional resistance does not permit saw logs to slide freely, horses or tractors are used to pull or push them in long strings or trails.

Chutes are also classified by the design or form of the trough. In Idaho the hewn V-shaped two-pole chute was widely used; probably 95% of all chutes used in Idaho were of this variety, though occasionally the three-pole chute was employed for saw logs. In the Northeast and eastern Canada, the semicircular or U-shaped chute constructed of three to eight or more small poles serves for the transportation of short- and long-length pulpwood.

Chutes may be classified according to their adaptability to:

1. Saw logs, piling, poles, and similar medium-size and heavy products.
2. Small products such as pulpwood, fuel wood, chemical wood, and bolts.

FIGURE 106. Newly laid trail chute, showing method of hewing faces on logs, Lolo National Forest, Mont. Logs are decked on the landing along the chute awaiting transportation. Most of the chutes used are the two-pole hewn-face type.

Chutes may be classified further as to portability—the Sykes plank slide used in the Adirondacks, the Tryon chute in the Black Rock Forest of New York—or as permanent or temporary chutes. They are seldom installed on a permanent basis. Some are constructed for only a few hundred cords of pulpwood. Others see service for several years. Earthen chutes iced in the winter are occasionally used for short distances in cold climates, as in Canada, New York, and infrequently in the Southern Appalachians. Sawed timber chutes fashioned from cants to form a trough and spiked poles or sawed cross ties are sometimes used if the location is accessible to a sawmill. Plank chutes are sometimes seen in Europe.

Recently, lightweight portable metal chutes have been used on several pulpwood operations in eastern Canada. They are laid directly

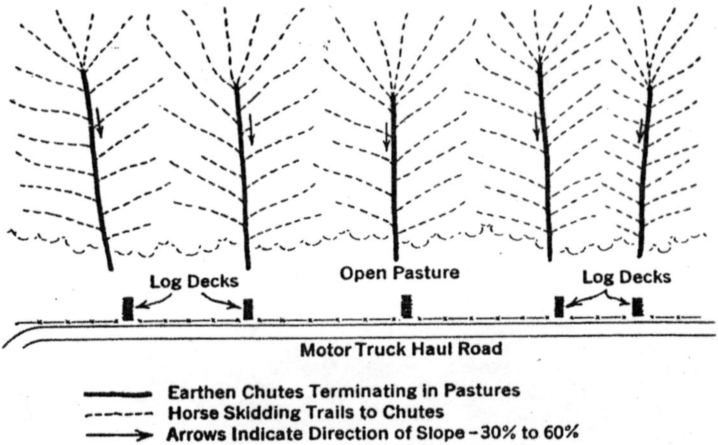

FIGURE 107. Sketch illustrating the use of earthen chutes on steep slopes, generally 30% to 60% or more. The chute is constructed by dragging two or three large logs when the ground is soft, preferably wet. During freezing temperatures logs slide readily to the base of the slope. Earthen chutes are constructed and operated very inexpensively for relatively small quantities of logs and for short distances. They are used on small-mill operations in steep topography in the East. Chutes on this operation were placed 400 to 1000 ft. apart and were 500′ to 1100′ long. Not drawn to scale.

on the surface of the snow and sometimes elevated on temporary supports or suspended by cables. In New York one company uses about 900 ft. of aluminum portable chutes on pulpwood operations. This is a variation of the portable Tryon chute.

FACTORS AND CONSIDERATIONS DETERMINING USE

Chutes are used to transport logs from skidways, located both at the intake and along its entire route, to a landing at or in a stream or to a flume, railroad landing, or log dump where they may be loaded on a truck, sled, or other vehicle. Chutes are generally an intermediate method of log transportation or a connecting link between the skidding operation and some other form of major transportation. They have played an important part in logging operations in the steep gulch topography of Idaho and western Montana, where for many years $500,000 to $1,000,000 has been expended annually for chute construction and probably $1,000,000 or $2,000,000 a year for operation. Chutes are also used extensively in British Columbia.

Neff estimates that an annual average of 250 miles of chutes was built in the Inland Empire where nearly every logging operation may install 1 to 20 or more chutes in a season. Chutes are being rapidly replaced, however, by tractors and trucks as transport devices.

Since the more accessible stands in eastern Canada and the Northeast have already been cut, operators use some chutes because logging jobs are now located in steeper and rougher topography.

The conditions which determine the use of chutes may be summarized as follows:

1. Chutes are generally initial means of major transportation and, therefore, constitute an intermediate step between skidding and some other form of transport to the mill. They are seldom an exclusive or principal method of transport from the woods to destination. They may, therefore, be considered one of the minor steps in log transport.

2. They are utilized under conditions of steep topography where construction costs for other transport facilities would be excessive and, therefore, impracticable. They are particularly adapted to short sharp drops in elevation such as down steep ravines, gulches, and from benches and terrace-like topography as in northern Idaho where slopes of 30% to 60% or more are frequently encountered. Chutes are usually located in depressions or ravines where logs may be skidded to them and rolled directly onto them from both sides.

3. They are adapted to relatively small or medium-size timber such as western white pine and ponderosa pine of the Northern Rocky Mountains (generally 5 to 16 logs per thousand board feet) and to pulpwood, fuel wood, and other small forms of forest products. When located in or adjacent to heavy stands of ponderosa pine, western white pine, or eastern spruce pulpwood, they may be operated most economically. They are also adapted for use in the heavy stands of hardwoods in the upper coves of the Southern Appalachian Mountains.

4. Availability of other logging equipment—tractors, trucks, flumes, and the like—may prove to be determining factors in the decision to install chutes or other means of transport. Chutes represent generally a small investment. Local materials are used, and relatively cheap and ordinary woods labor may be employed in their construction and operation.

CONSTRUCTION

Chutes are generally designed, located, and constructed by experienced local woodsmen. Careful surveys and advance computations of gradients, tractive resistance, and the like are seldom made. Two important considerations in chute construction are vertical and horizontal location. Grades should be maintained uniform to insure delivery of logs with the least additional assistance and the least losses possible. Sharp curves are inadvisable because of the likelihood of the logs being lost en route by "jumping," the possibility of damage to the chute itself, and the difficulty of operation. Straight chutes or those with few degrees of curvature are much preferred. However, chutes should fit the topography as effectively as possible because trestles are generally very expensive and susceptible to damage by storms, washouts, and falling timber.

Two-pole chutes for saw logs are generally laid directly on the ground and held in place by ties at regular or irregular intervals, depending on the size and weight of the logs, length of operation, gradient, and footage to be transported.

Construction generally starts at the bottom or terminus of the chute. When the chuting of logs is completed, the logs used in construction, if valuable, are themselves sent down. Owing to the frequent necessity for partial crosscutting on curves, chopping, and spiking, the more valuable species are seldom put into saw-log chutes. Local inferior species such as western hemlock, western larch, and white fir serve in the Northern Rocky Mountains, and hardwoods, cedar, and so on in the Northeast and in Canada. Spruce, however, is sometimes employed for chute construction in the latter regions and is salvaged on completion of the operation.

Practically all the pulpwood and fuel-wood chutes are gravity chutes. They must be built with sufficiently steep grades to carry the material unassisted from the intake or loading point to the terminus.

Pole chutes for saw logs are frequently trail chutes and may have grades varying from level to 60% or more. Idaho chutes are generally found on 10% to 35% grades. Trailing chutes are preferred because of the lower losses incurred in bringing in logs by this method.

Chute sticks (logs) are generally laid with the small end downhill to avoid slivering and excessive abrasion. Long logs 20′ to 40′ in length with relatively little taper are preferred for two-pole saw-log chutes. In Idaho, logs with tops 10″ or more in diameter and butts not over 16″ in diameter inside the bark are preferred. There are generally about 90,000 to 100,000 b.f., log scale, in a mile of two-pole chutes. This is equivalent to about 20 b.f. to the running foot. Small logs are spaced 1″ to 3″ apart and large ones 5″ to 7″. Where both small and

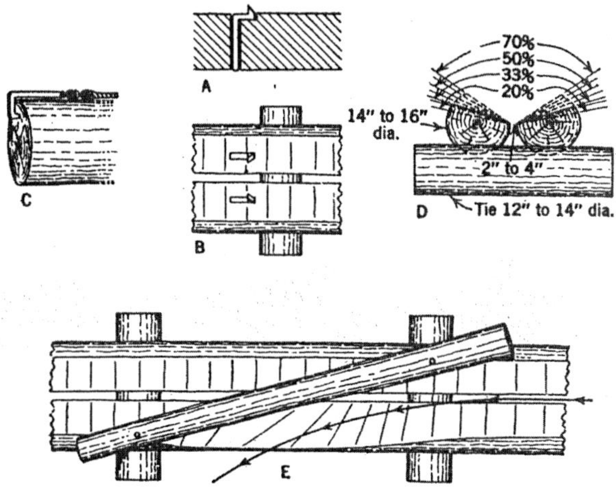

FIGURE 108. Details of the use of two-log chutes. *A* and *B* illustrate the use of goosenecks to retard the speed of logs in the chute. *C* illustrates the use of the L or J hook used in trailing a string of logs down the chute. *D* illustrates a cross section of a two-pole chute with varying degrees of facing depending on the size of logs chuted. The face is generally at an angle of 33°. *E* illustrates the operation of a whippoorwill at the terminus of the chute. Another method of unloading chutes is to depress the log on the landing side.

large logs are used, the chute logs are spaced 2″ to 4″ apart, or small logs are spaced between large ones.

In northern Idaho, $5\frac{1}{2}$ rods of two-pole chutes were built per day on a 25% to 30% grade with much rock work and down timber in a steep narrow gulch by the following crew:

1 foreman or chute builder
3 helpers
2 sawyers cutting chute poles
1 swamper
1 teamster hauling chute poles
4 men to clear right of way
Total 12 men

In addition, one team for skidding chute poles, 10 lb. of powder, and 22 drift bolts were required for this unit.

On an easier logging chance with 15% to 25% grade, only nine men and a team completed the same amount per day at a much lower cost. Fewer ties are required and construction is cheaper on the more favorable or lower gradients.

On a 30,000-cord job in the Adirondacks, there were 15 separate chutes of five different types on one area approximately 1 mile square.

These included the six- and eight-pole U-shaped chutes, the eight-pole square chute and three-, four-, and five-pole chutes.

One chute was 5040′ long, made up of 280 units of 18′ logs. The chute rested on the ground wherever posible, but, in order to maintain an even gradient, there were trestles varying from 1′ to 10′ or more in height. Five men constructed this eight-pole chute in 40 working

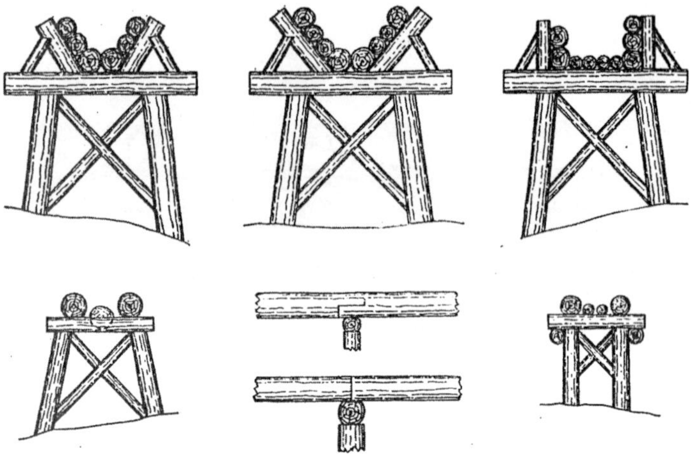

FIGURE 109. Five types of logging chutes used in transporting spruce pulpwood and two methods of making joints, as found on one operation. The two lower chutes are used for short steep pitches where only a few hundred cords are to be transported. The upper three are the more permanent types required for longer distances up to a mile where large quantities of pulpwood are transported. The upper right sketch illustrates a method of banking with an extra log to prevent logs from jumping on curves. The flat base of this chute proved to be unsatisfactory. Generally peeled spruce pulpwood logs 18′ long are used, the small sizes being 4″ to 6″ at the top, and the larger logs 6″ to 8″ at the top. Bents and braces are frequently cut from cedar and hardwoods.

days, or a total of 200 man-days. About 1800 cords were chuted, and the drop in elevation in this distance was 1400 ft. Only 4′ peeled wood was used. The larger bolts traveled with great velocity, whereas the smaller bolts often stopped or at least slowed down on the nearly level places. The bolts left the slide at a dump where they were loaded on sleds and hauled to the driving stream. During the winter a jobber's chute 3500 ft. long joined the main channel. This was an eight-pole rectangular or square chute with four small logs forming the base and two of larger size forming each side. The flat wide base permitted too much side sway, causing excessive vibration and springing. Many bolts were thrown out, resulting in great damage to the chute itself as well as in loss of wood. At the end of pulpwood chutes a special 16′ portable chute was put on the last section and elevated at the end in order to carry the bolts farther out over the log dump.

CAPACITY

The capacity of chutes is governed largely by the amounts that can be loaded into the chute by the crew at the intake or loading point or by the amount that can be handled at the terminus, or both. Gravity and trail chutes generally have similar capacities.

The average capacity per 8-hour day in the Northern Rocky Mountains is 400 to 600 logs or 40,000 to 100,000 b.f., depending on the size and length of the logs. The number of logs is a much more important controlling factor than their size.

Pulpwood chutes in the Northeast and Canada, which are almost exclusively gravity chutes, have nearly unlimited capacity from one loading point. From two to four men load the chute from adjacent skid piles. Pulpwood is loaded by hand. Chutes that end in a stream have unlimited capacity. Extensive landings are generally constructed for chutes that terminate at a logging railroad spur. Pulpwood chutes which end in a log dump are generally unrestricted in capacity.

OPERATION

Saw-log chutes are used in the following ways in the various logging chances in northern Idaho:

1. Chutes only.
2. Chutes and flumes only.
3. Chutes and railroads.
4. Chutes and motor trucks.
5. Chutes and sled or dray hauls.

Many more chutes are constructed and operated in places where they either alone or combined with flumes constitute the major means of log transportation. Motor trucks and tractors, more than any other improvements, have replaced chutes.

Operating methods vary widely between gravity and trail chutes. Gravity chutes are operated at the time the logs are skidded to them or later during the same day. They may also be operated after the logs have been cold decked along the chute for some time. When the logs are cold decked, the chuting crew generally consists of four to six cant-hook men with a horse or team, with additional men to regulate the speed along the chute. On trail chutes, a string of 10 to 20 logs is made up. Trailing is generally used on distances of ½ mile or more. Small logs are placed ahead and the larger ones behind, since the latter slide more easily in gravity chuting and serve as "pushers" in trailing. Permitting the logs to slide separately is known as "wild catting" and is generally regarded as poor practice unless the chute is short or the terminus of the chute is directly into the water.

The motive power in trail chuting is supplied by horses and tractors. For long-distance trailing, tractors are preferred. Tongs or J-hooks are set into the last log, the tongs about 3 ft. from the front end.

On a 1¼-mile two-pole saw-log chute in Idaho, 120,000 b.f. were handled by:

5 trail teams and drivers
3 horses and 3 men to make up log trail
2 men rolling the logs into the chute
4 oilers to facilitate movement of the logs
2 flagmen
3 spike tenders
1 watchman at terminal
2 helpers
Total 22 men and 13 horses

BRAKING AND ACCELERATION DEVICES

Owing to the varying grades found on the same chute, several methods are employed to retard or increase the velocity of the timber transported. Peeled logs slide much more readily than unpeeled ones. Logs used in the face of chute channels are always peeled to decrease the frictional resistance.

The speed of logs on two-pole saw-log chutes may be checked by the following devices:

1. Goosenecks or spikes. The former are made of 1½″ square steel and are used only on steep grades. They are very destructive to soft-textured woods, such as western white pine.

2. Sand, gravel, or dirt which is sprinkled on steep pitches and also on ice and snow to check the speed at curves or on sharp gradients.

3. Sharply hewn angles which provide more frictional resistance than more open or flatter angles.

The speed of logs or bolts on pulpwood chutes is generally checked by wolf brakes, also known as bear traps, made from one or more small 12′ to 18′ logs fastened at one end directly over the chute. The lower end rests on the bottom of the chute channel. As the log strikes the lower end of the brake, it is lifted by the impact, its speed thus being retarded. This is very effective on steep grades to retard the velocity of pulpwood.

Sliding facilities on saw-log chutes to speed up the movement of logs on low grades or on almost level places are provided by the use of:

1. Grease, tallow, or heavy oil. In winter light oils are used, and heavy oils or grease in the summer.

2. Water inducted into the chute from near-by streams. Where this is practiced, the chute is calked to make it water-tight or partially so. When the chute rests directly on the ground, water can be readily used in this fashion.

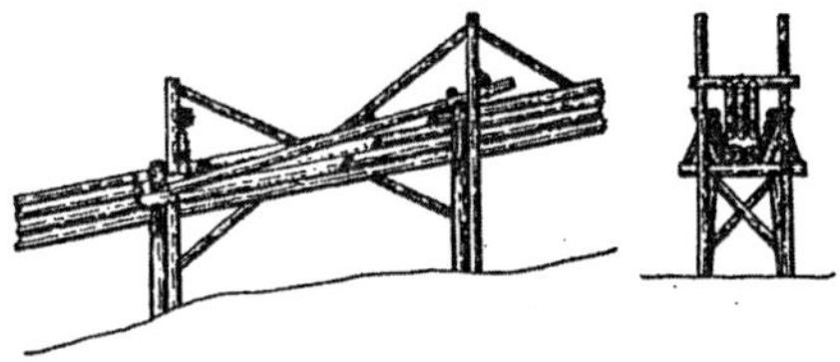

FIGURE 110. A side view and cross section of a trap or brake used to retard the speed of pulpwood. Logs go through the trap at high speeds striking the lower part of the three-pole brake, thus elevating the lower end and permitting passage of logs down the chute. Small-size bolts pass through uninterruptedly; large ones which travel at high speeds are retarded in their progress. These traps are placed at or just above the steepest pitches in the chute.

3. Ice. Water is sprinkled on the chute or turned in from a near-by stream. When this water freezes, the velocity of the logs is increased very materially. Rossed or peeled logs slide much more rapidly than logs with bark on them.

4. Snow shoveled into the chute during the winter. This is a common method.

ADVANTAGES AND DISADVANTAGES

The advantages of chutes may be summarized as follows:

1. They may be used on steep gradients and rough ground conditions where other methods of transport would be impracticable or prohibitively expensive.

2. They reduce skidding distances, thus lowering skidding costs and, therefore, the major transportation unit expense. The cost of operation and maintenance is generally low if conditions for construction and operation are favorable.

3. Local materials and labor may be used. When favorable log sizes are available and labor costs are low, this may be a very definite advantage.

4. No back haul or empty rehaul is required as is with tractors and trucks.

The disadvantages are:

1. Depreciation of log values due to damage and breakage.
2. Marks and cuts due to braking devices.
3. Holding logs on the rollways for indefinite periods awaiting weather conditions. This applies especially to western white pine. There is little, if any, depreciation on pulpwood transported on gravity chutes.

THE SYKES PORTABLE LOG CHUTE

This can be used efficiently for distances up to about ½ mile from the railroad track or truck road on areas suited to the use of the log chute and in timber producing logs that average 8 to 15 per thousand board feet. In general, such areas are moderate slopes, and experience on these slopes with the Sykes chute has been satisfactory in summer, fall, and winter logging. The log chute works best in two extremes of weather: (1) hot dry weather when a decidedly small amount of "slide oil" is daubed on the log chute, and (2) the freezing or frosty weather of winter or early winter when no oil is needed. Log chutes may be used on the level and at times are successful in moderate upgrade conditions.

FIGURE 111. Sykes log chute used for hardwood logs in the Northeast.

The first time the Sykes log chute is used on an operation, it should be in charge of men who have had previous satisfactory experience with it, since inexperienced men often are not successful with it and, therefore, become prejudiced against it.

Logs can readily be skidded into the chute and made up into trails of 3 to 12 logs coupled together with grabs from anywhere alongside the chute after it is properly laid. This is done by taking out of the chute one end of one side member which makes a place where logs can be readily skidded into it. The end of the side member removed from the chute is, of course, in the rear of the direction in which the logs are to be moved. In nearly all cases it is best not to have certain teams work merely hauling the trail of logs from where they are made up while

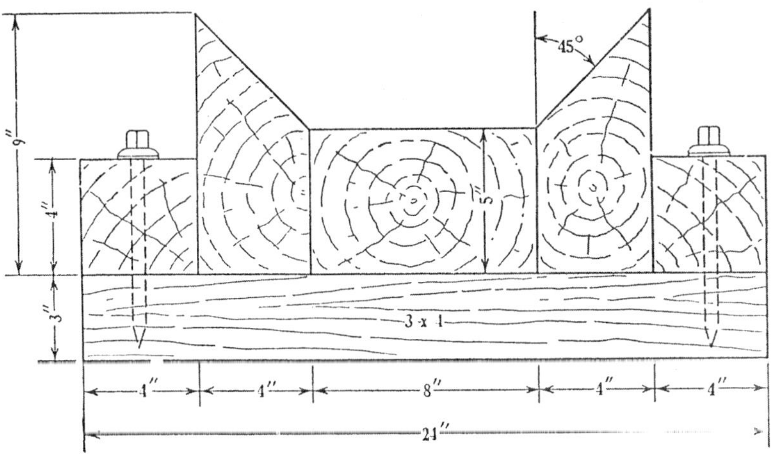

FIGURE 112. The upper view illustrates the winter operation of a Sykes chute. The lower sketch is a cross section showing the conventional method of holding the portable chute sections in place on a 3" × 4" cross tie at the joint.

other teams are making up the trails by skidding logs into the chute. This usually results in an unbalanced condition, and as a result the teams making up the trails get either way behind or way ahead in their work. The best method is to have each team work in both making up trails and taking trails through to the delivery point.

Thus when a team returns to the woods, if a trail is already made up, this team can haul the trail to the delivery point. If no trail has been made up, then it can skid logs into the chute and make up a trail.

It is likely that some very light tractors might be used efficiently on the log chute, especially for longer distances, but, just as horses still seem to be the most efficient for short-distance skidding in the East, they are generally also more efficient for moving logs in the log chute. The log chute is simply an efficient means of long-distance skidding which enables a team to handle from 3 to 12 logs, depending on the grade, weather conditions, size of logs, and so on.

The very sparing application of oil with a sort of swab or dauber is important not only because it saves oil, but also, because the logs thus slide much more easily than when much oil is applied. As an illustration of how this works, a few drops of rain on the rails causes a locomotive to slip its drivers very easily. The tendency to slip is much more pronounced when there are only a few drops of rain on the rails than after the rails have become wet all over. Scattered dabs of oil on the log chutes work the same way.

CHAPTER XIV

Wheeled Vehicles*—

Animal and Tractor Draft

SUMMARY OF WOODS OPERATIONS WHERE USED

Wheeled vehicles, generally known as log wagons, have almost disappeared as a general means of log transportation except in some restricted sections of the East and South and, to a lesser extent, in the Northeast and the Lake States. The wagons are found mostly on small-mill operations. They are used occasionally for both southern pine and hardwoods in the South and for miscellaneous hardwoods and white pine in the East and the Lake States. They have gradually been supplanted as a transport device by trucks. Since 1920 the motor truck has furnished the most effective competition against wagon transportation; wagons are now found in only a few regions and even in those regions are seldom used.

Where they are used, log wagons are frequently taken directly to the stump or felling area in both hill and bottomland hardwoods as well as in southern pine. Thus the logs are hauled short distances by wagon to the landings, and log assemblage is eliminated entirely. One improvement in log wagons has been the introduction of rubber-tired wheels.

The main requirements of a log wagon are:

1. Strength and stability to resist strains and stresses incident to the rough heavy work of log hauling. The lightest weight in construction consistent with the strength and stability to do the work with least tractive resistance is the important factor.

2. The practicability of making repairs on the wagon while in the woods.

3. Simplicity and ease of loading, unloading, and turning.

4. Low center of gravity so that rough and uneven roadways may be traversed with least danger of accident.

* Other types of wheeled vehicles such as bummers and high wheels are used for bunching (skidding) logs and are described under Minor Transportation.

FIGURE 113. Eight-wheel broad-tread wagons are sometimes used to haul Mississippi Delta hardwood logs over soft ground from stump to railroad. Note skids resting against the oak logs to assist in crosshaul loading, and also boxed wheels to reduce hauling resistance during muddy seasons.

TYPES OF WAGONS

Wheeled vehicles, which are generally of the four-, six-, and eight-wheel types, have mostly been superseded by motor trucks. The four-wheel wagon is the principal type of log wagon in use today.

The four-wheel wagon. Relatively small logs, averaging about 20 per thousand board feet in the pine sections of eastern Virginia and the Carolinas, are sometimes brought in from the felling areas, both with and without preliminary assembly, by four-wheel carts or wagons. Occasionally two to four or more logs are bunched together in the woods or along the haul roads by one man and a horse equipped with a short chain and a pair of skidding tongs. The wagon teamster drives the cart directly over the logs, which are underslung beneath the wagon. The wagon tongue is lifted, and a chain is passed around the logs and fastened to a trip device. When the tongue is lowered, the logs are lifted from the ground as in the high two-wheel carts. This trip device enables the teamster to unload the logs by dropping them without stopping at the log yard.

WAGON-HAUL ROADS

Relatively little investment for road construction is customary in the southern pine region. The topography is generally flat or gently rolling on the South Atlantic and Gulf Coasts so that little road construction is necessary. Bottomlands are generally favorable for wagon haul in dry weather. During wet or rainy seasons, however, wagon haul may be very difficult or impracticable. In Arkansas and some of the hill sections of the Piedmont region, some rough topography is encountered with fairly steep slopes requiring haul roads which may cost from $100 to $200 per mile. Generally a little surfacing, grading, and infrequently rock work may be required.

For hardwoods, both in the South, the Lake States, and the eastern wood-lot regions, unless the logs are skidded out to the haul roads, which is the prevailing practice, roadways must be cut through the woods.

On a large operation in North Carolina, logs averaging 50 b.f. each were usually hauled a distance of 600 ft. on four-wheel mule carts directly from the stump to the sawmill.

ADVANTAGES AND DISADVANTAGES

The advantages of wheeled vehicles are:

1. The initial investment for wagons and animals is small. Frequently these animals are used for other work during parts of the year and, therefore, may serve a dual purpose. Investment in large and expensive machinery such as tractors, or even motor trucks, may not be justified for the amount of logs to be hauled.

2. Fairly heavy logs may be hauled on soft or wet bottoms with wide-tired wagons under conditions in which motor trucks could not be profitably or satisfactorily operated, especially in rainy seasons.

3. For short hauls and small scattered units of timber, low-cost transportation can be furnished. Good motor-truck haul roads may be unavailable or too expensive to construct under the conditions. It is frequently unnecessary to build special roadways where wagons are used, especially in southern pine.

The disadvantages are:

1. Wagons are exceedingly slow; therefore, they are not adapted in themselves to supply any but small mills or to be more than an auxiliary to other means of major transportation. They are used occasionally as feeder units to logging railroads on large operations in the South or for short hauls to small mills in the East, South, and Central States, and occasionally in the Lake States. The wagon itself could not be relied on to furnish a large and constant flow of logs to supply sawmills of 50,000-b.f. capacity or more daily.

2. They are not recommended where motor trucks may successfully serve as the major method of log transportation.

CRAWLER WHEELS AND WAGONS *

The crawler wheel as illustrated in the Athey truss wheel was a pioneer for carrying loads on wheels of the crawler type generally behind or in connection with the tractor.

FIGURE 114. Two types of log wagons mounted on crawler or track-type wheels. The upper one is built in 10-, 15-, and 20-ton capacities, and the lower one is built for 20-, 30-, and 40-ton capacities. Both are for use with tractors. The upper type of log wagon is sometimes operated in trains. When the front ends of logs are loaded on a single bunk between a two-wheel crawler, it is generally known as a bummer.

These crawler wheels, sometimes referred to on logging operations as bummers, are used in pairs to elevate the front end of the log, thus permitting the rear end to drag. On four-wheel and two-wheel bummer wagons they support the entire load without dragging the end; also they serve to support the fair-lead arch known as the Willamette-Hyster arch. These bummer or crawler wheels are found in the South and on the West Coast, particularly in ponderosa pine, as well as in Idaho, the Southwest, and in the Lake States, and in Mexico, Central and South America, and on the west coast of Africa.

* These types of equipment may be used with both major and minor transportation, depending on local conditions.

FIGURE 115. Hauling gum logs from swamp on Athey wagon in Louisiana. This was a 3000-ft. haul, and the average load was 2800 b.f. About 36,000 b.f. were handled in a day. The tractor used about 3½ gal. of fuel per hour.

FIGURE 116. Hauling Engelmann spruce logs on 2-wheel cart, also known as bummer from felling area to landing where they are loaded onto trucks and hauled to sawmill at Pueblo, Colo.

Athey wheels are particularly suitable wherever it is necessary to haul heavy loads over soft roads or bottoms. The fair-lead-arch type of equipment mounted on crawler wheels is especially adapted to mountainous topography, because the trailing logs serve as a brake in descending the steep mountain grades. On level soft terrain the wagon type of equipment is preferable, since it requires less power to haul the load if the entire weight is carrid on the wheels than if part of the load is trailed or dragged.

Several hundred crawler-wheel wagons, crawler wheels used separately as such, and crawler wheels for the Willamette-Hyster arch are in use in American logging operations as well as in many countries abroad. They are extensively used in logging operations in Siberia. They are also employed for hauling products other than logs, notably in the cane fields of Cuba.

The Athey tractor wagons are made in 10-, 15-, and 20-ton models adapted for use with Diesel tractors. The Athey combination trailer is especially designed for long loads and long hauls, notably for transporting long logs, piling, and telephone poles as well as long pipe, derrick timbers, and other commodities. These combination trailers are made in 20-, 30-, and 40-ton models adapted for use with the Diesel tractor units. The Athey truss-wheel log trailer or bummer (two bummer wheels) is made in 3-, 6-, 10-, and 20-ton models, with capacities, respectively, of 1200, 1800, and 4000 b.f. of ponderosa pine logs.

LOGGING TRAILERS

Roads are sometimes impassable for rubber-tired equipment in wet weather, or a timber stand sometimes is too sparse to justify the expense of building roads. Logging trailers with track-equipped wheels and hauled by tractors, designed for use on hauls more than 1500 ft. long, are particularly applicable on roads which cannot readily be negotiated by log trucks. They are used principally in the southeastern logging areas, especially the hardwood bottoms of Mississippi, and also in many sections of Canada, the Northeast, and the Northwest.

Athey track-equipped logging trailers are designed to carry heavy full payloads of timber. Some of this trailer equipment is often held in readiness for emergency use so that sawmills can be continuously supplied with logs and keep operating in spite of adverse weather conditions in the woods. Broad tracks permit hauling through mud and over rough and rugged terrain. An important advantage is the freedom from tire trouble normally caused by stumps, rocks, or bumpy spots in the roads. Also, the heavy expense of building and maintaining roads under adverse conditions is eliminated when these track-type tractors and trailers are used. Moreover, logs hauled on these trailers arrive at the sawmills relatively free from dirt, imbedded rocks, or other undesirable material which dull or injure the saw teeth. It is especially important that timber cut for veneer logs or pulpwood arrives at the mills clear and in good condition.

FIGURE 117. Athey logging trailers averaging 1400 b.f. of hardwood per load at work in a Mississippi logging operation. Track-equipped trailers are especially suited to swampy areas.

Generally logs are bunched at a loading point by tractors pulling sulkies or arches or by ground skidding. This log assemblage at loading points rarely exceeds 1500 ft. in distance. Several methods of loading logs on trailers are used, including a winch-equipped tractor, an engine-powered skidder, a tractor equipped with a fixed or swinging boom, or even by a truck-mounted boom loader. Loading by the old crosshaul method using horses or mules is still done in several sections of the South. A good crew will load the smaller logs (up to 16″ top diameter) on the trailer bunks in about 1 minute and larger sizes in southern hardwoods and southern pine in 1½–2 minutes.

After the trailer is loaded, binder chains are used to prevent logs from rolling off the trailer bunks en route to the reloading point or sawmill.

These logging trailers are generally used in four sizes ranging in capacity from 6 to 20 tons. A Diesel D6 tractor with 65 drawbar horsepower will handle a 120-ton trailer or two 15-ton trailers in tandem.

CHAPTER XV

Sleds

Sleds, known also as sleighs and sledges, are among the oldest forms of log transport. Single sleds, drays, or bobs are used in summer, generally on steep gradients, as well as in winter for skidding saw logs,

FIGURE 118. Sled logging along Kankakee River, Ind.

cross ties, pulpwood, and other small products in the Rocky Mountains, the Southern Appalachians, Canada, and the East. In most hardwood regions, the Lake States and the Northeast, single sleds are used principally as a skidding device; they are described briefly under minor transportation.

Two single sleds, coupled together and known as double sleds or two-sleds, drawn by one horse, a team, or a tractor (in the case of several sleds coupled together in trains) are generally used for major transport of saw logs and pulpwood in the Lake States, the Northeast, and eastern Canada. These double sleds are also employed for hauling saw logs, to a limited extent, in the Northern and Central Rocky Mountains. Snow and iced roads furnish a minimum of frictional resistance in hauling logs from skidways in the woods to landings at the railroad, driving stream, or other method of major transportation. In eastern Canada three sleds are coupled together and drawn by a single horse or team.

Sleds are used to transport a good share of the several million cords of pulpwood cut annually in eastern Canada and a good part of the forest products in the Northeast and, to a lesser extent, in the Lake States. They are used for a very small percentage of the saw logs, cross ties, poles, and other products of the Rocky Mountains. Sleds are being replaced by motor trucks which go directly into the woods even in winter under deep snow conditions.

Sleds were early developed as a means of transporting logs to the driving streams in the Northeast. They are being supplanted by the motor truck in New England, notably in Maine, as well as in New York and, to some extent, in Canada. In Colorado, New York, and the Northern Rocky Mountains, sleds are sometimes used during the winter and motor trucks during the summer on the same operation. In northern Idaho sleds have been almost entirely abandoned in favor of motor trucks.

Motive power has for many years been principally supplied by horses, but recently tractors or trucks with various applications to snow conditions and tractor trucks have been introduced and found advantageous. Horses, however, are still used on many pulpwood operations of eastern Canada because of the prevailing short-length hauls, seldom exceeding 2 or 3 miles in distance. Horses are preferred because of the lower expense involved on short-distance hauls, although Diesel tractors are gradually replacing horses even for this purpose.

The development of sled transportation is due to the practical ingenuity of experienced woodsmen rather than to the skill of trained engineers. Sleds are usually homemade devices constructed in the camp blacksmith shop. Haul roads are laid out by the camp boss or foreman to bring in the logs at a minimum of expense. Great skill and enterprise are usually exhibited by these practical woodsmen in locating the roads so that skidding and sledding distances are properly balanced and continuously favorable grades for hauling out the product are maintained.

Advantage must be taken of a limited period of sustained cold weather to use the snow or iced roads for hauling purposes. Winter sled hauling in the northern United States is generally practiced between January 1 and March 1. In eastern Canada and Newfoundland, the period may be more extended, especially in the northern portions. In some sections of Canada, owing to the accumulation of deep snows in the latter part of the winter, sled hauling must be finished ordinarily by February 15. Occasionally, however, hauling may be continued to March 15 or even later.

There are generally five periods of pulpwood logging in connection with winter sled hauling in the northeast and eastern Canada as follows:

1. Felling in the summer. Peeling bark, if done, is usually confined to the period from May 1 to about August 1. Peeling and cutting of short-length pulpwood (4′) is the common practice in New England, but long lengths (10′, 12′, 14′, and 16′) unpeeled are the prevailing practice in eastern Canada and 100″ in the Lake States. Felling and peeling are done together when the bark is peeled. Felling of long-length pulpwood, not peeled, may commence any time between August 1 and November 1 in Canada and in some parts of the Northeast.

2. Skidding, bucking, and piling in late summer and early fall. Felling and piling are sometimes done in one operation, and no skidding is necessary, as hauling is direct from the stump to the landing. On one operation in northern Quebec, 75% of the 12′ logs are hauled directly from the stump to the landing. Where sap peeling is practiced, the trees are felled, then skidded to landings and bucked into 4′ or longer lengths, and piled along the haul roads.

3. Haul roads constructed during the fall. This work may be done between August 1 and November 1, depending on the season. Frequently, there is an interim period of inactivity from about December 1 to January 1, depending on the weather conditions, the amount of work to be done, availability of labor supply, and similar factors. In Quebec, many farmers who are employed on their own places during the summer work at road construction in the fall and winter sled haul during the cold weather, a small number continuing in the spring drive. Branch roads leading to the main haul roads are sometimes made by the cutters as a part of their contract.

4. Winter hauling from about January 1 to March 1 as described above.

5. Spring driving season from April or early May to June 15 to July 20. In northern Quebec driving ordinarily starts about May 20 and continues to July 31. In New Brunswick it generally starts about May 10 and continues to about June 15. In New England and New York, the drive takes place during May and June.

In some parts of Canada and the Northeast there is a strong tendency toward "hot" logging or the continual flow of logs from the stump to the stream landing. In this case, logging starts late in the fall or as soon as freezing temperatures are available, and cutting, skidding, and hauling continue as long as the haul roads are open or available. With motor-truck logging as practiced in Maine, New York, and New Hampshire, hot logging continues throughout the fall and winter.

TYPES OF SLEDS

There is wide variation in the type of sled construction, materials used, gage, length of life in service, and hauling capacities. These

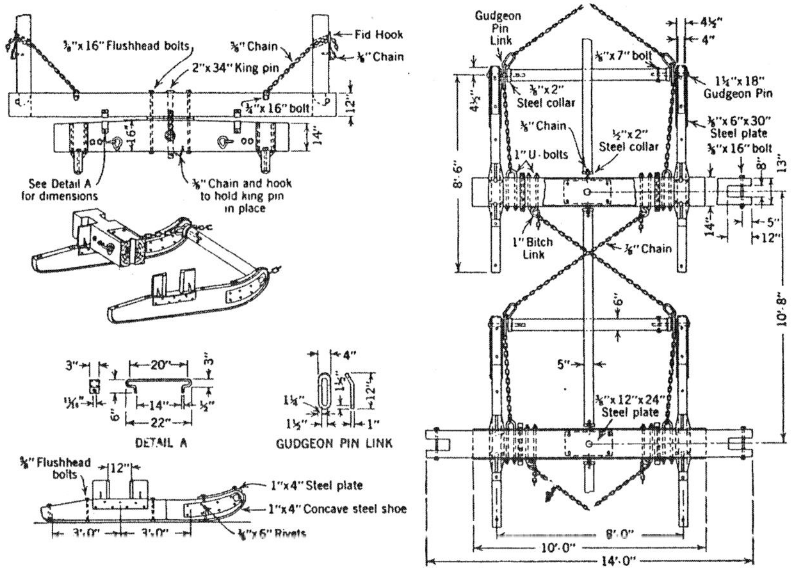

FIGURE 119. Sketch of logging sled in common use for hauling logs on iced roads with tractor. Sleds are built proportionate in strength to match larger or smaller tractors.

types in turn depend on local customs, length of haul, whether tractor- or horse-drawn, size of loads to be carried, and especially the type of woods product. Both the construction and size of sleds vary markedly in the several regions as well as within each region. Sleds with single racks for short-length spruce pulpwood in the Northeast do not require the heavy construction, large size, or wide gage necessary for hauling heavy hardwood logs in the Lake States. Double sleds drawn by tractors are of much heavier and stronger construction than those hauled by horses.

The gage of the sleds, or distance between the runners, varies from 3′ to 9′, generally being 4′ to 7′. When horse teams or tractors are used for hauling saw logs, the gage is usually wide. With a single horse, the gage is uniformly narrow for single, double, or triple sleds. For tractor hauling, the sled runners are generally 12″ to 18″ outside the tracks of a tractor, but on iced roads a 9′ gage is recommended, especially in hauling hardwoods and white pine in the Lake States and Ontario. For a small tractor, the 8′ gage is preferred. The advantage

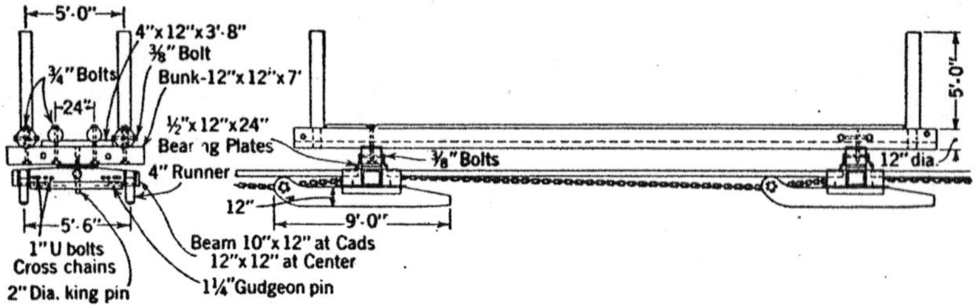

FIGURE 120. Long-racked pulpwood sled in common use. Will hold 9 cords of 8′ pulpwood when loaded crosswise to top of stakes. Stakes are 5′ apart at ends and 29′ 8″ apart lengthwise.

of wide-gage sleds lies in the provision for bunks 12′ to 18′ in length, thus permitting easy loading and unloading and large loads. With wide bunks, loads are not generally piled very high; therefore, quick and easy loading and unloading are possible. No standard gage is recognized for hauling either long saw logs and pulpwood or short pulpwood such as 4′ or 8′.

The number of sleds required on a given operation depends on its size, the length of haul, the form of product, and the hauling schedule. Generally sleds are hauled down from branch roads to a make-up point, and, in long hauls, trains of sleds are joined together for hauling to the landing. Horses or tractors should not be kept waiting at the make-up point on long hauls where a train of four or more sleds may be hauled at one time.

Sleds are usually made in the logging camp or a local blacksmith shop. They are generally constructed of well-seasoned birch and last about 5 years without major repairs. With repairs they give service indefinitely. Owing to the generally uneven character of the roads, especially the steep grades and sharp turns, and to the heavy loads transported, they are subjected to rough usage.

Hardwood saw-log loads of 1000 to 2800 b.f. or more are used, depending on the size of the bunks and road conditions, favorable or adverse grades on the haul road, and the motive power. With tractors, and especially on the longer hauls, trains of 3 to 20 or more sets of sleds are possible, which thus increase materially the volume of logs hauled at one trip. As many as 29 double sleds of cross ties, pulpwood, and saw logs, coupled together, have been hauled at one time

FIGURE 121. Side view of double-rack two-sleds. The racks are 28′ long between 6′ stakes. The runners are 8′ long, the bunks are 7½′ long by 7″ by 10″ in cross section. These double-rack sleds generally carry six to ten cords of 4′ spruce pulpwood per trip, depending on the grade and condition of the road.

by a tractor for a distance of 5 miles, thus making a total weight of 1100 tons. The bunks or crosspieces on which the loads rest are free to swing on the turns, regardless of the direction of the runners. The bunks rest on crossbeams connecting the runners. A king bolt holds the bunk to the crossbeam. For the heavy-duty and largest sleds for hauling hardwoods, eastern hemlock and northern white pine, bunks may be up to 18′ long. For pulpwood logs and small loads, bunks may be 7′ to 12′ long. For 4′ pulpwood bolts in the Northeast and, to a limited extent, in eastern Canada, either single or double racks 16′ to 28′ long are used to connect the forward and rear sleds and support the load. The 4′ bolts are laid crosswise on these racks up to 6′ or more in height and are held in place at the two ends by stakes. Bunks for single racks for 4′ wood are 44″ to 48″ long. One company uses 7′ to 9′ bunks for single-horse sleds and 10′ to 12′ bunks for team sleds for long pulpwood.

Runners are made of yellow birch, maple, or white oak, and may be 3′ to 9′ long. Approximately three fourths of the length of the runner constitutes the bearing surface. Runners are curved in front and rear in order to provide for backing when required. They are generally shod with flat, convex, or flat with keeled iron strips: runners may be used on iced, rutted, or snow roads. Convex runners (4″ wide and ¾″ thick at center) operate smoothly on iced roads and very successfully on snow. They are preferred to others also because sleds equipped with them are easy to start, control, and steer. The flat runner with keel is used to decrease the frictional resistance on iced roads and also helps to prevent the loaded sled from slipping off the road on short turns, especially on steep and sidehill slopes. Flat runners are not generally used.

Saw logs, long pulpwood (12′ and 16′), poles, and similar long-length products are loaded parallel with the road. Cross ties, posts, 4′ and 8′ pulpwood, fuel wood, and other products which generally do not exceed 10′ in length, are loaded crosswise on racks which are bolted or fastened to the bunks. With 4′ pulpwood, double racks are commonly used. These racks may be 16′ to 36′ long and vary in width with the product to be hauled. Stakes set in each end of the racks hold the short-length products in place. Saw logs, poles, and long pulpwood are held on the sleds by means of stakes at the ends of the bunks.

CONSTRUCTION AND OPERATION OF WINTER-SLED-HAUL ROADS

Haul roads for winter sledding are generally constructed during the summer and fall, while saw logs, pulpwood, and other material are being felled and skidded to the haul roads. The main and branch roads must traverse the entire area to be logged, a proper balance being maintained between skidding and hauling distances. The volume of timber to be moved generally determines the character of the road to be constructed. For branch or strip roads, a little brushing out may be sufficient, whereas on the main haul road a fairly uniform grade must be maintained. Stumps may be cut low or blasted out, cuts and fills must be made, and bridges built. Haul roads should generally follow the natural drainage. Adverse grades should be avoided, as they may limit the size of the maximum load to be hauled. A 1% increase in adverse grade may decrease the hauling capacity by 15% to 25%. Advantage may be taken of long level

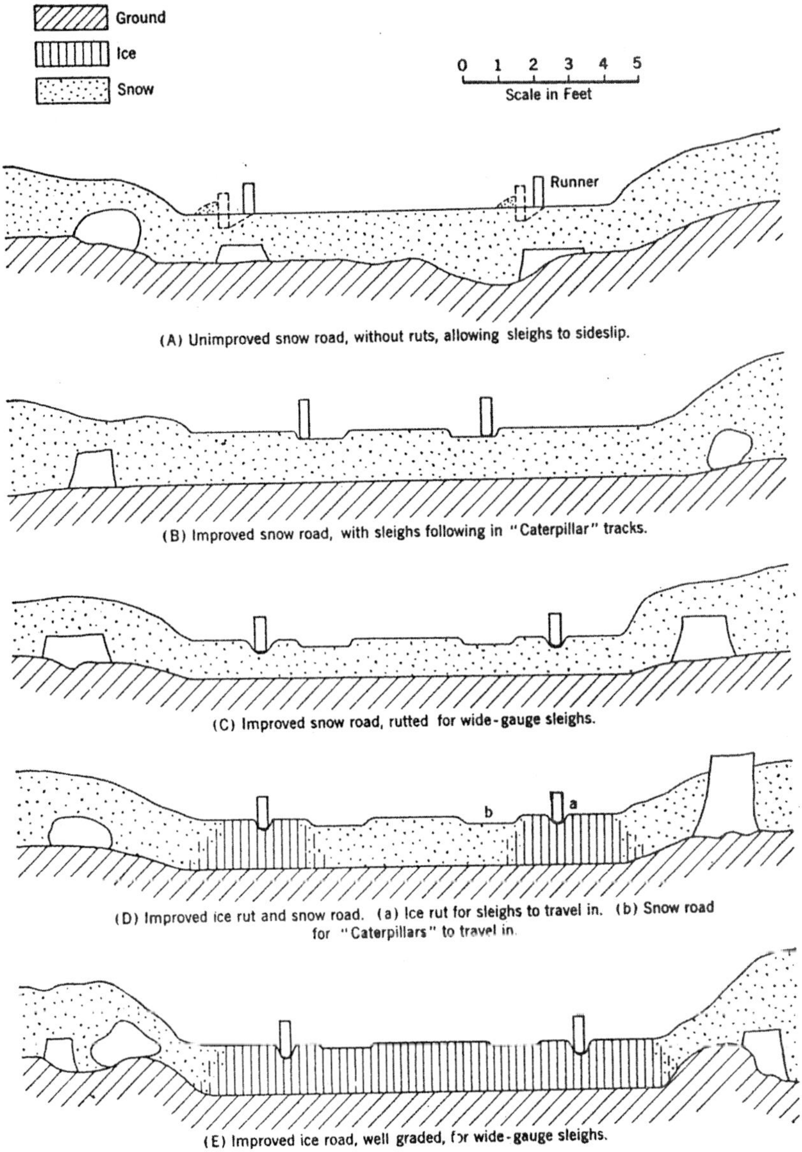

FIGURE 122. Types of sled roads.

places, such as swamps, ponds, and even rivers, because when frozen they may serve as good hauling locations. They are generally the last to freeze, however.

A right of way is cleared 10′ to 24′ wide on the main roads and 8′ to 20′ or less on the branch roads. When the road surface is covered by a depth of about 12″ of snow, many inequalities in it are eliminated.

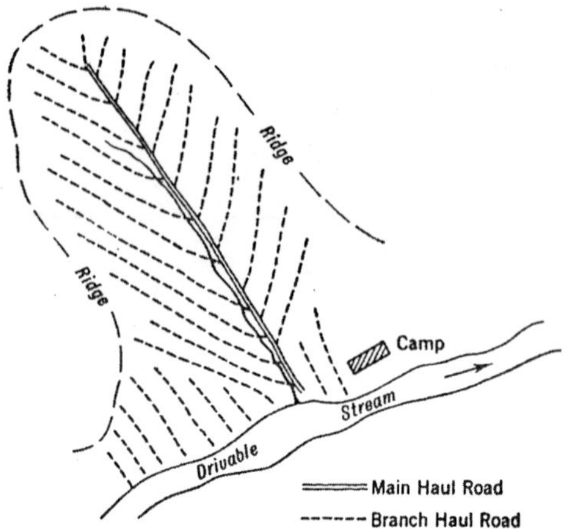

FIGURE 123. Sketch illustrating a typical pulpwood operation in the Northeast and eastern Canada. Approximately 10,000 cords of spruce pulpwood are produced each year from a 100-man camp operating on five main haul roads. This illustrates one of these main haul roads. The branch roads illustrated in dotted lines are made by the cutters and spaced generally 60 to 100 ft. apart, and approximately parallel to each other. The main haul road is located along a dry gulch or nondrivable stream, and approximately 2000 cords were hauled from this drainage unit as illustrated. The pulpwood was unloaded on landings along drivable streams and floated to the mill during the spring. Not drawn to scale.

Low or wet places may be corduroyed. When deep snows occur, several devices may be used to remove it or pack it down; they may be briefly described as follows:

1. Snow plows. These are generally of three different types: (*a*) the V-type hauled by team or pushed by tractor; (*b*) the rotary type operated by tractor; and (*c*) combination snow plow and rutter. All except the rotary type are homemade affffairs.

2. Rollers 6′ to 8′ or more in diameter and covered with small lumber or poles. These are generally made in the local blacksmith shop and hauled over the road by a team. They are especially advantageous in remote regions where the snowfall is light and soft.

3. Horses equipped with snowshoes. These are customary in Newfoundland where the snowfall usually is very light and deep. One company employs as many as 300 horses so fitted out.

Roads may be snow- or ice-covered; the latter may be used with or without ruts. Iced roads reduce the frictional resistance and increase the possible size of the load to be hauled. The ruts accentuate these favorable conditions. However, satisfactory ruts are difficult to construct and maintain and are not generally used, in spite of their theoretical advantages.

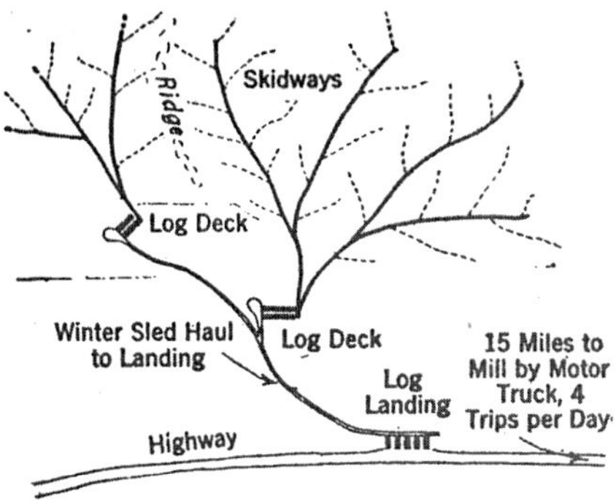

FIGURE 124. Diagram illustrating methods of main and branch skid roads to log decks. Both drays and direct horse skidding were used to bring logs to the decks, the latter for small logs and short distances and the former for the larger logs hauled for longer distances. The logs were hauled 1 to 3 miles on the winter-sled road to the landing and then loaded on motor trucks and hauled 15 miles to the mill. Skidways were placed generally every 300 to 500 ft. apart, depending on the number of logs tributary to the main haul road. Not drawn to scale.

A sprinkler for icing roads and ruts consists of a tank holding 100 to 500 cu. ft. of water and mounted on sleds. The tank is filled generally with water from an adjoining stream by the cross-haul method, a barrel being elevated on inclined skids. Sprinkling is usually done at night or during the early morning hours in order to take advantage of freezing temperatures and not to interfere with log hauling.

In some sections of eastern Canada, the Lake States, and the West, "hot" logging is used; that is, trees are felled, skidded, and hauled continuously throughout the winter as long as hauling conditions permit. This eliminates decking, which may be expensive.

Pulpwood is brought to the haul roads by means of single racks loaded on bobs, as well as by chutes, chaining, trailing, and hand logging, as described elsewhere. On one operation in the Lake States where cross ties and pulpwood are cut, three 30-hp. tractors are used for switching and hauling in the loads over the strip or branch roads to the main haul road. This is known as the make-up point, generally at a skidway. Jobbers cut and deliver the wood to the strip roads (which they also cut out, about 200 to 250 ft. apart). Horses skid logs to the strip roads; tractors bring in the empties and spot them at the log deck to be loaded. Six men load about ten sleds per day, then the 30-hp. tractor hauls out one, two, or three sleds at a time, for an average distance of ¾ mile to the make-up point. Here the sleds are hooked up in trains of 10 to 20 or more and hauled 10 miles by a tractor to a landing along the stream or in other cases to a railroad. Three swings or round trips are made on a 10-mile haul with a tractor, which means a total distance of 60 miles per day. The number of trips depends on the time consumed in loading and unloading rather than on that actually devoted to hauling.

For the longer distances, tractors, tractor trucks, gasoline log haulers and steam log haulers are used. Throughout eastern Canada and Newfoundland, hauling distances are seldom more than 2 to 3 miles; therefore, horses are usually adequate for hauling purposes. In New England, New York, and Canada, teams are preferred on the short-distance hauls. Tractor trucks occasionally carry loads on the truck itself and also haul double sleds in trains.

Where hauling involves adverse grades and long distances, tractors are required. The average speed of horses on the level is about 2½ miles per hour as against 3 or more miles per hour for tractors. On steep downhill grades, horses with various means of retarding velocity are preferred to tractors.

DEVICES TO RETARD SPEED

As sled logging is being practiced on more difficult topography and frequently on very steep slopes, especially in the mountainous sections of New England and eastern Canada, the importance of methods of controlling the speed of heavily loaded sleds on downhill grades is at once apparent. These methods may be summarized as follows. The adoption of any one of several possibilities may depend on the severity and length of grade, weight of the loads carried, condition and direction of the roads (curves), and local materials available.

1. On gentle slopes, ice roads may be left as snow roads, which offer more tractive resistance than ice roads. This method, however, is seldom sufficiently effective.

2. Local material, such as hay and straw, is spread with a fork along the path of the runners where only a slight check in speed is required. For steeper and longer grades, sand, gravel, or dirt, usually taken from an adjoining bank, is used. This method is feasible on grades of 10% to 18%. Occasionally hot sand is employed, but it is expensive to prepare and transport, and it wears out the runners too quickly.

3. Chains may be wrapped around the sled runners to impede the progress of the sleds. This is not generally very effective because they tear up the road and are not easily removed. They serve as an additional safeguard in connection with other methods of retarding speed.

4. A snubber made of steel cable or heavy rope is occasionally used, especially on short steep pitches. The snubber is generally wound around a stump and "payed" out by a man stationed at this point for the purpose.

5. The Barienger brake is common in the Adirondacks and other mountainous sections for long and exceedingly steep slopes, from 20% to 60%, and for distances up to 1 mile. It is most efficient, however, for distances not greater than ½ mile. On one slope ½ mile in length, with an 850-ft. drop in elevation, these brakes were efficiently used. This brake has four, six, or eight drums weighing 3000 to 9000 lb. each. A ⅝″ cable is used for the two smaller varieties and ¾″ cable for the largest size. They are anchored to a stump and operated by one man. When one load is being lowered, the return cable is hauled up the slope automatically. Seven of these brakes were used on one pulpwood operation. On long steep slopes they are cheaper than labor for sanding or graveling the steep pitches. On one haul road, which varied from almost level to 60% grade, four separate Barienger brakes were successfully used, one each being employed for distances of 1900, 1600, 1300 and 850 ft.

6. Chaining, described under minor transportation, is employed to haul 4′ pulpwood bolts crosswise behind loaded sleds and also to retard the speed of sleds on steep slopes when they are connected to bobsleds or drays. It has been very effective in the Adirondacks.

The back-haul road for returning empty sleds is constructed in steep topography along a different route, to avoid the heavy adverse grades, especially when these have been sanded or graveled. It would be exceedingly difficult to haul even empty sleds up many of these roads with or without sanding.

TOTING (PORTAGING) SUPPLIES AND MATERIALS

Since winter-sled hauling is widely practiced in the North and particularly in eastern Canada in connection with stream driving where the method of major transport does not provide a means of return haul with men, camp supplies, and so on, as railroad and motor-truck hauls do, the question of furnishing an adequate and inexpensive means of toting in supplies is of great importance. Especially is this true when the principal means of motive power is horses, which require large quantities of feed for the arduous work is cold climates. Toting, like logging, is a problem of transportation: that is, one of moving great weights and loads at a minimum of expense.

Experience has shown that 6 lb. of dry weight of food per man and 55 lb. of feed per horse per day or a total of 61 lb. per day for a horse and driver are required in the Northeast and Canada. For a camp of 100 men and 20 horses, this means $[(100 \times 6) + (20 \times 55)] =$ 1700 lb. per day for human and animal sustenance.

Supplies are toted to camp in the North and Canada by motor truck, winter-sled haul with horses and tractors, canoe, shallow boats pulled by horses up a stream, boats propelled by tug or oars, wagons, and other vehicles, or various combinations of these methods. Generally, if good highways are available, supplies are toted directly into the camps by truck during the summer and fall before the snow flies. As many provisions as can reasonably be stored and cared for are toted to the camp in advance of deep snows. Tractor- and horse-drawn sleds are used in the winter. Secure packaging is important for the protection of the contents as well as for easy handling. At Shelter Bay on the St. Lawrence River, camp supplies are first shipped via boat from Montreal and then unloaded on scows to be taken to shore. Here they are loaded onto wagons and hauled to a warehouse. Again they are loaded onto trucks, hauled 12 miles to a lake, and loaded onto scows to be transported several miles. Once more they are trucked 12 miles to the camp stores. Thus, it is evident that the containers must withstand rough and frequent handling and yet not be too heavy or bulky.

CHAPTER XVI

Motor Trucks*

Since 1940 the motor truck has been greatly improved in power, efficiency, and size, and, with better and more frequent hard-surfaced roads, it has become the most important means of major log transporta-

FIGURE 125. Usual method of loading truck from elevated skidway, commonly found on small operations. These are hardwood veneer logs cut in the Green Mountain National Forest of Vermont. These loads carry generally 1200 to 1700 b.f. *Courtesy U. S. Forest Service.*

tion. The truck was first used for logging in 1913 in the Douglas fir region where it has replaced logging railroads. It is extensively used in the pine forests of the West and South and, to an increasing degree, in all forest regions of the United States and Canada. It has widely supplanted the logging railroad, chutes, flumes, stream driving, sled hauling, and other methods.

* For recent and detailed data on motor trucks used under West Coast conditions see *Cost of Hauling Logs by Motor Truck and Trailer*, by J. J. Byrne, R. Nelson, and P. H. Googins, U. S. Forest Service, Portland, Ore., October 1947.

There has been an increasing trend toward motor-truck transportation in logging throughout the country ever since World War I and

FIGURE 126. Log trucks have displaced, in some cases, skidding as well as several phases of major log transportation. In this case, the truck is driven directly into the Southern pine woods without advance road preparation, and so at least some of the skidding is eliminated. Logs are hauled directly to a mill, thus reducing costs. Trucks are also used to go directly into the woods in eastern Canada and the Northeast to replace skidding and sled haul, even in winters with heavy snows. View taken in southern pine where Hyster winch loads up to 20,000 b.f. per 8-hour day and uses less than 1 gal. of fuel oil per hour. Tractor with winch shown in background. Single center tongs used in loading. *Photo taken near Meridian, Ga.*

especially since 1945 as a result of the development of inexpensive, light, powerful, pneumatic-tired trucks and detached trailers, the latter being easily loaded and unloaded, and the former capable of traveling 30 to 40 m.p.h. or more on the "go back." The motor truck is frequently used to transport logs directly from the felling areas to the sawmill or other destination, thus partially or completely eliminating the skidding or assembling operations ordinarily necessary in general logging procedure. This is notably so in the southern pine and the ponderosa pine forests in the West, especially in Oregon and the Southwest. Trucks are frequently equipped with a winch to draw logs

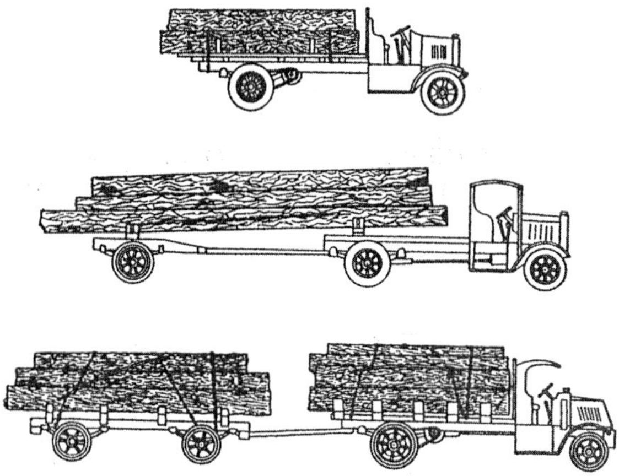

FIGURE 127. Motor trucks are the principal means of log transportation in most forest regions of the country. Here are three types used for logging. For small and short-length logs, the top type is generally used. For long lengths, especially in the Douglas fir region, the trailer type as shown in the center is widely used with dual and tandem wheels. On good highways with favorable grades and for long hauls, the bottom type is preferred. Motor trucks are used to haul logs for distances up to 70 to 90 miles or more.

to the truck and load them when it is impracticable to drive directly to the logs. The winches powered by the truck engine are thus used for skidding logs short distances.

Motor trucks are used chiefly to replace wagons, standard-gage railroads, and the various forms of animal haul, and have been successfully operated under conditions where tractors and power skidding have been used. They have especially replaced spur and main-line railroads in the South and the Pacific Coast, and also chutes and flumes, and have reduced skidding distances as in the western white and ponderosa pine types of the West. They are widely used in regions of second-growth timber such as the Southeast, the Lake States, and the Northeast, where small sawmills often depend on the open log market, purchasing logs as needed. Thus, many mills are entirely or partially independent of reserve timber holdings as the motor trucks offer a wide flexibility in source of log supplies. This flexibility means that trucks may be widely used in connection with selective logging.

In eastern and southern logging for small mills, the same crew is frequently employed in felling, skidding, and trucking logs as well as in sawmill operations.

Many states have legal restrictions which limit the weight and size of logs carried over public highways. These restrictions may be left to the state highway commission or similar agencies, or they may be matters of legislative enactment, specifying the length, width, height, and weight of loads and size of logs transported over roads constructed at public expense. Many regulations of this kind are now in the process of enactment, and changes are being constantly made. Fees for motor-truck licenses are based generally on the weight of the truck; they vary widely among the various states and in Canada. In the Northwest, some operators find it more economical to build their own roads for trucking logs rather than use the public highways.

Motor trucks are used in connection with every other type of logging. For example, one company in the South which operates the largest paper mill in the world, using 2500 cords of pulpwood per day, employs 250 motor trucks and 30 tractors in order to carry sufficient wood to maintain capacity operation of the mill. This wood is brought in from distances up to 350 miles by common-carrier rail, 250 miles by barge, and 60 miles by truck.* Jeeps which were developed during the last war for so many purposes are even used to skid logs, especially on small wood-lot jobs.

DISTANCES

Logs are hauled by motor trucks for varying distances up to 70 to 100 miles or more. Generally, however, logs are hauled 3 to 15 miles under conditions of rough topography; longer hauls are possible on hard-surface highways and on level or favorable topography. Other conditions determining the length of haul are weight, length, and character of the logs hauled. Short hauls generally prevail for small sawmills. The very longest hauls are generally in the southern pine region and in the Southern Appalachians where high-value logs are hauled 10 to 70 miles or more. Where logs are exceedingly large and heavy and the topography is rough and steep as in the Northwest, short distances of 4 to 10 miles or less are the prevailing rule. Long hauls are sometimes necessary in all forest regions, depending on local

* Data supplied by J. E. McCaffrey of Georgetown, S. C.

conditions and requirements. Short hauls are generally made on private roads built for truck transportation; long hauls, only on improved public highways. Walnut veneer logs and high-value hard maple, oak, and birch have been transported for distances greater than 100 miles.

The number of daily round trips possible for a truck depends on the distance; type of road; load; grades, both favorable and adverse; and the loading and dumping facilities. On a comparatively level 10-mile road of good gravel, loading with logs located about 500 ft. from the road and dumped in deep water or on a good clear landing, the truck should average four round trips in an 8-hour day. In the Lake States, a contract trucker makes four round trips per 10-hour day on a 22-mile haul (total, 176 miles per day). A loading jammer is mounted on the truck with a power take-off. Loading and unloading of pulpwood into gondola cars are done by driver and helper.

Motor trucks are used to carry logs directly as well as to haul semi-trailers and four-wheel, six-wheel, and eight-wheel trailers. Double loads are thus commonly hauled where road conditions and grades permit. In one example, loads of 3300 b.f., log scale, of Douglas fir with an average weight of 9 lb. per scale of board foot, or a total of 29,970 lb., were hauled a distance of 60 miles in the Olympic Peninsula of Washington. The trucks made two round trips per day, or a total distance of approximately 240 miles. One of the largest pulp mills in the East receives 20% of its wood supply by truck from distances up to 35 miles. At this maximum distance, three round trips are made daily. Pulpwood is being hauled up to 65 miles in Maine, the deliveries being made both at the pulp mills and in driving streams for further transport to pulp mills. One company hauls about 40,000 cords of pulpwood, annually, direct to the mill and another 40,000 cords to the landings at the river. Loads vary between 1½ and 3 cords of spruce. Trailers on sleds and wheels are both used during the winter in Maine. Company trucks in Wisconsin generally carry about 5 cords of pulpwood, whereas contractors ordinarily use Ford and Chevrolet light trucks which carry 4 cords each and up to 6 to 7 cords with trailers.

Trucks can carry log loads up adverse grades as great as 10% to 25% without additional power, depending on condition of the highway, weight of the load, power of the truck, and length of the grade. On steep inclines trucks are let down or hauled up by means of steel cables and winches or by tractors on grades up to 50% to 60% or more.

Trucks have even replaced tractor-drawn sled trains on iced roads in the Lake States. In one case International K.R. 11-F. 10-wheel 6-

and 8-ton trucks have hauled a train of four sleds, each of which was loaded with four cords of pulpwood a distance of 20 miles on winter roads from January 1 to March 1.

DEPRECIATION

Depreciation on motor trucks is generally figured at 25%. This means an average life of 4 years, although trucks often last 3½ to 7 years. Motor-truck operators generally estimate 22 to 25 days as the operating time per month, the remainder being accounted for by Sundays, repairs, and changing the loading locations. In some regions, trucks are operated 23 to 26 days per month, repairs and overhauling being done at night, while separate crews arrange for loading facilities.

Truck tires generally last 8000 to 15,000 miles. Life expectancy is usually estimated at about 10,000 miles. Actual life depends on loads, character of road, mileage covered, and kind of tires.

NUMBER OF TRUCKS IN USE

Probably more than 300,000 trucks are used in the several branches of the lumber industry. No survey has been made to determine the number employed in the logging phases of the business, but there are probably 100,000 or more. Some companies own and operate 5 to 30 trucks or more for logging purposes alone. One company in Arkansas used 39 trucks on its operations. Probably between 5000 and 5600 trucks are utilized for logging in Oregon and Washington. The largest volume of logs in the East and South is carried by companies or individuals owning and operating one truck. Trucks are used very largely to get out the 9 million cords of pulpwood produced annually in the South. Trucks are the principal means of supplying small sawmills with logs. There are said to be more than 15,000 small mills in the South. One pulp mill in Wisconsin has 20 active logging operations, all of which use trucks for major transportation.

CONTRACT HAULING

Contract hauling, known as the gyppo system in the West, is the usual practice in motor-truck logging. Many of these gyppo contractors do not accurately determine the costs of trucking, including depreciation, taxes, interest charges, gasoline, oil, and repairs, and they engage in contracts for hauling logs below actual costs. This uneconomic competition has contributed to occasional demoralization of the log market in some parts of the country. Logging companies let contracts for hauling logs generally from skidways, cold decks, landings, or directly from the woods to destination at a flat price per thousand feet. Truck contractors often operate 10 to 12 hours or more per day and in this manner cut hauling costs below those of companies which observe the hours of the usual working day. Many farmers and truck owners devote part time to log transportation whenever an opportunity is afforded, especially on small operations. Large mills throughout the

FIGURE 128. Truck loaded with 4′ spruce pulpwood in eastern Canada with Montague loader. From 10 to 15 cords of 4′ wood (128 cu. ft.) per hour are loaded with a five-man crew, depending on the size of the wood and the organization of the operation. With 12′ wood, 16 to 20 cords are cut and loaded per hour, including lost time due to moving, etc. *Courtesy of I. F. Fogh.*

country receive at least a part of their log receipts from small contract jobbers.

Contract hauling is sometimes done by those who undertake to fell and buck, skid and haul logs to the mill. Frequently these contractors sublet various portions of the job to others. The timber is generally owned by the company which saws the logs, especially in the South. The contract price for hauling varies widely with local conditions such as length of haul, stand per acre, character and grade of roads available, and whether hauling is during the winter or summer.

CAPACITY

The rated capacity of motor trucks used for long haul is generally 1½ to 15 tons. The smaller and medium-size trucks predominate, especially in supplying logs to portable and small sawmills of the southern pine region and sections of the East where a large portion of the lumber output is produced from small mills. These mills cut second-growth timber which generally runs from 10 to 25 logs of 16′ length per thousand board feet, log scale. The largest-capacity trucks are used in the redwood and Douglas fir regions. The capacity of

FIGURE 129. Douglas fir logs arriving on double trailer at sawmill in Oregon.

truck and trailer is generally 40% to 80% more than the truck alone, especially with dual and tandem wheels.

The 1½-, 2-, and 3-ton trucks have proved to be very satisfactory and are predominantly used because the investment in them is relatively small. On good improved highways they have given excellent results. They are easily handled, and operating and repair costs are low. Much time is lost in maneuvering large trucks and trailers at loading points and at unloading landings.

The great majority of motor trucks are of the 1½-ton size with many 2-, 2½-, and 3-ton sizes in use. In handling the large logs of the Pacific Coast forests and the Northern Rocky Mountain forests, 3½- and 5-ton trucks and even 15-ton and larger-size trucks are used. The 5-ton truck carries up to 3500 b.f. of logs and with a trailer may haul

4000 to 6000 b.f., log scale, or more. Road conditions must be exceptionally favorable, however, for loading trucks to this extent.

The average hardwood truck load is 1000 to 2000 b.f. for the larger trucks, and still smaller amounts for the lower-capacity trucks. Hardwood logs weigh very much more than softwood logs. Douglas fir generally weighs about 6 to 9 lb. to the board foot in the log, depending on size. Western red cedar and Sitka spruce weigh about 6½ lb. per foot.

FIGURE 130. A gooseneck type of semitrailer used to haul southern pine pulpwood to large newsprint mill at Lufkin, Tex. This type truck hauls approximately six cords of 4′ pine pulpwood, two tiers to the load, for distances up to 50 miles directly to the mill. Great improvements have been made in truck design, operation, and maintenance to haul the several million cords of pulpwood required each year in the South.

An operator with a 1½- to 2-ton truck and two-wheel log trailers can handle satisfactorily about 1500 b.f. of Douglas fir or 1800 b.f. of Sitka spruce or cedar. A 3- to 4-ton truck with two-wheel trailer will ordinarily handle about 2000 b.f. of Douglas fir; a 4- to 5-ton truck and trailer will handle about 2500 b.f.; a 6- to 7-ton truck and trailer approximately 3000 b.f.; and a 7- to 9-ton truck with a heavy-duty trailer 3500 to 4000 b.f. or more.

Operators are generally inclined to overload to some extent, as it is not always possible to load logs to make the footages comply exactly with the theoretically maximum capacity of each truck and trailer. Contractors paid on the basis of thousand board feet delivered usually carry overloads. In southern pine, 500 to 1200 b.f. of logs are carried on the small trucks commonly used, depending on the roads, weight of logs, capacity of truck, and so on. During dry weather 10% to 50% more footage can be carried than in wet weather on private roads.

The average board footage logs hauled by trucks of various capacities with and without trailers is shown in Table XXXIV. These pertain to the smaller trucks used in the South and East where most of the footage is logged.

TABLE XXXIV

	By Truck		*By Truck and Trailer*	
Manufacturer's Rated Capacity	*b.f.*	*Weight, lb.*	*b.f.*	*Weight, lb.*
1½ tons	1,200	6,450	1,855	10,910
2 and 2½ tons	2,059	12,080	3,060	19,810
3, 3½, and 4 tons	2,436	13,040	3,442	22,070
5 tons and larger	3,325	18,850	4,167	25,670

Average loads of ponderosa and western white pine carried in the Inland Empire are generally greater than those shown in the table.

Motor trucks and trailers have replaced tractors to some extent for hauling trains of sleds on winter haul roads in the Northeast. Two-and-one-half ton trucks can haul eight and nine cords of 4′ spruce pulpwood on double racks or trailers on hard-surfaced roads. Trucks are operated throughout the winter on short hauls on woods roads costing about $300 per mile and frequently less. It would be impossible to haul logs by wagons on these roads in the summer, but, when the ground is frozen and the right of way has been roughly cleared and brushed out, trucks are used successfully on the packed snow.

THE PETERBILT MOTOR TRUCKS

The truck built for the heaviest loads in logging is the Peterbilt truck. There are two types, one especially designed for redwood operations and the other for Douglas fir, ponderosa pine, sugar pine, and associated species. Redwood logging is regarded as the more difficult. In the use of logging trucks, some of the largest operating companies have learned that it is more economical to purchase and operate heavy-duty equipment than the lighter and medium-size equipment which was used largely before World War II. Experience has proved that equipment of proper sizes to handle the larger loads will operate throughout the logging season with a minimum of failures. Trucks are generally put in the shop during the rainy winter season when driving conditions are not always favorable, and, if the trucks are put in shape, they are capable of operating throughout the next season with few if any failures.

The Peterbilt Company has recently delivered trucks equipped with 12″ × 24″ tires in front and 8–14″ × 24″ tires in the rear. The logging

FIGURE 131. One of the newer models of the Peterbilt motor trucks. It carries loads up to 27,000 b.f., an average weight of about 10 lb. per foot. This means a net load of 135 tons. Rear wheels have 14 × 24 tires, and front wheels are equipped with 12 × 24 tires. It is used only on private roads because heavy loads of this type are not permitted on public highways.

dolly also is equipped with 8–14″ × 24″ tires. For this exceptionally heavy type of equipment and the very ponderous loads, trucks are only allowed on private roads. Loads up to 27,000 b.f. have been carried with this truck. These loads average 10 lb. per foot, making the net load about 135 tons. The truck is powered by a 1090 cu. in. engine and burns Butane fuel.

OPERATING COSTS

Operating costs may be classified for purposes of accounting into three general groups, namely:

1. Investment costs, including cost of truck delivered, trailer, or special equipment.
2. Fixed costs, including such items as interest, insurance, taxes, wages, garage expense.

3. Variable costs, including repairs and maintenance, depreciation, fuel, lubricants, tires, and parts.

Operating costs are figured on the basis of thousand board feet per mile or total costs of mileage covered expressed in per mile charge.

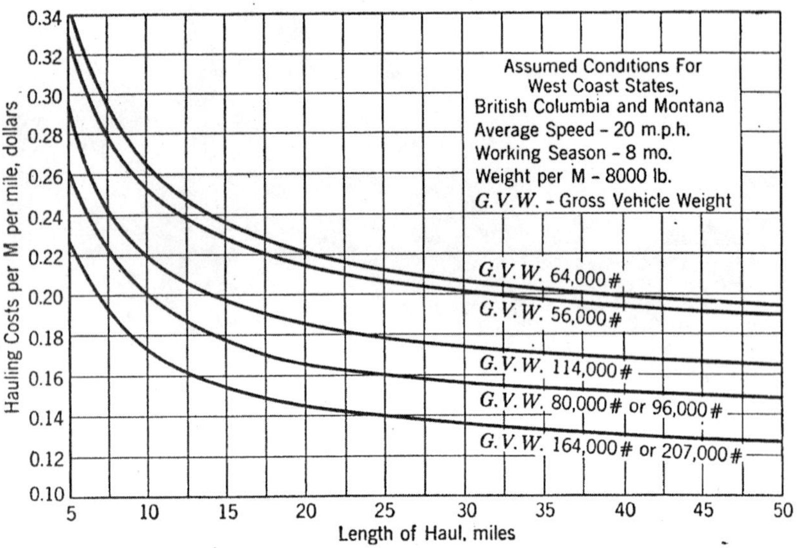

FIGURE 132. Log-hauling costs for various lengths of haul on gravel-surfaced roads to show that costs per unit are reduced with length of haul for all types of trucks. *From Cost of Hauling Logs by Tractor, Truck, and Trailer, by J. J. Byrne, R. J. Nelson, and P. H. Googins, U.S. Forest Service, 1947.*

The costs vary considerably with local conditions such as distance covered, character and condition of roads, adverse and favorable grades, local costs of fuel, lubricants, and repairs.

Costs of operation for the various items vary widely. A summary of many operating costs is as follows:

Items of Expense	*Percentage of Total Cost*
Gasoline or Diesel fuel	22
Depreciation	25
Repairs	15
Tires	12
Interest, insurance, taxes	13
Miscellaneous, including garage, lubricants, parts	13
Total	100

The General Motors Truck Company compiled the operating records * of 46,000 trucks. In this study 1264 trucks were used for logging, lumber, and mill work. Of this number 598 were light-duty trucks of ½ to 1½ tons, 525 medium trucks of 1½ to 3 tons, and 141 heavy-duty trucks of 3½ tons or more. From these data Table XXXV was compiled.

TABLE XXXV

Items	*Light Trucks Average ½ to 1½ tons*	*Medium Trucks Average 1½ to 3 tons*	*Heavy Trucks Average 3½ tons up*
Miles per gallon gasoline	12.5	9.4	5.9
Miles per quart oil	86.8	80.0	71.7
Average payload (pounds)	2645	5307	9612
Average number of stops per day	19.6	14.9	8.2
Average length of haul (miles)	11.5	16.8	30.8
Total operating cost (including maintenance and depreciation) per running mile, cents	15.0	20.6	29.0

It is a notable fact that neither the fixed nor variable charges previously described under Operating Costs increased with the truck capacity. Table XXXV points out that the operating cost, including maintenance and depreciation per running mile, is far lower with the light trucks than with the heavy trucks, but it is cheaper per ton-mile to operate a 3-ton truck than one of 1½-ton capacity, provided conditions justify the use of both trucks and are comparable.

REASONS FOR INCREASED USE OF TRUCKS

The reasons for the introduction and extended use of motor trucks may be summarized as follows:

1. Marked improvement in motor-truck design, durability, efficiency, construction, and ability to carry heavy loads at a rapid rate of speed as well as on adverse grades.

2. Improved highways for the main haul. Before and since World War II vast sums have been expended by both the Federal and state governments for main route and secondary systems of highways throughout the United States. These highway systems have penetrated to most forest regions, notably throughout the Southeast, the

* As only a relatively small number of these trucks were used under the rather severe and difficult conditions of log hauling, it is not believed that these figures are directly applicable for average conditions encountered in log transportation.

Lake States, and the East. Many sawmills located in communities along the main and subsidiary highways can readily take advantage of these facilities to lower transportation costs by motor truck. Estimates of costs being lowered 10% to 50% below previous methods of transportation as a result of improved highways and the use of the motor truck are frequent. Improved highways have been a vital

FIGURE 133. Ford truck converted by a Thornton tandem drive to a 4×6 truck with a 209″ wheel base and 20′ wood-carrying space. The load carried is approximately 5½ cords (128 cu. ft.) of 5′ 3″ southern pine pulpwood. This type of hauling unit has proved very satisfactory where extreme driver care and good preventive maintenance are followed. The tandem-drive unit gives good flotation and improved traction for boggy and difficult woods operations.

factor in the haulage of logs to thousands of small-capacity sawmills, especially in the second-growth timber throughout the eastern forest districts of the country. Probably 90% of all logs hauled to small sawmills a distance of more than 2 miles are hauled by motor truck throughout the East and South.

3. The time element and the relatively low initial installation and depreciation costs. Sawmills in which production is started after shutdowns or to take advantage of improved market conditions can secure logs promptly and speedily by motor-truck delivery from both distant and near-by timber. Motor trucks provide the most rapid method of log transportation yet devised from the stump to the mill.

4. Reduced time required for loading and unloading. Within recent years, marked improvements have been made to facilitate loading,

especially pulpwood in the South. Unloading facilities have been greatly improved in the big timber of the West Coast.

CONDITIONS TO WHICH MOTOR TRUCKS ARE ESPECIALLY ADAPTED

These may be briefly summarized as follows:

1. Scattered bodies of timber not reached by other methods of log transportation can be felled and brought to the market at low cost.

FIGURE 134. One of a fleet of nine International trucks carrying loads averaging about 17,000 b.f. with Idaho white pine logs 32′ long. Up to 75 logs are carried per load. The trailer has 35,000 lb.-capacity axles and 12′ bunks and 4′ stakes. Logs are cut from selecting-logging operation on a National Forest in northern Idaho and hauled 40 to 45 miles to a log dump in a lake and thence floated to the sawmill. Two round trips are made each day. The truck and trailer has 18 wheels with tandem and dual wheels.

Trucks are especially adapted for use in salvage logging as well as prelogging and relogging. The extension of railroad spurs to bring in timber from isolated spots or even from remote corners or separate watersheds may be prohibitively expensive for the amount of timber to be hauled. Other logging improvements may result in much greater expense than truck haulage. Truck haulage is commonly used where the distance precludes the use of tractors or teams or where the installation of power skidders or railroad spurs would undoubtedly be expensive. In the East, motor trucks can transport 10 to 20 logs from one small wood lot or several thousand logs from another wood lot at equal expense per thousand board feet. No other transport device could be used with the same minimum expense. Integrated logging of scattered bodies of timber is well adapted to trucks, especially on small operations.

2. Short hauls, especially on rough ground in mountainous regions, can frequently be handled by motor truck more cheaply than by other

FIGURE 135. Load of about 5000 b.f. of ponderosa pine near Deer Park, Wash.

forms of transport. Road construction for motor-truck haul is generally low. All roads for trucks in Maine have been constructed at a cost of $200 to $1000 per mile and in the Southern Appalachian region for around $100 to $400 per mile in the few places where roads are required.

3. Trucks are widely used as an auxiliary or supplement to existing transport facilities, especially stream driving, winter-sled haul, rafting, and logging railroads. For some of these purposes, they have successfully competed with horse- and tractor-powered vehicles.

4. Where improved state or county highways are available for at least 60% and preferably 80% of the total length of haul, log transportation is especially cheap. If truck-haul roads must be built, they should be constructed at low cost except for large volumes of standing timber or for high-value species.

5. For selective logging, that is, removal of only a portion of the stand—by groups, sizes, or species—trucks may be especially adapted.

The removal of only those trees which have high conversion value requires a low-cost flexible method of log transport because of the low value of the timber stand per acre and the scattered nature of the logging. In the South, only 1000 to 3000 b.f. per acre may be removed under a selective system of logging. No other device could handle the low yield per acre so successfully as the truck.

6. Prompt action is necessary in the salvaging of timber damaged by wind, fire, or other natural agencies. Trucks can be used to log isolated trees or groups which more expensive forms of logging improvement would render prohibitive. In Oregon trucks were used to haul fire-killed timber 7 miles and bettle-killed timber 10 miles to a railroad, to haul logs to a portable sawmill established to cut salvaged logs, and haul lumber 17 miles to a seasoning and shipping plant.

ADVANTAGES

1. **Flexibility.** The truck is serviceable under a great variety of conditions including adverse as well as favorable grades; it can be backed or turned with great facility and started and speeded up quickly. The truck lends itself as readily to hauling in supplies and men as to hauling out logs, moving donkey engines, tractors, trailers, machinery, and similar items. It can be used on lateral highways while logging machinery is being moved to another location so there may be none of the shutdowns or delays that occur frequently in other methods of log transportation. Truck logging permits wide flexibility in sources of log supply. Therefore, it is particularly adapted to selective logging.

2. **Economy of labor.** Most of the methods used on major operations require many men, some of whom must be of the skilled type who command high wages. The logging railroad and power skidding require many men for operation. In contrast, trucks operate with one driver. In many operations in the East and South two trucks are operated together, the two drivers helping each other to load and unload.

3. **Low initial investment with consequent low maintenance,** depreciation, and interest charges per thousand board feet compared with the logging railroad, power skidding and other systems which require heavy initial investment and, therefore, high interest and depreciation charges.

4. **Speed and reliability in delivery.** Loaded trucks may travel a mile in 2 to 6 minutes on the level on good highways. Return trips can

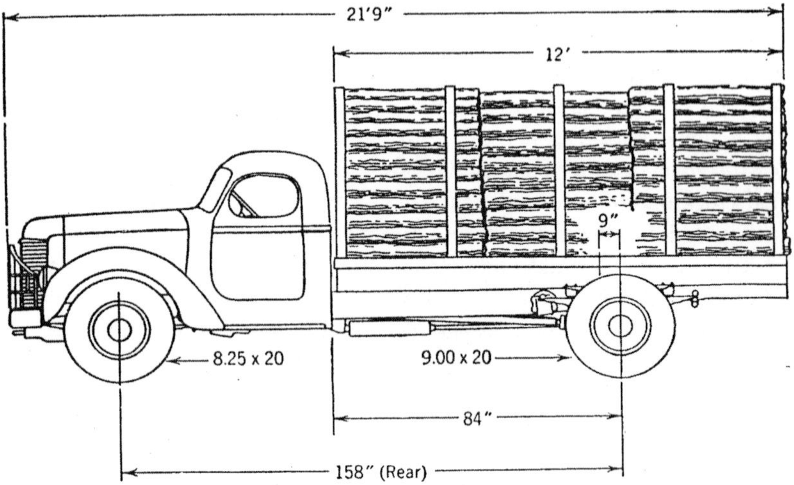

FIGURE 136. Pulpwood load carrying up to 12,000 lb. payload as used with the KBS-6 International 4-wheel truck especially in the South and Northeast. It has a cruising speed of 45 m.p.h. and a maximum speed of 55 m.p.h. Long-length pulpwood is loaded in this way; short-length pulpwood and cross ties may be loaded in this way or crosswise.

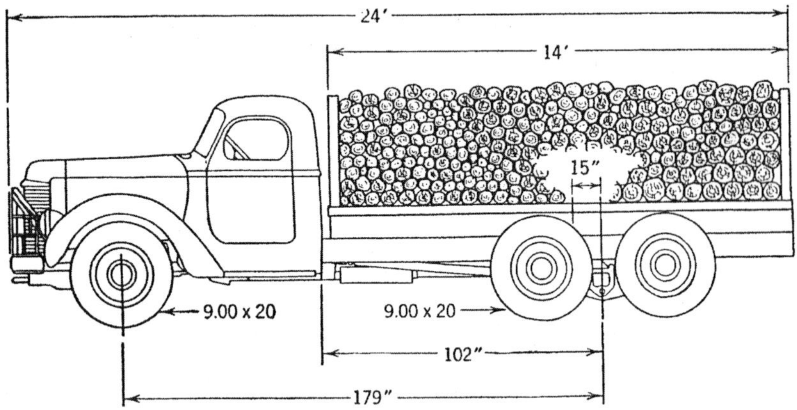

FIGURE 137. Method of hauling pulpwood up to 20,000 lb. payload on KB-8-F 6-wheel International truck as widely used in the South, Northeast, and Canada. This has a cruising speed of 35 m.p.h. and a maximum speed of 40 m.p.h. The 9.00 × 20 tires give satisfactory results.

be made in much less time. The increased volume of deliveries in a given length of time due to the greater number of round trips per day has been a potent argument for the replacement of other forms of log transportation by trucks.

5. Economy in road construction. Roads built for truck operation are generally far lower in cost than spur lines in logging railroads. However, in the Pacific Northwest, truck roads suitable for year-round (winter) logging may cost nearly as much as railway spurs.

6. Speed in loading and unloading trucks and trailers. Occasionally trucks can be loaded in 5 to 15 minutes and unloaded in 2 to 7 minutes. This depends, however, on size, weight, and length of logs; number of logs per truck; methods of and facilities for loading and unloading, and similar factors. Since 1945, mechanized loading devices, especially in southern pulpwood operations, have decreased costs and, therefore, assisted in increasing the efficiency of motor-truck logging.

LIMITATIONS OF MOTOR-TRUCK HAULING

These may not be of great importance but should be carefully considered in determining the method of logging extensive tracts of timber and the largest trees.

1. Motor trucks cannot handle the very largest and longest logs felled and used in redwood, Douglas fir, Sitka spruce, and other large species on the Pacific Coast, under some of the forest and topographic conditions where they must be hauled, owing to their excessive weight. Douglas fir logs produced by one company average 84′ in length.

2. Motor trucks require reasonably good roads and fairly level topography. In the pine regions of the Sierra Mountains of California, motor-truck logging is frequently not adapted to the rough topography with deep ravines, sharp curves, steep slopes, and so on, where road construction is excessively expensive. However, trucks are now used for hauling logs out of very rough and difficult places, and with auxiliary devices such as cables and tractors on adverse grades they are overcoming the handicaps of rough topography.

3. Trucks cannot be used on large operations requiring the delivery of 100,000 to 300,000 b.f. or more per day for mills located at long distances from the logging jobs. Railroads must be used where long hauls and large daily output of logs are required.

4. There is a definite limit of distance beyond which log haul may not be profitable. Load and road conditions determine this.

5. In some regions trucks may be operated only 8 to 10 months of the year on account of wet weather and the condition of lateral highways.

6. Restrictions regarding weight, width, length, and height of loads on public highways in some states have resulted in the building of private roads in the Northwest.

RAILROAD VERSUS MOTOR-TRUCK LOGGING

Motor trucks have several advantages over railroads in logging. Except on the very largest operations especially where a large volume of logs is required, the motor truck has made definite inroads on the railroad. This is notably true in the South, the Southeast, and the Lake States. The following shows the comparative advantages of truck hauling over railroad logging under usual conditions encountered in the woods, especially on those operations involving a sufficient outturn of logs for the use of railroads to be considered:

FIGURE 138. An old landing along a railroad spur in the Menomonee Indian reservation in Wisconsin where the forest has been cut selectively. About 75% of the merchantable stand of white pine, red oak, hard maple, and hemlock has been removed, cut to a flexible diameter limit around 18″. This spur track was taken up and replaced with a motor-truck highway.

1. The initial investment for right of way, grading, equipment, and operation of railroads is exceedingly heavy. They must be operated at full capacity in order to minimize depreciation charges per thousand board feet. The depreciation factor may be exceedingly heavy if only partial capacity is utilized or the mill is shut down for an extended period of time.

2. Comparatively large storage facilities must be provided at the sawmill to take care of long railroad trains delivering logs once or twice a day. This means a large and expensive log pond for softwoods or extensive yards with costly handling facilities for hardwoods. When trucks can deliver

logs directly at the base of the jack ladder at the sawmill, sinkage from "dead heads" is entirely eliminated. This is especially important with hardwoods.

3. Railroad grades must be favorable, curves must be wide (not over 20° generally), and the roadbed must be smooth and properly maintained in order for logs to be hauled economically. Many cuts and fills are necessary together with heavily constructed bridges and culverts. Trucks can more readily lend themselves to the conditions of topography. Road construction for motor trucks is simpler and less expensive, and bridges and culverts may be much lighter in construction. Furthermore, motor trucks may be used on more severe adverse grades than railroads, which are generally limited to grades of about 6%.

4. In railroad logging, there is a strong tendency to clear-cut all available timber because of the large initial investment and the concurrent desire to remove all the timber in order to justify the investment, depreciation, and other carrying charges. Trucks are much more mobile and lend themselves more readily to selective logging procedure, as a few trees per acre may be cut. Motor-truck logging tends toward better silvicultural management.

5. In the hardwood forests in the South and in the Lake States, motor-truck roads can be constructed for approximately one-fifth the cost of railroads under similar conditions and for similar purposes. For example, on one location, truck roads cost about $2100 per mile contrasted with about $10,200 per mile for railroad construction. Both types of roads accomplish the same purpose.

6. The cost of decking logs for railroad haul is much greater than for truck hauls. Large, expensive landings are generally required for railroad loading facilities. For truck hauls high banks adjacent to highways are generally used for decking logs from which they may be rolled directly on the trucks at a minimum of cost.

7. Public highways are frequently available for truck hauls, eliminating the necessity of part or all of the road-construction costs. Frequently, highways parallel both common carriers and private logging railroads. Within recent years, public-highway facilities have been extended into many forest regions formerly considered too remote and inaccessible for logging.

8. A large area can be used as a source of log supply for truck hauls because of their greater speed. This flexibility of source of supply, coupled with low-cost transportation units, greater speed, and quick deliveries, means frequently better selection of logs, a quick shifting

of log supplies to suit market conditions, and the elimination of the necessity for forecasting market requirements.

9. Smaller camps operating at a lower unit cost may be operated with truck logging than with railroad logging, as men may be carried greater distances to work in a minimum of time. Furthermore, labor may be recruited from near-by communities with motor trucks. Workers frequently live at home or on neighboring farms or with relatives.

FEATURES OF IMPROVED LOGGING TRUCK DESIGN

During World War II and since, many improvements have been made in the design of motor trucks to withstand the tremendous

FIGURE 139. An 18-wheel truck and trailer used for carrying the heaviest loads as used on the West Coast. The largest types have a gross vehicle weight of 207,000 lb.

stresses and strains incident to hauling heavy green logs out of the woods over rough terrain and very rough and uneven roads. Under no comparable circumstances are trucks subjected to such strains as in hauling heavy logs, especially on the West Coast and in the rough topography of the Southern Appalachians, the Northern and Central Rocky Mountains, and the Northeast. The following is a summary of some of these features:

1. The smaller trucks and trailers are made with 4, 6, and 8 wheels and the largest trucks and trailers with 12, 14, 16, and 18 wheels. The largest trucks, carrying the heaviest loads and particularly long heavy logs on the West Coast, use dual and tandem wheels.

2. Oversize brakes, used to assure greater safety, have longer life and efficiency in operation.

FIGURE 140. Heavy-duty log truck used on the West Coast. Of the 18 tires used when loaded, eight on the trailer are pulled onto the main truck bunk. This provides for better maneuverability and saves wear and tear on the trailer rubber on the return run. *Courtesy American Forest Products Industries.*

3. Heavy-duty and oversized front and rear springs with two speed axles are provided.

4. Very large and heavy-tread tires are used, especially on dual and tandem wheels for the heavy and long Douglas fir, redwood, and ponderosa pine logs trucked on the West Coast. There the sizes vary from $10'' \times 20''$ up to $16'' \times 24''$.

5. Auxiliary transmission provides extra low-gear ratios and vastly increases the pulling ability, especially on adverse grades. The increased power provided is often up to 100% above previous designs.

6. Five or more speed transmissions are provided plus ideal gear splits when used with overdrive transmissions.

7. Reinforced frames provide adequate strength for maximum loads on extremely rough woods road.

The correct size of motor trucks for carrying loads is an important feature of truck logging. The capacity of the structural materials used in a motor truck has been very carefully estimated by truck designers. Trucks are manufactured that will operate over a longer period of time, keep the factor of depreciation as low as possible, and observe the proper factors of safety. When loads above the recommended ones are imposed, the factor of safety is definitely reduced. Therefore, overloading trucks means (1) accelerated depreciation, (2) higher maintenance costs, and (3) lost time which should be unnecessary.

All of this means higher costs for trucking. For example, a truck designed to cost about $3000 and carrying the same load as a heavily overloaded $2000 truck can very readily be operated for $600 less per year and earn increased revenue equivalent to perhaps $1000 of additional pulpwood delivered to destination because of less time lost due to breakdowns. In a fleet of 10 trucks this would mean $6000 savings per year plus an additional 10,000 cords of pulpwood hauled.

According to J. J. Byrne of the U. S. Forest Service, in reporting to the Pacific Logging Congress, the advantages of using larger equipment instead of the smaller trucks are more apparent where:

1. Road construction and maintenance costs are low.
2. Speed of travel is limited by conditions of terrain, climate, or other causes.
3. Length of haul is short.
4. Traffic density on single-lane roads causes excessive delay time at turnouts.
5. The number of qualified drivers is limited.
6. Labor costs are high.
7. Volume of logs to be hauled is great.

There is no doubt but that hauling costs are lower with the heavier units. The big problem is whether the savings in handling costs would pay for the additional road construction and maintenance required with the larger units. In most instances this will depend on the volume of logs to be carried over the road involved.

LOADING

Loading is generally done by one of the following operations:

1. By hand from elevated skidways or ramps with small and short logs. This method is used with second-growth southern pine and

FIGURE 141. Manual method of loading truck. Generally advantage is taken of high banks or built-up skidways so that the logs may be rolled and dropped by gravity onto the truck-bunks. These are black cherry logs in Pennsylvania. *Courtesy U. S. Forest Service.*

FIGURE 142 Semitrailer being loaded from conveyor, shown at left, of a portable pulpwood mill. Load capacity is about 3½ units (168 cu. ft.). About 9 million cords are produced yearly for the 50 pulp mills of the South. Trees are felled with power saws, tractor-skidded to pulpwood mills, crosscut to length, and loaded on trucks. This is one of the best examples of efficient mechanized logging. *Photo by G. E. Knapp.*

spruce pulpwood, and often with eastern hardwoods and northern pine.

2. By crosshaul by horsepower or power supplied by winch or winding drum on the truck itself or trailer. This system is commonly employed with long lengths, poles, piling, and the like. Tractors are also used for crosshauling.

3. By gin pole and drum operated by power supplied by the truck itself, by separate motor, or by tractor.

4. By power loaders operated at the landing or deck in connection with railroad operations or by donkeys from cold decks in the Northwest.

5. By mechanized-conveyer loaders, especially for loading pulpwood in the South, Northeast, Lake States, and eastern Canada. See chapter on Loading for more extended treatment of this subject.

UNLOADING

Unloading is accomplished by rolling the logs off the truck with a cant hook or peavy down an incline at the mill landing or log pond.

FIGURE 143. Hydraulic unloader used to dump truck load into Willapa River, near Raymond, Wash. Time required to unload—2 minutes. *Courtesy White Motor Car Co.*

Tilted roads to depress one side of the truck facilitate unloading at the landing. Logs may be forced from the truck bunks by a steam or electric-power "kicker" similar to that used on the log deck of a sawmill. Winches operated by power supplied by the truck may also be used, as well as the crosshaul operated by a winch or team. Unloading problems are most difficult on the West Coast where the huge weights of large and long logs require mechanical rather than manual methods of unloading.

TYPES OF ROADS

Many types of roads are constructed to bring logs to the landing or to main or improved highways. These secondary or lateral roads are generally crudely constructed and may be ½ or ¼ to 5 or more miles in length. On the West Coast, they may be many miles long and often parallel public highways. Generally they are built for short distances and over wet places, ravines, and streams. In the southern pine region and ponderosa pine forests of the West, roads are occasionally not necessary, owing to the level character of the ground, lack of underbrush, open growth of the timber, and small number of trees per acre.

Some of the principal types of roads are:

1. **The lengthwise plank type** in which two planks are laid on each side of the road after the right of way has been cleared of tree growth or brush and leveled by hand labor, scraper, or bulldozer. Sometimes a plank flange is placed on the inside of the wheels of the truck to keep the truck on the road. On steep inclines, loads are lowered by means of a steel cable and winch operated by a donkey engine. This type is found in the Lake States, the South, and on the Pacific Coast.

2. **Ordinary rough woods road.** Frequently the right of way is cleared of trees, and wet places are covered with corduroy made of small timber. Temporary bridges of round timbers or planks are used to cross rivers, creeks, and low places. This type is common in the Rocky Mountains and in the East.

3. **Wooden stringer roads** with round timbers or planks as cross ties. Trucks equipped with double-flanged wheels are used on this type of road. It is found in the Douglas fir forests of the Northwest.

4. **Pole roads** where the truck wheels roll on logs 12″ to 16″ in diameter faced or flattened on the top surface. The truck is prevented from rolling off by two logs placed on either side of the faced log and above the level of it. The three logs are spiked to round cross ties at the joints and as frequently between the joints as may be necessary. This type is constructed in the southern pine and hardwood forests for short distances where the cost of building more usual types of roads

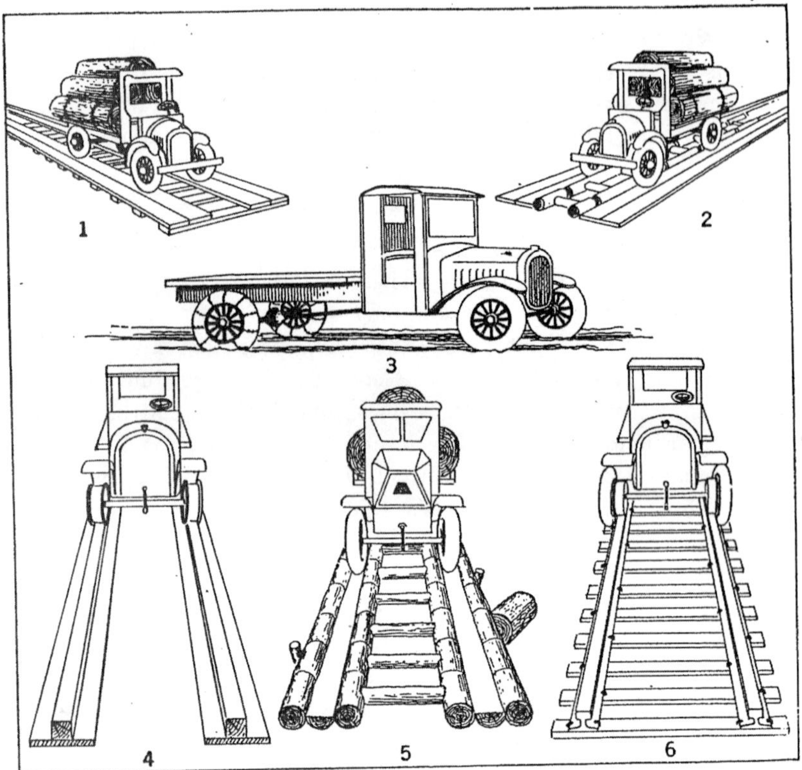

FIGURE 144. Some types of roads used in motor-truck transportation, especially for short distances.

1. Illustrates the longitudinal-plank-road type construction with portable plank sections. This is used to cross low, wet, or swampy places for short distances.

2. Practically the same as 1 but may be portable or permanent. It has poles or stringers on the inside of each track to insure greater safety. Some roads of type 1 have sections of type 2 over dangerous places, as across streams or ravines or low swampy places.

3. Ordinary woods road swamped out on dry well-drained topography. This type is built at relatively low cost. Types 1, 2, and 5 are often used in connection with type 3.

4. Trucks equipped with flanged wheels are sometimes run on wooden rails. They are not common, however, because the type of wheel is not adapted to use on ordinary roads.

5. A common type of lateral log road used in crossing streams, swamps, etc. Logs placed on both the inside and outside of each track give added protection.

6. Special types of wheels are sometimes designed to run on steel rails as well as on ordinary roads. Sometimes wheels designed to operate on steel rails are used. The type illustrated is especially serviceable on steep inclines and can be run on ordinary roads as well as on rails. On one operation in Washington trucks were let down by cable a distance of 4700 ft. with a maximum grade of 64%.

would be exceedingly expensive. Pole roads are extensively used in the Northern Rocky Mountains on steep grades and over wet places.

5. Steel railroads on which trucks equipped with especially constructed flanged wheels are used. Five-ton trucks are run on some roads of this type. One is 4700 ft. long with a maximum grade of 64% down which the loaded trucks are lowered by means of a cable and winch.

FIGURE 145. Mats of 1″ lumber 30″ wide laid cn woods roads prior to hauling pulpwood by motor truck in the South. Note large pulpwood truck at left and loader shown in upper center.

6. Portable plank roads. Sometimes planks, laid laterally, are placed end to end on the ground and spiked to cross planks or round timbers. Two planks are in each section, which varies from 16′ to 20′ or more in length. They are adequate only for short distances.

7. Snow roads. With steering skis substituted for steering wheels and a tractor drive in place of the rear wheels, trucks known as snowmobiles have been built for winter haul in eastern Canada.

In the South, low-grade dimension stock of 2″ × 4″, 2″ × 6″, 2″ × 10″, 2″ × 12″ is laid longitudinally and tied in by crosspieces of 2″ × 4″ or 2″ × 6″, forming a road serviceable during wet weather and over low swampy places.

Another method sometimes adopted in the South is to plow two furrows wide enough to fit the truck wheels and fill them with gravel and stones. They cost only $50 to $120 per mile on the relatively flat topography of the southern pine region. They are used only during wet weather.

TRUCK ROADS FOR ALL-WEATHER LOGGING

With the increase in the use of motor trucks as the principal means of major transportation from the woods to the mill or to another method of transportation, the construction and operation of all-weather roads has become a matter of major importance in logging. For the relatively level topography and open timber stands of southern pine and the frequently flat areas of ponderosa pine, truck roads can be built at little expense, with a little brushing out and perhaps bridging and corduroying low swampy spots. On the other hand, in mountainous or rugged terrain and in heavy timber, the construction of roads is a matter of major expense. Unfortunately, because of restrictions on the use of public highways, many companies, especially on the West Coast, have been forced to build private highways, which sometimes parallel public highways, in order to get the logs to the mill or other destination.

Various forms of trail builders, bulldozers, and angledozzers are used to build these truck roads. Generally woods managers or logging bosses, rather than civil engineers, do the necessary planning in connection with a topographic contour map. Then the roads leading from the felling and skidding areas to landings or sawmills must be built, together with the necessary clearance of right of way, excavation, drainage, surfacing, culverts, bridges, and other necessary phases of road construction. Sometimes expensive rock work, cribbing, and retaining walls must be constructed, especially if the road is of a permanent or semipermanent nature. Frequently the cost of road building must be weighed against the cost of a tramway or logging railroad in order to get the logs to destination at a minimum cost.

The U. S. Forest Service * has built a good many roads in connection with the logging of National Forest timber, especially in the West. This was done to open up remote areas otherwise inaccessible. These roads have furnished a pattern for many truck roads built by private companies.

* The subject of truck roads, although an important phase of logging, is too large to be completely covered in this elementary book on logging. For further details, see *Northeastern Loggers' Handbook,* by F. C. Simmons, U. S. Forest Service, and engineering textbooks and manuals dealing with the construction of woods roads. The *Truck–Trail Handbook* of the U. S. Forest Service is especially helpful.

The maximum permissible adverse grades, width of highway, degree of curves, and other items must be carefully planned so that truck loads may be carried at reasonable expense. Sometimes the charges for blasting and removal of stumps, work incident to cuts and fills, especially in connection with rock work, may be exceedingly expensive although these roads frequently are only of temporary use.

FIGURE 146. Loading Douglas fir logs on White 6-wheel, tandem dual-drive truck near Enumclaw, Wash. Average load 4500 to 5000 b.f.

During winter logging in the Northeast, the Lake States, and eastern Canada truck roads have been built and used to replace winter-sled hauls. With improved types of snowplows and large-sized dual and tandem wheels, trucks go directly to the felling areas, where previously only sleds could be successfully employed; load out saw logs and pulpwood; and haul to the railroad, stream, or mill, thus eliminating greater charges incident to methods previously used.

PRIVATE MOTOR-TRUCK ROADS

Because of state restrictions regarding the weight and sometimes the length, width, and height of loads (especially at certain seasons of the year when public highways are susceptible to injury from excessive loads), many logging companies and even subcontractors have been forced to construct their own highways. These sometimes parallel public highways, since they can be built in such locations at a minimum of expense. During World War II emergency highways known as access roads were built in many National Forests in order to get out logs to meet the emergency log situation. These excellent woods highways have been adopted as a standard by many private companies building similar highways, especially with respect to grades, curves, ditching (drainage), and surfacing. Although construction of private roads has greatly increased logging expenses, the existence of large volumes of 50 million to 800 million or more board feet of stumpage has justified their construction for removing timber in places where public highways cannot be used. This is especially true in the heavy timber of the West Coast, where trucking has replaced railroads (previously for more than 50 years the principal method log transport), and where "on highway" and "off highway" truck hauling are a problem confronting the logging industry. This problem is assuming critical importance in many sections from California to British Columbia and east to Idaho and Montana.

TABLE XXXVI

	Weight per cubic foot, Green
Western Softwoods	
Cedar, Port Orford	36
Cedar, western red	27
Fir, Douglas	38
Fir, lowland white	44
Fir, noble	30
Hemlock, western	41
Larch, western	48
Pine, sugar	51
Pine, western white	35
Pine, ponderosa	45
Spruce, Engelmann	39
Spruce, Sitka	33
Eastern and Northern Softwoods	
Fir, balsam	45
Hemlock, eastern	50
Pine, northern	36
Pine, Norway	42
Spruce, red	34
Southern Softwoods	
Cypress, southern	50
Pine, loblolly	54
Pine, longleaf	50
Pine, shortleaf	51
Pine, slash	56
Hardwoods	
Ash, white	48
Aspen	43
Aspen, largetooth	43
Basswood	41
Beech	54
Birch, yellow	57
Cherry, black	46
Chestnut	55
Cottonwood, black	46
Elm, American	54
Gum, black	45
Gum, red	50
Gum, tupelo	56
Hickory, shagbark	64
Maple, sugar	56
Oak, red	63
Oak, white	62
Oak, willow	67
Poplar, yellow	38
Walnut, black	58

WEIGHTS OF GREEN LOGS

Weights are an important consideration in all log transportation but particularly with motor trucks and sleds, where the entire weight of the log must be carried. Table XXXVI * shows the weights of wood when green. It is based on an average of heartwood and sapwood as represented by test specimens, taken from the pith to the bark, from the top 4′ of 16′ butt logs. Variations frequently occur up to 20%, owing to wide differences in moisture content. This is particularly true of specimens which have a high moisture content in the sapwood. Young softwood trees contain relatively more sapwood than old trees, and some hardwoods such as maple and beech may have a very large proportion of sapwood. Logs from trees of this type generally average heavier when green than logs cut from virgin or old-growth timber. The table includes 42 different species which are frequently transported by motor truck.

Rapraeger has determined weights of various size logs expressed in thousand board feet for the Pacific Northwest. Table XXXVII credited to Rapraeger, shows the weights of the principal species used in motor-truck logging in Washington and Oregon.

TABLE XXXVII

WEIGHTS OF VARIOUS SIZE WESTERN LOGS AND SPECIES IN M.B.F.

	Weight in pounds per 1000 b.f., Scribner Decimal C Log Scale		
Species	*Small green logs* 16″ diameter	*Medium green logs* 30″ diameter	*Large green logs* 50″ diameter
Softwoods			
Port Orford cedar	7,050	5,411	4,997
Western red cedar	5,289	4,058	3,748
Douglas fir	7,446	5,711	5,276
Lowland white fir	8,621	6,613	6,108
Noble fir	5,878	4,509	4,165
Western hemlock	8,033	6,162	5,691
Lodgepole pine	7,641	5,862	5,414
Ponderosa pine	8,817	6,864	6,247
Sitka spruce	6,465	4,960	4,582
Hardwoods			
Red alder	9,013	6,914	.....
Oregon ash	9,013	6,914	.....
Black cottonwood	9,013	6,914	.....
Bigleaf maple	9,209	7,064	.....
Oregon white oak	13,519	10,370	.....

* From Technical Note 218, U. S. Forest Products Laboratory, Madison, Wis.

Table XXXVIII, based on actual experience, shows the weights and sizes of pulpwood as usually cut in the East and South.

TABLE XXXVIII

LENGTHS OF PULPWOOD LOGS

Lake States	8′
Pennsylvania, New York, New England	4′
Solid South	5′
Log lengths for trucking (generally)	16′

APPROXIMATE SHIPPING WEIGHTS OF WOOD IN STANDARD CORD OF 128 CU. FT.

Northern softwood—peeled	3200 lb.
Northern softwood—rough	4300 lb.
Southern softwood	5000 lb.
Hardwood—peeled	3800 lb.
Hardwood—rough	4500 lb.

CHAPTER XVII

Forest Railroads

HISTORICAL

On the basis of volume of board footage, railroads were formerly the most important single medium of major log transportation in the United States. They are particularly applicable for and the chief means of delivering logs from large and scattered bodies of timber for long distances to the sawmill or market. It is assumed that sufficient board footage must be available for transportation over the logging railroad in order to justify the large expense incident to the construction and operation of a system of main and spur lines with motive equipment. A considerable portion of the nation's lumber was once produced by large sawmills supplied by logging railroads. Motor trucks have replaced many logging railroads in all forest regions.

Forest railroads came into prominence for logging purposes during the great period of railroad building and westward extension of our common-carrier systems. The first forest railroad * was in Steuben County in New York, for logging white pine in 1852. Railroads were widely developed in Pennsylvania and the South during the last half of the nineteenth century. They were extensively constructed for the exploitation of the large virgin white pine forests in Michigan, Minnesota, and Wisconsin. Still later, railroads were widely developed and used on the Pacific Coast where the timber of great size and weight could not be profitably handled by other means of transport until the development of heavy motor trucks.

At first, logging railroads were exceedingly crude in their construction and methods of operation. Round logs or poles and sawed stringers served as rails. The motive force was sometimes supplied by horses and mules as well as by locomotives, chiefly of the geared type which were adaptable to adverse grades and sharp curves.

* This railroad, using wooden stringer rails and a locomotive, hauled logs to the Tioga River. See *History of the Lumber Industry in New York*, by W. F. Fox, Bulletin 34, Bureau of Forestry, Washington, D. C., 1902.

Types of logging railroads as developed in the United States may be classified as:

1. Pole roads.
2. Stringer or tram roads.
3. Narrow-gage railroads—generally 3′ gage.
4. Standard-gage railroads—4′ 8½″ gage.

The last named now carries at least 98% of the total volume of logs hauled by railroads in the country.

FIGURE 147. Log train crossing a trestle in the Northern Rocky Mountains. Railroads are used to supply a constant flow of logs to large sawmills located at considerable distances from logging operations.

The construction and operation of logging railroads often represent 20% to 47% or more of the total cost of delivering logs to the mill pond (including felling, bucking, skidding, camps).

There were, at one time approximately 6000 miles of logging railroads in the three Pacific Coast States. Probably more than 2000 miles of spur lines were replaced every year in this region. These spur lines may be used only about 6 months to 2 years, whereas the main lines may be used from 5 to 20 years or more, depending on the length of the operation and whether sustained yield management is practiced. It is estimated that 25% of all spur lines are replaced every year. Spur lines often represent 20% to 60% of the total logging railroad mileage. Spur lines are sometimes served and replaced by motor trucks.

Logging railroads are frequently operated in connection with existing common carriers. Other forms of major transportation are sometimes used to "feed" them along both main and spur lines. Chief among these auxiliaries are motor trucks, tractors, chutes, flumes, stream driving, and rafting. Most of them precede the railroad haul; some, such as rafting and booming in the Pacific Northwest, may follow it.

The locomotives and log cars are subjected to very severe operating conditions. The tracks are usually uneven and poorly ballasted; the grades may be steep and the curves sharp. Railway equipment is generally worked heavily with relatively little attention and repairs. Effective spark arresters are required to guard against forest fires, especially if wood is used as fuel.

CONDITIONS DETERMINING USE

The installation of large-capacity sawmills necessitated a continuous, large, and reliable flow of logs daily throughout the year. This made the logging railroad the most important single medium of major transportation for timber until the introduction of the motor truck.

Logging railroads are especially adapted to long hauls and to large tracts of timber to supply sawmills with a daily capacity of 30,000 to 40,000 b.f. or more in hardwoods and 50,000 b.f. or more in softwoods. It is assumed that a capital outlay of at least $50,000 is required as the minimum amount for initial investment. Some forest railroads have cost up to $1,000,000. Logging railroads generally signify a fairly long-time investment. There must be a large steady flow of logs to justify the installation of a logging railroad with the relatively large depreciation, interest, and maintenance charges incident to them. The period of logging generally continues 280 to 300 days per year, which means literally year-round operation. Railroads must be planned so that the mill will be supplied with a constantly stocked log pond. As it is essential to keep woods operations as well as the logging-railroad operations in continuous use, so must the sawmill be in continuous operation, and markets must be found for the product. Therefore, successful marketing and selling may justify railroad installations. The important factors in determining the amount of capital to invest in a logging-railroad system are the volume of standing timber and the distance of haul to the mill. Frequently this haul may be 40 to 150 miles long. Common-carrier systems are often used in connection with the hauls, particularly in the Pacific Northwest, the ponderosa pine operations of the West, and the southern pine operations of the South and Southeast.

Logging railroads are especially adapted to heavy and long-length logs usually cut in logging Douglas fir, ponderosa pine, and virgin southern yellow pine. They are also an important medium of transportation for redwood, cypress, and hardwoods on extensive tracts.

FIGURE 148. One of the fundamental differences between American and European logging practice is that, in the former, logging railroads, flumes, chutes, splash dams, and other forest improvements have usually been regarded as temporary in their installation and frequently abandoned when the logs are hauled from the forest. In Europe, there is a greater sense of permanence in constructing forest highways, tram roads, and stream improvements. Overhead and other depreciation charges in European logging operations are much lower because the flow of forest products is constant. The illustration shows a narrow-gage logging railroad in the forest of Sihlwald in Switzerland where forest products have been cut under a system of forest management for a thousand years.

As a minimum, about 50 million b.f. should be available for the installation of a short-distance logging railroad. Some operations using railroads cut 50 million to 200 million b.f. or more per year. On a few operations, from 1 billion to 12 billion b.f. are available to a single railroad. The larger the flow of footage over a railroad, the lower will be the charges per thousand board feet for construction, operation, interest, depreciation, and maintenance. For example, if the construction of a railroad involves an expense of $1,000,000, and 1 billion b.f. are chargeable against this construction cost, the expense will be $1 per thousand board feet, whereas, if the footage is only one half, or 500 million b.f., the expense per thousand board feet would be double, or $2.

Although the stand per acre is important in determining the advisability of the use of railroads, the total volume on a given tract is a more important factor, provided construction and operating costs are not prohibitively high. Forest railroads have been constructed and operated successfully on large properties in the South that yielded an average of only 2000 to 2500 b.f. per acre. It is advisable, however, that stands on large tracts should produce at least a minimum of 5000 b.f. per acre of merchantable timber in order to justify the expense incident to the construction, operation, and maintenance of forest railroads.

LOCATION AND GRADE

The proper location of the roadbed is of vital importance. An expensively located and constructed road may be inexpensive to operate over a period of years, or vice versa. There is a proper balance in road location between construction and operating costs. These two costs are generally separated in logging accounting.

Railroad location must be based on and include plans for the minimum cost of yarding the tract, together with available sites for landings along the track as well as probable camp sites, water facilities, switchbacks, and spar-tree locations (in power skidding). A logging railroad must be so located that it can be built and operated at minimum expense. Rail hauls are becoming longer and more difficult to construct, especially in virgin-timber regions on the West Coast. Therefore, their location is increasingly important. Excessive grades, curves, cuts, fills, and trestlework should be avoided if possible. However, the cost of cuts and of moving earth seldom exceeds 20% of the total railroad construction costs, and it is generally considered cheaper to make cuts or fills than to build trestles or indulge in excessive grades.

In the Southeast and Southwest, grades seldom exceed 2%. There railroad location and operation are relatively simple problems. This holds true of other regions where there are long level stretches or undulating topography such as parts of eastern Oregon. In the rugged steep topography of the Pacific Coast, the Southern Appalachians, and the Northern Rocky Mountains, 9% grades are generally the absolute maximum, and grades are seldom found over 6% except for short sharp "pitches." Above these grades considered as a maximum for any given condition or region, inclines are used. On the main railroad lines of the Pacific Coast region, grades seldom exceed 5% to 6%. The steepest grades are found on spur lines. Generally 30° to 35° curves are the maximum for logging railroads. Geared locomotives are required on severe adverse grades, either for loaded or empty cars and for the sharpest curves. Therefore, geared locomotives are generally used for switching cars and for bringing loaded cars from the spur lines to the make-up point on the main line. Thence the log trains are frequently hauled to the sawmills by rod or direct-connected locomotives.

In location work, the object is not, as with a common carrier, to provide a road at minimum grades through the property to connect two distant points, but rather to provide a network of main and spur lines to move all the merchantable logs from a given area to a distant point at the least expense for construction and operation. Thus it must be located with reference to the skidding or hauling devices adapted for each particular operation. With power and tractor skidding, distances generally vary up to 600 and less frequently to 2000 ft. (in southern pine up to 660 ft.); therefore spur tracks must be spaced approximately ¼ mile and, less frequently, ½ or 1 mile apart.

An accurate survey including timber estimates and contour maps must be provided in units of 40 acres in legal subdivisions, or by drainage basins. Rates of curvature or grade are governed by conditions found in the timber stands. If the topography (slope) and density of timber were uniform, railroad location would indeed be simple. Frequently main lines must cut across the topography, whereas the spur lines generally follow the creeks and draws. In other logging units main lines follow the main river or creek levels, and spur lines follow each subsidiary drainage basin in order to procure the timber at minimum expense.

Main lines must be more carefully located than spur lines. The number of carloads per train that can be hauled has a definite relationship to grades; it varies also with the power, type, and condition of the engines used. The more loads per train the main-line engine can haul at fair speed, the lower the cost per thousand board feet. Spur-line location is dependent on the topography and the quantity of timber per acre. Expert judgment must be exercised to determine whether spurs should be built or the timber should be transported to the existing railroad by sleds, motor trucks, tractors, or cable lines.

CONSTRUCTION

The methods and costs of construction vary widely in accordance with local conditions. Railroad construction is a relatively simple and inexpensive operation in flat topography, especially in open timber stands as found in Florida and many parts of the Gulf and South Atlantic States. In some of the ponderosa pine stands of the Southwest and in eastern Oregon, railroad construction is also comparatively inexpensive. It is most expensive and difficult in the Pacific Northwest, California, Northern Rocky Mountains, Southern Appalachians, and other mountainous regions. Costs are generally figured on the basis of stations of 100 ft. or distances of a mile. For purposes of accounting, construction costs are based on the charge per thousand board feet. Railroad construction generally involves the following in chronological order:

1. Right-of-way clearance. The cost of clearance depends on the character of the timber, slope, amount and kind of windfalls, presence of rock, and width to be cleared. The width varies from 15′ to 50′ on spur lines and up to 70′ on main lines, depending on the height of adjacent timber and other local conditions. In Wisconsin, the grade is brushed and stumped to a width of 14′ to 18′ for narrow-gage lines and 18′ to 24′ for standard-gage lines. The cost varies from practically a few dollars per acre, or relatively little expense, in open ponderosa stands in Oregon, the Southwest, and southern pine, to $600 per acre or more in Douglas fir, redwood, and northern Idaho. The cost may be as high as $700 to $900 per acre in unusually dense timber stands with many windfalls and similar problems. Tractors and, still more, the bulldozer have entered prominently into clearance work.

2. Building roadbed. The cost depends on (*a*) the amount and character of material to be moved for cuts and fills (such as rock, gravel, loam, and sand), (*b*) the efficiency of labor and machinery (equipment), and (*c*) the season of the year. Many kinds of Diesel bulldozers and scrapers are used in making cuts and fills. Sometimes, as with bulldozers, clearance and road building are carried on simultaneously. Many laborsaving devices have been introduced within recent years. The completed grade is 8′ for narrow-gage lines and 10′ to 12′ for standard-gage track.

3. Laying ties and steel. In the South and Southeast, ties and rails on spur lines are sometimes laid directly on the ground without preliminary grading or surfacing. Track-laying and lifting machines are used, as described later in this chapter. Crooked rails, low joints, and battered ends may have a very serious effect on the equipment.

4. Ballasting roadbed. The cost depends on the kind of material used and the expense in hauling it to the roadbed, as well as the character of the ballast itself. Surfacing, including raising track and its alignment, and tamping ballast on main-line tracks, may cost up to $1000 per mile. Frequently spur-line tracks are not ballasted at all. On main lines used for a number of years, the work is done almost as efficiently and as substantially as on common carriers.

5. Picking up ties and steel rails. When no longer needed, especially on spur lines, both ties and rails are generally taken up, and this is considered a part of the construction expense. When a track-laying machine is operated, it is used to pick up ties and steel rails and convey them to the next roadbed.

The total cost of construction of main-line logging railroads may vary from $2000 to $70,000 per mile, generally being about $20,000 to

$45,000 per mile, depending on the afore-mentioned factors. Spur lines cost very much less, owing to their temporary nature and, therefore, to the lower grading expense, narrower clearance for right of way, and other construction factors such as lighter rails, smaller and fewer ties, little if any ballast.

Many types of bridges and trestles are built on logging railroads. Streams, from the smallest creeks to large rivers, must be crossed in order to obtain favorable grades and to avoid excessive rock cuts, fills, and other construction work. The cost of trestles varies with permanence and local requirements. Wooden trestles, even though they may be cheaply constructed, are often avoided in favor of fills in regions of serious fire hazard such as in the Pacific Northwest. Forest fires are a constant source of concern and expense on the Pacific Coast and in the Northern Rocky Mountains.

Generally trestles are of the following types of construction:

1. Crib type. This is elementary in design and low in cost. Earlier trestles were almost entirely of this type, but they have generally been superseded or abandoned on the larger and longer-lived oprations. They are susceptible to settling, and, where labor costs are high, the expense may be excessive. They generally have a solid deck, and the timber in them is not salvaged. Only short low trestles of this type can be built successfully.

2. Framed trestles of sawed or round timbers. These are widely used in many parts of the country where a long low trestle is required. The usual engineering specifications * are followed in their design.

3. Log trestles. Logs serve for girders or spans up to 50′ without any trusses. Spans up to 100′ have been built of logs in the Pacific Northwest by first placing them in a solid deck on falsework. Cross ties up to 20′ long are laid on top of this solid log deck, and then a long large log is placed on top of the ties at either side of the bridge. The whole structure is laced together with wire cable. It is especially adapted to high trestles where long large logs are available to give sufficient clearance, especially over navigable streams, railroads, highways, and other obstructions. Frame bents are erected to carry the deck during the falsework construction.

* For further descriptions and methods, consult engineering textbooks devoted to railway design, construction, and operation, such as *Railroad Construction—Theory and Practice,* revised 1932, by W. L. Webb; *The Elements of Railroad Engineering,* by W. G. Raymond; Railroad Structures and Estimates, by J. W. Orrock (all John Wiley & Sons, New York); and *Railroad Construction,* by C. L Candall and B. A. Barnes, McGraw-Hill Book Co., New York, 1913.

4. Piling trestles. These are built of piles up to 128′ in length with top diameters of 14″, especially in the large timber of the Pacific Coast where long straight piling is readily available. When the ground or rock surface permits insufficient penetration, the piles are held on place, butt down, by the driver until capped and braced. Some spur tracks have been built entirely of piling without recourse to trestling.

5. Miscellaneous. Wood and steel Howe trusses and steel-arch trusses, and some other types are infrequently used. Their design and construction generally follow standard railroad practice. Culverts to provide adequate drainage are built of logs from local timber stands.

Inferior species such as white fir in California and hardwoods of low value in the East and South are utilized for trestles, culverts, and the like, as well as for cross ties of sawn, hewn, and round types. Tunnels have been necessary in the construction of some logging railroads in the West. Higher and wider clearance than with standard railroads is generally required. Tunnels are usually lined with timber.

In the Northwest, construction costs on some of the logging railroads are approximately as shown in Table XXXIX (these are offered as exemplary of average conditions; there is wide local variation).

TABLE XXXIX

Items	*Percentage of Total Construction Cost*
Location and survey	4
Right-of-way clearance	5
Roadbed construction (grading)	14
Laying ties and rail	7
Ballasting and surfacing	10
Picking up ties and rail	7
Cost of ties and steel	53
Total	100

The charge for the last item is the greatest on the West Coast because of the heavy logs carried and the fact that heavy skidders and loaders are often transported and rest on the rails, thus requiring the largest sizes of ties and steel found in American logging practice.

EQUIPMENT, OPERATION, AND MAINTENANCE

The equipment and major materials, including locomotives, logging cars, and steel rails, generally depend on:

1. Gage of track—usually standard.
2. Length of haul to sawmill or destination.
3. Maximum adverse grades and sharpest curves.
4. Weight of loads or number of cars of stated capacity or logs to be hauled (or number of empties to be hauled to the woods).
5. Kind of fuel used.
6. Weight of rail for type of locomotives and weight of loads to be hauled.

FIGURE 149. Long lengths are customary in logging Douglas fir. There are generally three types of logging cars: (1) flat cars, (2) skeleton cars and (3) disconnected trucks. The last are used largely for loading logs 50′ or more in length. This illustrates the last type.

Early locomotives with eight wheels weighed only about 12 tons. Modern logging engines with 12 wheels weigh as much as 100 tons or more. Steel rails weighing 12 lb. per yard were used in the early types of logging railroads as contrasted with 60- and 80-lb. rails and even greater weights on modern logging railroads.

Steel rails vary in weight generally from 40 to 70 lb. On main lines 60- to 70-lb. rails are required for the heavy timber of the Pacific Coast, but 40- to 70-lb. rails are adequate for spur tracks. In the southeastern region, as well as in cypress and hardwood timber, 30- to 60-lb. rails are used, with 25- to 40-lb. rails on some of the spur tracks. Frequently rails are rerolled stock or second-hand equipment from standard railroads. Where heavy skidding machinery is mounted on cars, heavy rails are necessary on spur tracks in logging the heavy timber of the West Coast.

Logging cars are (1) the usual flat-car type used by common carriers in the East and South; (2) various skeleton-type cars with steel bunks, common throughout the country; and (3) disconnected trucks. With the last type, the ends of the logs are placed on disconnected trucks, and the loads are hooked together to make up the logging train.

FIGURE 150. Skeleton logging car loaded with more than 18 M.b.f. as scaled in three large 32′ logs, each of which is 57″ to 72″ in top diameter. In rear is shown a 32′ long, 90″ in diameter at the small end, which scaled 14,320 b.f. on a single log car. These logs are unusually large even for the Douglas fir region.

With southern pine, hardwood and lightweight small logs, loads must be fastened to the car by chains, various trip devices, car stakes, and so on. On the West Coast, single logs or even two or three logs frequently comprise the entire carload. Log cars are generally 40′ to 42′ long. On the West Coast they are constructed in lengths up to 70′. Disconnected trucks carry logs up to 100′ and even up to 150′ in length.* Speeds of 20 m.p.h., when the logging car is loaded, are customary, but on good straight track this speed is sometimes exceeded. Logging cars must be strongly built because of the heavy loads, the irregular and uneven condition of the track, and the fact that logs are

* One mill specializes in long lengths and fills orders which other sawmills having insufficiently long carriages cannot manufacture.

frequently loosened from the loading tongs and drop directly on them. Furthermore, derailment is fairly frequent, and cars are subject to severe shocks and strains.

Railroad cars of 42′ length on the Pacific Coast generally carry 5000 to 8000 b.f. The number of board feet carried on disconnected trucks is frequently 10,000 to 12,000 per load. Generally five to ten flat cars are hauled together on spur lines, whereas on the main lines trains of 20 to 30 cars may be hauled. On the common-carrier lines where the grades and roadbed are generally favorable, 30- to 40-car trains are used in both the West and the South.

Locomotives used for short hauls, for the steepest grades and sharpest curves, mainly on spurs, and for switching are generally of the geared type.* They weigh 40 to 90 tons each, and the weight is almost entirely on the driving wheels. The demand for geared lecomotives has been declining for a number of years. The horsepower varies from 55 for the smallest types up to 1200 for the largest ones. The horsepower is based on the drawbar pull or tractive power rather than the engine horsepower. The cost of geared locomotives per ton is usually higher than for other types.

Oil is favored for fuel in the West and especially in and near the oil fields of Texas. Coal and wood are burned in other regions. Electric locomotives have not been widely adopted. Some gasoline locomotives are used for light service.

Cross ties are hewn from local timber or sawed at some local mill. Generally inferior or low-value species serve this purpose, and the largest number are often required for spur tracks. Round logs are sometimes used as ties on spur lines in the South in connection with southern pine, hardwood, and cypress operations. Ties generally cost \$.75 to \$1.70 apiece either sawn or hewn, delivered to the tracks. Ties are used again and again on spur tracks. They are usually placed 2′ apart, center to center, or 2640 per mile. For the heaviest rails and service 7″ × 9″ × 9′ or 7″ × 8″ × 8′ ties are used. For lighter service 6″ × 6″ × 8′, 6″ × 7″ × 8′, and 5″ × 7″ × 8′ ties and round poles 8′ or 9′ long are used. Ties generally last 3 to 10 years, depending on the species. They are usually untreated with preservatives except on some of the larger and longer railroad operations.

On a typical operation in the Pacific Northwest the logging equipment required on the logging railroad, consisting of 25 miles of stand-

* Geared locomotives are generally of the Shay, Climax, or Heisler makes. "Saddle tank" locomotives are used on spur tracks in the Northwest.

ard gage with 60-lb. rails operating on a maximum of 8% grade and maximum curves of 32° with fuel oil, was as follows: 1 rod locomotive chiefly used for the main haul, 5 gear locomotives used principally on the spur lines, 72 sets of disconected trucks, 4 flat cars, 2 locomotive cranes, 1 steam shovel, 2 speeders, and 2 gasoline shovels.

On large forest railways, the operating division may be grouped under the following departments and officials: (1) bridge building, which has to do with the repair, maintenance, and building of bridges, trestles, culverts, and such; (2) the roadmaster, who has the responsibility of keeping the tracks in proper condition (roads are subject to constant sinking and creeping, causing derailments and time losses); (3) the mechanical department, which has to do with the maintenance, repair, purchase, storage, and inspection of equipment; (4) the trainmaster, who supervises the work of the train crews; (5) the dispatcher, who controls the operation of trains; (6) the car service department, which supplies logging cars as needed.

On most of the operations, however, the responsibilities may be combined into two or three departments.

TRACK-LAYING AND LIFTING MACHINES

In practically all forest regions with railroads, track-laying machines are sometimes used to lay ties and steel rails in building logging railroads either of the standard- or narrow-gage type and also to pick up steel rails and ties from roadbeds, especially on spur lines which are commonly changed every 6 months to a year. This machine is loaded on an ordinary flat car and attached to it with a U-bolt at each corner. The ties and rails are lifted and passed through the machine in either direction on a horizontal boom extending forward and to the rear by means of a two-drum unit driven by steam from the locomotive in attendance or with a gasoline motor mounted on the machine itself. All heavy lifting and carrying are done by the machine itself, thus relieving the crew of fatiguing manual labor that slows up this kind of heavy work and increases the labor turnover.

A full crew of 13 to 15 men is generally used. A crew for laying track under usual circumstances consists of:

1 foreman	4 spikers
1 engine operator	2 nippers
1 rail man on the car	2 bolters
1 tie man on the car	1 spike peddler

The loading capacity of these track-laying machines is 4000 lb. at the extreme end of the boom. The normal hoisting speed is 100 f.p.m. and the traversing speed along the boom 300 f.p.m. There is sufficient space within the machine to carry one hundred and fifty 60-lb. rails and ample lifting room to pass a bundle of 15 ties to or from the car at the rear. With two cars, one for the machine and rails and the other for ties, material can be carried at least ¼ mile of track. Thus,

FIGURE 151. A track-laying machine operating in ponderosa pine timber in Oregon.

the machine is used for laying steel and ties as well as for taking up these materials.

There are generally three types of track-laying and lifting machines. They are widely employed in the flat level Southeast, where ties on spur lines are laid directly on the ground with little if any preliminary grading or ballast. They have been especially used in Alabama, Florida, North Carolina, Louisiana, and Texas, and to a lesser extent in Minnesota, Wisconsin, Oregon, Washington, and California.

FIRE-FIGHTING APPARATUS

A fire hazard exists in the operation of forest railroads, especially in regions such as the Pacific Coast and Northern Rocky Mountains.

Many companies retain tank cars full of water in constant readiness at strategic points with pumps and up to 500′ or more of 1½″ or 2″ hose. All locomotive stacks, especially those burning wood or coal, should be screened to prevent sparks igniting the slash. Flues should be sanded only at specified safe places. Locomotives frequently have 50′ to 100′ of hose piped to the siphon for fires along the right of way.

Tracks are frequently sprayed with water to prevent sparks from igniting inflammable material.

When the humidity falls below a certain fixed minimum (30% in the Northwest) logging operations are sometimes shut down in summer. During dry seasons, loggers frequently work from 4:00 to 11:00 a.m. or until the humidity reaches the danger point.

ADVANTAGES

The following is a summary of advantages of forest railroads for the transportation of saw logs and other forest products:

1. Railroads provide a medium for the continuous and assured flow of a large volume of logs daily to sawmills. The log supply is generally delivered irrespective and independent of climatic conditions and at a favorable cost in comparison to other methods of major transportation. This constitutes the outstanding advantage of logging railroads and the reason for their installation in practically all sections of the country. The log supply is not subject to conditions beyond the control of the operator. No logs are left in the woods from year to year as with other forms of transportation, such as winter-sled haul and stream driving.

2. All species of logs can be utilized. With other forms of transportation, some logs may be left in the woods on account of nonfloatability, or because decay has set in as a result of delay in moving the logs during the warm season. Railroad logging is especially valuable for hardwoods, since many hardwoods will not float, and many species are susceptible to decay when left on landings or log decks awaiting transportation.

3. Special orders for lumber and timber can be cut, hauled, manufactured, and placed on the market within a minimum length of time. Orders have been accepted, timber logged and sawn, and the lumber delivered within a week or even less in extreme cases.

4. Products not requiring further manufacture or preparation such as pulpwood, hemlock bark, cross ties, posts, poles, fuel wood, and acid wood may be loaded on standard freight cars in the woods and hauled

directly to destination, thus eliminating rehandling or reloading charges incident to other forms of major transportation.

5. Long, heavy, and large logs can be more successfully and economically transported by logging railroads when the timber is available in sufficient volume. Motor trucks cannot economically haul the biggest logs of redwood, sugar pine, ponderosa pine, Douglas fir, Sitka spruce, and western hemlock, in large quantities daily.

6. Railroads can be used in connection with the existing roads of common carriers offsetting the necessity of building a transport medium the entire distance to the sawmill. Lumber manufacturing plants may be located 50 to 150 miles or more distant, and, if existing means of transportation can be used, it may materially reduce the cost of log transportation.

7. The losses in transportation are small or negligible. Losses or damage may be considerable with stream driving, fluming, chuting, and the like. Relatively few logs are lost en route on logging railroads except over especially rough roads on some of the more poorly constructed and ballasted spur lines.

8. Clean logs, free of grit and dirt, abrasions, and other defects are generally delivered to the sawmill unless they have been skidded under unfavorable circumstances prior to the railroad haul.

9. Labor and materials for camp construction and operation, as well as food supplies and logging equipment, can be quickly and inexpensively transported from the source of supply to the woods operation. This is an important factor on some operations. Stream driving, rafting, chutes, flumes, and such are only one-way transportation systems. However, motor trucks have this same advantage as railroads.

DISADVANTAGES

Among the disadvantage of forest railroads may be mentioned:

1. Railroads require a relatively large initial investment which normally requires a long period of time to liquidate. Logging railroads require capital investments of $50,000 to $1,000,000 or more, depending on the length of the railroad, character of construction, type of equipment, and similar factors. However, although a large capital outlay is necessary, there may be a very low operating charge owing to the large volume of logs hauled over the railroad. Where conditions are favorable, logging railroads may be the very cheapest kind of transportation. Generally, motor-truck installation and operating

costs are lower than railroads and, therefore, have been widely introduced and successfully used.

2. In some regions such as California and the Southwest a wet bottom may prevent railroad operation during wet periods in the spring. Deep snows may also prohibit it during the winter especially in the higher Sierras of California and in portions of Idaho and the Lake States. In the southern bottomland hardwoods and in the cypress forests, floods lasting several weeks may interrupt or interfere with railroad operations.

COMPETITION WITH OTHER METHODS

During recent years, motor trucks have supplanted railroads, particularly spur lines, to a considerable extent in the South, the Lake States, and the East and, to a large extent, in the West. Trucks are used largely, however, in an auxiliary capacity to "feed" both main and spur lines of railroads. Tractors are also used to eliminate spur-line construction partially or entirely. The lower operating costs of trucks and tractors have had an important influence in this respect in some of the prominent logging regions of the country. In addition to those regions previously mentioned, this has been true of the Appalachian hardwood region, the Northern Rocky Mountains, and the southwest ponderosa pine stands. Railroads formerly had little competition from other methods of log transportation, especially in heavy virgin-timber stands of the Pacific Coast States. When the virgin timber has been removed on the extensive tracts, the steel rails and cross ties are generally taken up and the road abandoned. Over 26 large sawmills and railroad-logging operations were abandoned within a 5-year period in one section of the southern pine region. The total cost of removing logs by truck, tractor, or railroad can be carefully estimated. The cost of overhead charges and interest for logging railroads is considerable and may be much lower for tractors and trucks. However, with a large volume of logs to be hauled, the charge per thousand board feet may be so low as to justify the operation of railroads rather than the adoption of competing methods. On large operations, it would be exceedingly difficult to supply enough logs exclusively by motor truck. There is a general tendency to retain main railroad lines but to replace the spur lines with tractors and trucks in many sections of the country. On long-established railroad operations involving a large investment in railroads already constructed, it may be inadvisable to replace the tracks with motor trucks as a medium of transportation.

PART VI
MAJOR WATER TRANSPORTATION

CHAPTER XVIII
General Considerations

The utilization of natural waterways constitutes the oldest and generally the least expensive form of major log transportation. The streams and rivers in nearly all forested districts throughout the civilized world serve to bring logs from the woods to the mill or market. In parts of Europe, notably in Sweden, Finland, Austria, Germany, and Switzerland, the streams have been used for many hundred years, in fact, since the very beginning of the utilization of those forests. On many of these streams, permanent improvements have been constructed of concrete, masonry and piling—crib piers, dams, and the like—that are used year after year. Forests managed under a permanent form of sustained yield may produce saw logs, cross ties, poles, posts, and pulpwood annually. The unit cost of log transportation is greatly minimized where stream improvements are maintained on a permanent basis.

Log transportation by water is generally the cheapest form when available and practicable. Since the earliest colonial days in the United States it has been widely practiced throughout the East and Northeast. It has been and is customary throughout the South, particularly along the streams, bayous, and estuaries of the South Atlantic and Gulf Coasts. Water transportation is probably more extensively used in the Pacific Northwest and eastern Canada than anywhere else, owing to the frequency and availability of streams and bays adjacent to the important timber districts, notably Puget Sound, the Columbia River, the Willamette River, Grays Harbor, Willapa Harbor, and Coos Bay in the Northwest, and the St. Lawrence River and its many tributaries in eastern Canada. Water is almost the only highway of commerce along the forested Alaskan Coast and is extensively utilized in logging.

For the purpose of analyzing and describing the subject, water transportation may be divided into the following:

1. Floating and driving.
2. River and lake rafts and booms.

3. Ocean rafts.
4. Flumes.
5. Barges and steamers.

Some mistakes of judgment have been made in attempting to use streams unadapted for log transportation. They have resulted in failures and great expense to the lumber companies involved. Attempts to drive combined hardwoods and softwoods in the Menominee Indian Reservation near Neopit, Wis., resulted in considerable loss of logs since most of the hardwoods did not float. Experienced Michigan stream drivers were sent to California to drive the sugar pine and ponderosa pine logs down the narrow, steep, and rocky canyons of the south fork of the American River in the Sierra Mountains. This turned out very disastrously. The company which attempted to drive Douglas fir down the Wind River in southern Washington to the Columbia River suffered very serious losses because the logs were too large and too long and there was insufficient water to carry them around and over the obstructions in the narrow rocky canyon.

FLOATABLE SPECIES AND RELATIVE FLOATABILITY

Fortunately a large proportion of our lumber, pulp, and other products is manufactured from species that float readily. Practically all softwoods float successfully. Floatability is essentially a matter of specific gravity. It also involves the hygroscopic and absorptive qualities of the different species. If logs are left in water for long periods of time, many of our leading species which ordinarily float may become "waterlogged" and sink. Some species continue to float better than others; others waterlog more quickly. The buoyancy of wood depends on the amount and character of cell cavities in the wood. It is the complete filling of these cavities by water that causes the waterlogged condition. In all streams which have been regularly driven for years, the stream bottoms contain many sunken logs which may become a menace to navigation or to logs floating in the same stream. Even white pine and eastern spruce, which possess excellent floating qualities, become saturated and may sink after prolonged periods in water.

Lumber from logs floated in water is said to season better. There is no present scientific explanation for it; no doubt many physical and chemical changes take place in wood that has been left for considerable periods of time in water. These may have an important bearing on the seasoning qualities of lumber or other forest products. Logs left in the water for a considerable length of time are likely to check readily when removed, although the floating of logs may itself prevent checking. Timbers that have been floated in water are often preferred in the export trade. Peeled logs, after floating, often check on the upper surfaces when exposed to the sun and wind for extended periods of time.

Hardwoods, on account of their greater weight, generally float with difficulty. Softwoods are generally light in weight and float readily.

FIGURE 152. Breaking up a log jam with dynamite. This is on one of the many rivers of the Adirondacks in New York. Several streams in the Northeast have been regularly driven for the past 100 to 300 years and will continue to be used for stream driving for many years to come. Along some of these streams, virgin timber was removed more than 200 years ago. The regrowth, however, has kept these industries alive and active.

Considerable water is retained in the heartwood of hardwoods, whereas conifers or softwoods contain relatively little moisture.

The species occurring along our streams during the early periods of settlement in the United States were fortunately of floatable softwoods such as white pine and later hemlock and spruce in the Northeast and Lake States and southern pine in the South and Southeast.

Many of the leading hardwoods such as oak, birch, and maple are of high density and do not float. The lighter-weight hardwoods such as yellow poplar, basswood, and cucumber float readily.

Western larch and redwood butts float poorly. Western larch is generally long butted, and the butt logs are left in the woods because of their shaky condition rather than their poor floating qualities.

Pitchy butt logs of longleaf yellow pine float poorly. Some species are barked or left to season partially before driving. There is danger, however, from severe checking if they are left too long in the woods, especially in summer, before floating. Southern cypress is sometimes girdled * before being felled to reduce the water content and increase floatability. Hardwoods are sometimes subject to serious fungus attack if they are left in the woods too long during the warm seasons, awaiting rafting in connection with lighter softwoods.

Table XL is based on average floating conditions. Many of the hardwoods listed are seldom if ever rafted or floated. Douglas fir, western hemlock, western white pine, southern pine, and eastern spruce

TABLE XL

FLOATABLE SPECIES AND RELATIVE FLOATABILITY

Species of Good Floatability	*Species of Medium Floatability*	*Species of Poor Floatability*
Softwoods	*Softwoods*	*Softwoods*
White pine	Southern yellow pine	Redwood (butt logs)
Western white pine	Douglas fir	Western larch (butts)
Sugar pine	Western hemlock	Eastern larch (tamarack)
Lodgepole pine	Western larch	
Ponderosa pine	Eastern hemlock (green)	*Hardwoods*
Norway pine		Oak (all species)
Spruce (all species)		Birch (all species)
Western red cedar	*Hardwoods*	Beech
Port Orford cedar	Red gum	Elm (all species)
Northern white cedar	Chestnut	Ash (all species)
Southern white cedar	Sycamore	Cherry
Incense cedar	Cottonwood	Hickory (all species)
True firs (all species)		Black walnut
Cypress (after girdling)		Hackberry
Eastern hemlock (seasoned)		Black gum
		Hard maple
		Soft maple
Hardwoods		Holly
Yellow poplar		Black locust
Cucumber		
Basswood		
Aspen		
Butternut		
Willow		

* When cypress logs are floated to destination, trees are girdled before logging. Cypress is largely logged by power skidders and railroad haul which eliminate the necessity of girdling. However, many companies prefer to use logs from girdled trees, since they float better in log ponds.

probably comprise 90% of all species rafted or floated in the United States. Probably 80% of all pulpwood in eastern Canada is stream-driven.

Studies made of relative floatability of freshly cut logs frequently rafted along the South Atlantic Seaboard indicate the following, the first-named exhibiting the poorest floatability and the last-named the best: (1) beech; (2) oak (water and post); (3) red gum; (4) black and tupelo gum; (5) heavy southern pine (butt logs and those of slow growth); (6) soft maple; (7) water ash; (8) cypress (not girdled); (9) southern pine (rapid growth and top logs); (10) yellow poplar.

Hardwoods in Nova Scotia are driven for several weeks in the spring with a sinkage loss of only about 10% of the total volume. This is accomplished by felling the trees in midsummer and leaving the tops and branches on the stems until late fall. Then they are topped, lopped, bucked, and skidded to landings and hauled during winter to the driving streams. Gradual evaporation through the leaves after felling reduces the moisture content of the logs sufficiently to float most of them to destination.

In Finland where rafting is commonly practiced, the sinkage of the valuable birch is 5% compared with 3% for the cheaper softwoods. Birch requires the absorption of little moisture in transit to cause sinking. Drying to reduce the moisture content is done by felling the trees in late summer and leaving branches on the trunks. Evaporation through the leaves consequently reduces the moisture from 80% to 50% in 2 weeks. Furthermore, to prevent reabsorption of moisture, the ends and side knots are treated with bitumen to insure longer floatage. Blue staining of logs while in storage is prevented by storage in water and by a moistening method to prevent the wood reaching a degree of dryness favorable to the attack and development of stain fungi.

CHAPTER XIX

Floating and Driving

Stream driving as applied to logging may be defined as the process of transporting logs by floating them in loose aggregations in water with the motive power supplied by the natural or flushed stream flow. Floating logs loosely downstream is frequently supplemented by raising the level of the streams from reservoirs of water stored in splash dams. If logs would float unassisted by flushing from splash dams, or by a follow-up crew at the rear of the drive, the operation would be exceedingly low in cost. Floating is possible on the large streams, especially in eastern Canada and in Maine, where logs and pulpwood are frequently loosely driven down the larger rivers. A combination of floating and driving is necessary on the smaller streams.

The logs in the North are usually hauled from the woods to the stream bank during winter and yarded in piles up to 50′ in height to wait for the "break up" of the ice in the spring. When the spring freshets nearly reach their peak, the log piles along the banks of the streams are broken down and the logs, cross ties, pulpwood, or other material rolled into the water. These logs, cross ties, and so on are generally banked along the streams above or below a splash dam or on the ice in a lake during the wintertime, or wherever they may gain the benefit of rising water in the spring. Thus the drive proceeds in a series of steps. Various types of dams, rollways, banking grounds, booms, and sluiceways are employed to put the drive through in the shortest length of time with the least expense.

Stream driving is substantially the same in practice as it has been for the past 100 years, but much new technique is being developed on individual operations. Eastern Canada is the principal location for floating and driving activities, which are devoted almost exclusively to short and long spruce and balsam fir pulpwood, that is, logs cut either in 4′ or in 12′ and 16′ lengths.

Stream driving was once widely practiced throughout the Northeast, the Southern Appalachian Mountains, and the Lake States. It has recently been limited very largely to the following regions:

1. Eastern Canada and the Northeast, including New York and New England. It is the principal method in eastern Canada.

2. Southern and western Wyoming for the large volume of lodgepole pine ties and a few other species (Douglas fir and Engelmann spruce) that were driven to points along the Union Pacific Railroad system. Most of the ties used on this system from Omaha west to Idaho are supplied by timber from southern and western Wyoming,

FIGURE 153. Western white pine and western red cedar logs banked along stream in northern Idaho. Driving streams with relatively straight channels and fairly steep banks are preferred to those with low banks, winding channels, and frequent sloughs, swamps, and meadows where the logs may be stranded or lost during the spring drive.

northern Utah, and Colorado. These lodgepole pine, Douglas fir, and spruce ties were taken from the streams to the treating plants at Laramie, Wyo., and Pocatello, Idaho. Recently much of this driving has been replaced by trucks taking sawn ties from small mills to the railroad.

3. Northern Idaho, where western white pine and the other species associated with it are driven to the mills. The availability of good drivable streams, swift currents, floatable species, and favorable precipitation make this region especially suitable for stream driving.

CONDITIONS NECESSARY OR ADVISABLE FOR FLOATING AND DRIVING

Briefly summarized, the conditions which should be present for successful floating and stream driving are:

1. **An ample and reliable water flow,** preferably throughout the season, but at least a sufficient seasonal supply in the spring. Advantage is generally taken of spring freshets and heavy rains which add to the water level, and also of the additional supply from the melting of the winter snows. Many sawmill and pulpwood operations are based on the expectation of the company's receiving its yearly supply during the spring and early summer drives.

2. **A stream sufficiently wide to take the longest logs.** Generally only short logs can be successfully floated and driven on the smaller

FIGURE 154. Part of a 20,000-cord spruce and balsam fir pulpwood drive in the Adirondack Mountains of New York.

streams. Frequently small streams are materially widened during the process of driving, as a result of the frequent flushing of the stream from splash dams and the impact of the logs along the stream bank. Some streams that have scarcely any water in them during the late summer are driven during the spring. Cross ties and 4′ pulpwood are the ideal sizes for stream driving.

3. **Steep high banks along the watercourse.** Sloughs, marshes, and low banks subject to flooding or overflow are not desirable because of the possibility of loss of logs. Logs, pulpwood, cross ties, and such can be more readily banked and rolled into the stream from steep regular banks than from flat low places along the streams.

4. **A sufficient volume of floatable species.** Eastern Canada and, to a less extent, the northeastern part of the United States are exceedingly fortunate in having an abundance of spruce, which is a very lightweight drivable species, and also a supply of available streams and abundant rainfall.

5. **Swift currents with few obstructions and curves to cause log jams.**

ADVANTAGES

Following are the principal advantages of log transportation by floating and driving:

1. They are the cheapest forms of log transportation, provided favorable conditions of water flow, precipitation, and floatable species are present.

2. The initial investment and operation costs are relatively low. Unskilled labor may be employed, although because of long hours and arduous work high wages are generally paid.

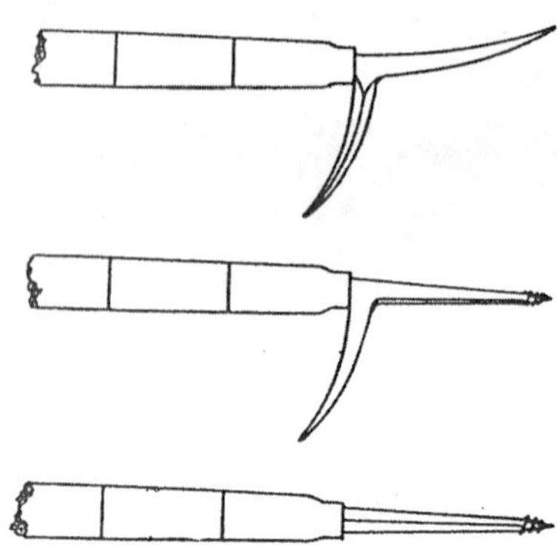

FIGURE 155. Three types of pike poles. Top to bottom: pike pole with curved pike and hook, with straight pike and hook, with pike only. They are commonly used in rafting and log driving as well as in moving logs in the storage pond at sawmills to the foot of the jack ladder. Handles are generally made of second-growth white ash or spruce and vary in length from 4½′ to 20′.

3. The time consumed to bring in the year's supply is relatively short. This concentration of time and effect may be very advantageous for certain financial and management policies as well as in the employment of labor. (Labor employed on the drive is generally occupied the remainder of the year in woods operations or at the sawmills.)

4. Logs stored and transported in water are much less susceptible to end checking than when transported by some other methods. Logs left on skidways too long in the summer, fall, and winter, however, awaiting the spring drive may affect this theoretical advantage.

5. These methods are especially adapted to remote, inaccessible, and primitive conditions and require little equipment of an expensive nature.

DISADVANTAGES

Although floating and driving are common methods of log transportation in the regions described, they involve certain hazards which must be carefully reviewed in deciding on transportation policies. Among the disadvantages are:

1. Losses due to sinkage, strayage along the banks and in low places, and breakage in jams and on obstructions during unfavorable seasons

may be considerable and may vary from 2% to 26% or more of the total volume.

2. These methods are adapted only to floatable species and to relatively small short logs.

3. Investment in stumpage, logging costs, and the like is large and can be liquidated generally only once a year on arrival of the annual spring drive.

4. Driving costs vary widely and unpredictably, depending on conditions beyond the control of the logging company. Unusually light snowfall during the winter, cloudbursts, and floods, or unevenly distributed rainfall may have an important bearing on costs. Floods have caused boom breakage or loss of other protective improvements resulting in partial or complete loss of the drive or unusually difficult and expensive log jams.

FIGURE 156. Attempting to break up the advance part of a log jam on the Moose River at Lyonsdale in the Adirondacks. This stream has been driven for more than 125 years and will continue to be driven indefinitely. The timber is principally spruce, with some balsam fir, hemlock, and white pine.

5. Grade depreciation may result in saw logs of some species owing to blue stain and loss in overrun as a result of heavier edging and trimming of driven logs. In a study in northern Idaho there was a difference of $6.05 per thousand board feet in favor of Idaho white pine logs delivered by railroad, compared with those driven by stream to the sawmill.

6. Stream-driven logs accumulate sand, grit, and dirt which may be very harmful to band saws or pulpwood grinders or chippers.

7. Miscellaneous disadvantages might be mentioned, such as (*a*) danger of fire destroying logs banked along streams; (*b*) necessity of building tote roads for hauling in men and supplies which could be avoided on a motor-truck or railroad haul; (*c*) the leaving of valuable logs in the woods as the result of failure to deliver logs to the stream in time for an early spring "breakup" or freshet.

FIGURE 157. Mine props and saw logs banked along stream in a lodgepole pine forest of Wyoming. Sometimes, when advantage is not taken of high water from spring rains and melting snows, logs, cross ties, poles, and mine props are left until the following season. One of the hazards of stream driving is the possibility of not being able to get out all the products logged during the season.

SPLASH DAMS

At or near the outlet of lakes and on the headwaters of streams used for driving, one to several dams are often constructed to furnish a reservoir of water. Concurrently with the melting of the winter accumulation of snow and the spring rains, these dams are opened at their gates to flush the streams below. On this elevated and swifter current of water, the logs, bolts, and so on are floated to the mill or other means of transport. These dams may be filled and opened several times during the driving season.

Splash dams are of several types of construction. Local materials such as logs, rock, and earth are the chief sources of construction. Experienced local woodsmen rather than trained engineers are employed to design, construct, and operate them. If they are for one or two seasons' use, they are of very simple, light, and temporary construction. In Europe they are frequently built of masonry and concrete * and designed as permanent features of the woods operations. On several western National Forests in the Rocky Mountains, as well as in parts

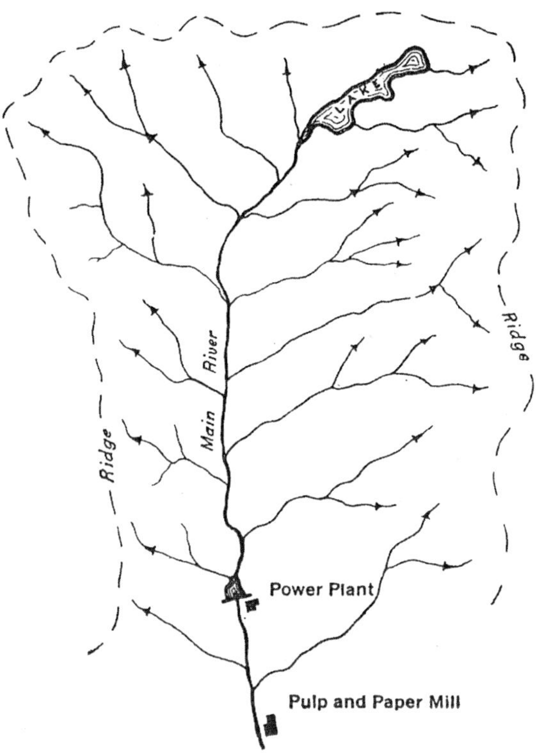

FIGURE 158. Diagrammatic sketch illustrating a system of splash dams at the headwaters of tributary streams used in flushing or driving pulpwood approximately 80 miles along a main river from a lake to a pulp and paper mill. There are approximately 800 square miles in this drainage unit. The main river and the lower reaches of the principal tributary streams do not require splash dams because of sufficient water flowage available. Not drawn to scale.

of Idaho, eastern Canada, and the Adirondacks, semipermanent dams are built.

The usual types of splash dams are:

1. The crib-type dam. This is the most expensive and permanent form used on driving operations. It is built of round logs and filled

* A splash dam constructed of concrete at a cost of $200,000 was built in West Virginia.

with rock, stones, and local earth, preferably clay. It is faced on the upper side with plank sheathing or occasionally with round logs, the seams being carefully calked. This type of construction takes more wood than other types, and the timbers need more adjustment. Large long spikes are required, which are not so necessary in the other types.

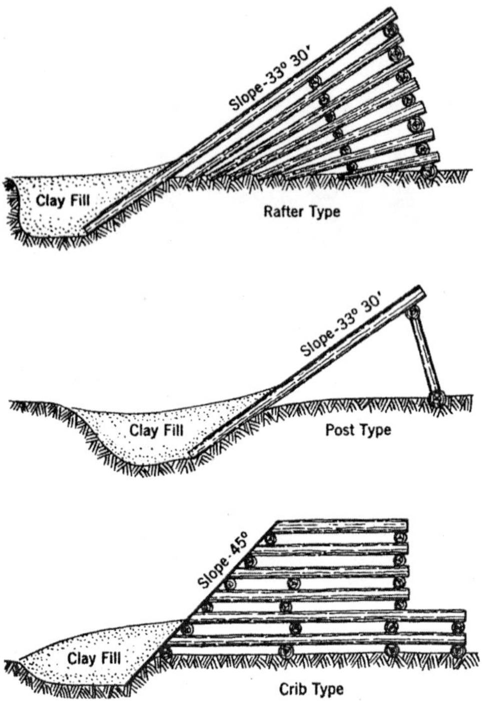

FIGURE 159. Cross sections illustrating three usual types of splash dams used in the Northeast and eastern Canada. The crib type is generally most permanent in its construction and also most expensive. The least expensive type is the post dam which is generally constructed for only 1 year's operation.

The crib dam will last longer and resist the frost better than others. The upstream face is generally placed at an angle of about 45° and may be perpendicularly covered with 2″ planks if a local lumber sawmill is available, or, in Canada and the Northeast, with round spruce, cedar, or hardwood logs. The plank sheathing (or logs) is generally inserted into a rock or hardpan bottom and then covered with dirt, preferably clay, for one fourth or more of the front of the dam.

Crib dams generally last 4 to 16 years, the average being about 8 years. They may vary 6′ to 18′ or more in height and cost about twice as much as the other types. Crib-dam gates generally vary from 10′ to 24′ in width, depending on local conditions such as volume of flow, head of water, and whether saw logs, pulpwood, or other timber are to be driven through the dam. A study of six different crib dams in the Northeast reveals some of the principal factors involved in the construction and operation of typical splash dams (Table XLI).

TABLE XLI

COST AND TIME OF FILLING AND EMPTYING CRIB-TYPE SPLASH DAMS OF VARIOUS SIZES

Dam	*Cost*	*Length, feet*	*Width, feet*	*Width of Gate, feet*	*Maximum Head, feet*	*Time for Filling, days*	*Time to Empty, hours*
1	$7000	210	40	24	14	8	36
2	7000	225	25	22	14	15	40
3	3500	100	20	20	8	6	30
4	2500	90	20	20	10	4	36
5	1000	100	20	14	9	30	48
6	650	125	20	10	6	30	48

2. The rafter-type dam. This type shown in Figure 160 is used for smaller dams for more temporary purposes, and especially for 1- or 2-year driving operations. Not so much timber or labor is required in this type. Spiking is not necessary. The face of the dam is at a smaller angle than in the crib dam, the weight of the water itself being used to hold the dam in place. Thus it is frequently referred to as a self-loading as well as a rafter type of dam. Dams of low head do not require earth or rockfill in the body of the dam. This type can be constructed for approximately one-half the cost of the crib dam of the same size (width, height, and length).

3. Post-type dam. This is probably the cheapest of all, the sheathing or face acting as the principal part of the dam which is supported at the top and bottom and occasionally between. One of these types is illustrated in the accompanying diagram. Red cedar logs, generally 8″ at the small end, are used as sheathing in eastern Canada. When properly built, these dams serve the purpose, especially on poor foundations, as they are likely to settle considerably. No spikes or drift bolts are required in the wings. Where several years of service are required, this type of dam is not recommended, as there is considerable leakage, and the timbers in the sheathing tend to fail after extended use.

4. Miscellaneous types. Occasionally pile, earthen, timber, and combinations of several types of splash dams are constructed.

The kind of dam to be installed depends on the materials available; the size of the dam, including height, width, and length; acreage feet to be impounded; number of years the dam is to be continued in service; and other local conditions.

FIGURE 160. Downstream view of rafter or self-loading dam used in driving pulpwood in eastern Canada. The height from the brook bottom to top of gate was 21′, from the sluicing platform to top of gate 11′; the gate is 6′ wide, and the length of the top of the dam is 140′ from bank to bank. The dam required 550 man-days and 65 horse-days to construct. Local materials were used almost entirely, the only cost being 10 barrels of 10″ and 12″ nails and 1 barrel of 5″ nails. Calking is generally done with rags and old bags. The drainage area above the dam is approximately 6000 acres, and the acreage of the reservoir when full about 3 acres. Given an 8-minute flood with the gate half open, 2500 logs 10′ long can be sluiced in 15 to 18 minutes. There were 3000 cords of pulpwood landed in the stream bed below the dam. Another dam was located at a distance 2¾ miles above this one to assist in flushing out the logs. The distance from the dam shown in picture to main river was about 30 miles. *Photo by V. E. Johnson.*

The calking on all dams is done with the best local materials available. In eastern Canada, sphagnum moss, old bags, rags, and oakum are used. Sometimes the lower portion of the dam is faced, and occasionally the entire front is covered with earth, preferably clay. When the dam is constructed of planks, seams are often covered with battens; when it is constructed of round logs, the calking presents a much more difficult problem.

Prior to construction, the front of the dam site should be well grubbed out of all stumps, stones, and vegetative material. These may cover a fissure which in turn may cause a serious leak. Grubbing is generally limited, however, to the lower part of the face of the dam so that planks or poles may be inserted for some distance, generally 2 to 10 ft. or more in a diagonal direction. The excavated material is thrown inside or below the dam.

Framing is constructed of local timbers, generally not less than 8″ at the top. Peeled timbers last longer. Drift bolts when used are ½″ or ⅝″ square and 10″ or more in length. Every piece of timber crossing another should be drift-bolted to it.

Gates are generally of the lift type operated by hand. Vertically placed planks are lifted out individually. They are preferably located in the center of the dam where maximum depth and rate of water flow are found. Gates must be sufficiently wide to permit a rapid outpouring of the water as well as of logs in case the logs are driven through the gates. "Barn door" gates similar to those at canal locks are occasionally used. The greater the pressure of water against the gate and the smaller the water supply, the tighter the gates must be constructed. The frequency with which dams must be filled and opened depends on the volume of water stored; the nature and size of the material to be floated; the height, depth, and gradient of the driving stream; and the volume of logs, pulpwood, and other timber to be driven.

Splash dams are generally constructed during the summer or autumn preceding the drive or repaired during the summer if used perennially.

Below the gate an apron or sluiceway is located to conduct the water properly downstream. This should be made long enough so that any undermining caused by the falling water will not affect the dam adversely. Occasionally a floating platform is attached to the lower end of the apron by chains so as to receive the water and to prevent undue undermining.

In crib- and rafter-type dams 10 to 13 lin. ft. of timber per cubic yard of crib is necessary for the framework.

Some of these splash dams have been in constant use for 50 years or more. When the dam is filled with water from May until January, the timbers will remain sound. Greater permanence is assured by peeling the logs and by tarring or creosoting the timbers or planks in the sheathing and gate sections.

The following considerations should be carefully reviewed in selecting the site of splash dams:

1. The dam should be so placed that it will drain all the stored water. The gate sills should be located below the normal low water level of the lake or stream to be impounded. The dam must always be located

downstream from the headwater and preferably below the outlet of a lake.

2. The foundations are of great importance and are directly related to the cost, permanence, and efficiency of the dam. Rock outcrops are preferred but are not necessary, especially for temporary types of dams.

3. The narrowest site may not always be the best location, because there must be sufficient space to pass the freshet water. However, a narrow high dam is preferred to a wide low dam and can be constructed at a much lower cost.

In figuring flow data, it is assumed that about 50% of the precipitation is absorbed by evaporation, and 40% if retained in the watershed will flow naturally down the stream.

A splash dam in order to remain in place requires resistance to three ever-present conditions:

1. Sliding.
2. Overturning.
3. Crushing.

It is most important that water be prevented from getting underneath the dam. This will undermine it, causing it to slide or overturn.

The danger of crushing dams is not considered important because the height of splash dams is not generally great.

The time required for filling splash dams generally varies from 4 to 40 days, depending on the size of the storage and drainage basin, precipitation, and amount of snow stored through the winter. The time required to empty splash dams varies from 30 to 50 hours, depending on the head of water, width of gate, and acre feet contained in the reservoir.

A dam known locally as a rafter and strut-framed type faced with planking and earth was constructed by a lumber company in the Rocky Mountains for the purpose of driving approximately 150,000 cross ties per year down the creek to a river. The creek carried insufficient normal flowage to float cross ties. No flushing was required on the main river itself, although the flushing of the lateral creek was of great assistance in transporting the logs to destination.

The dam was 340′ in length, 18′ in height at the center, and 20′ in thickness at the base at the center, and it tapered to 6′ thick at the edges. It required 210 man-days to construct it. The clearing of the reservoir site to highwater level required 50 man-days. The team hire on the dam construction and the clearing extended over a period of 82 days. Seventeen thousand feet of sawn lumber were used in sheathing the face of the dam. The cost of this material delivered on the dam site was $15 per thousand board feet. Incidentals such as iron, bolts, and nails used in dam construction cost $150.

The following summarizes the construction costs before World War II:

210 man-days for dam construction at $5.50 per day	$1155
50 man-days to clear reservoir site at $5.50 per day	275
82 team-days at $4 per day	328
17,000 b.f. of lumber at $15 per M.b.f.	255
Bolts, iron, nails, spikes, etc.	150
Total	$2163

If 150,000 ties were driven each year and the dam lasted 12 years (or there was sufficient timber to continue driving for that period), the annual charge-off for depreciation (2163 ÷ 12) would be $180.25. Interest on the initial investment per year at 6% would be $129.78. If it cost approximately $200 per year for upkeep, repairs, and operation, the total annual charge would be $510.03 for construction and operation of the dam, or $0.0034 per tie.

RIPARIAN AND BED IMPROVEMENTS

There is a great variation in the quantity and quality of stream improvements. The expense may vary from practically nothing per thousand board feet up to 5 cents per cross tie, 10 to 20 cents per cord, or 30 cents or more per thousand board feet. The unit charge for riparian and stream-bed improvements is lowest under conditions of European log-driving practice where permanent improvements make it possible to continue driving the streams year after year at a minimum cost. In southern Wyoming, chiefly from the Medicine Bow, the Wyoming, and the Washakie National Forests, and from the Wasatch National Forest in northern Utah, the lodgepole pine forests have been and will probably continue to be the principal source of cross ties for a portion of the Union Pacific Railroad system. Riparian and bed improvements are often on a permanent basis where the forests are managed on a sustained-yield principle, as in these National Forests.

The principal forms of stream improvements are:

1. Blocking off slough, swamps, low meadows, and banks along wider parts of the streams by means of log cribbing to keep the logs and other timber in the main stream channel. This generally represents the lowest unit cost in stream improvements. Cribbing must be erected

up to elevations above the highest probable level of the spring freshets and may extend from a few yards to ½ mile or more in length and from 4′ to 16′ in height. Local logs of durable species are used to face and reinforce the cribbing which is held in place by rockfill. Log cribwork on sharp turns serves to prevent jams and piling up of logs and cross ties. The outside turns on stream curves carry the preponderant share of the drive, on account of the swifter currents in that part of the stream. These log cribs must be more smoothly faced than the cribs described previously. Often floating booms are placed along the face of the crib dams of this type in order to shunt off the logs and prevent them from piling up in one place. Frequently a number of men are required to prevent log jams at these points.

2. Blasting out or removing boulders, large rocks, leaning trees, sunken logs or obstructions of any kind in the main bed during periods of low water. Any obstruction or accumulation of debris, such as floating trees, brush, and rocks, often causes serious and expensive log jams during the driving season. In some streams, the cost of improvements of this nature has proved to be prohibitive. In other streams practically no such expense has been necessary.

3. Placement of booms to hold up the drive until proper stream levels are reached at the lower parts of the same stream or to divert logs, cross ties, and other timber into the proper channel. Where lakes occur on a driving stream, logs are sometimes collected in booms and towed across to the outlet to prevent scattering and subsequent losses.

4. Extensive systems and combinations of piling, cribbing, or other means of supporting booms to divert or sort logs, especially on the larger rivers, as in Canada. Frequently several companies organize a subsidiary organization to operate a stream during the spring drive and to build and operate such improvements as piling, cribs, and sorting booms as may be necessary to sort and divert the product of the several companies involved.

5. Miscellaneous improvements such as timber, plank, crib, or rock aprons to conduct logs properly under bridges and avoid jams. Narrow places in a stream, islands, or rocky outcrops may require special devices to prevent logs from piling up. Special weirs and sluiceways are sometimes required to carry logs over narrow or shallow places in a stream. High overhanging banks on curves must be protected.

The improvements necessary on small streams in eastern Canada may cost from about $200 to $1000 or more per mile. The unit charge per cord depends on the number of cords of pulpwood driven. Streams driven year after year require relatively little expense for upkeep after the initial investment for improvements. Floods or unusually high stream levels, however, may destroy or seriously impair the improvements.

The cost of improvements on a typical Canadian pulpwood operation, compared with other charges for driving, are shown in Table

TABLE XLII

Operation	*Cost per Cord of 4′ Spruce Pulpwood*
Actual driving	\$0.46
Sinkage and losses in driving	0.39
River and stream improvements	0.10
Total charges	\$0.95

XLII. A small brook was driven with 6000 cords of 4′ wood, and then the bolts were boomed and towed across a lake 8 miles and finally driven or floated down a river for about 60 miles.

OPERATION

Operation of drives has continued along similar lines for many years. At first all timber within easy access of the stream was cut and loosely

FIGURE 161. A typical driving camp, which houses 35 men for 3 to 4 weeks, while they are starting the main drive of 26,000 cords during the spring on the upper Hudson River, N. Y. It generally requires about 2 months to drive this quantity approximately 100 miles. After this permanent-type camp is left, tent camps are moved to suit the progress of the drive, while the men "sweep" the logs into the main stream.

floated down the adjacent river. If the timber was too far to be profitably hauled by oxen to the mill or stream, the logger moved to another location. Then it became necessary to go greater distances for the timber. This involved the introduction and use of river landings, log yards, log driving, rafting, towing, and booming. Still later the more distant timber required the use of splash dams and sluiceways; expensive stream improvements; the use of canals, tramways, trestles, log chutes and slides, motor trucking and railroads in connection with floating and driving.

Some rivers have carried 20,000 to 200,000 cords of pulpwood or up to 500 million b.f. of saw logs at one spring drive.

Probably log jams constitute the greatest hazard and item of expense in stream driving. Jams are caused by low water or by piling up of logs in narrow places or at eddies and curves or against some obstruction. Breaking jams is exceedingly dangerous and requires sustained strength, great skill, and much experience. Generally one or more "key" logs are located and released by men using peavies and pike poles. Frequently dynamite is necessary to break apart the entangled logs, especially if ice forms to aggravate the condition. The increasing use of 4′ bolts of pulpwood in stream driving in eastern Canada, Maine, and New York in place of the 16′ lengths has reduced the number of jams during river driving; long poles and logs in Idaho greatly increase their likelihood. Cedar poles in Idaho are generally driven after and separately from the logs.

When two or more companies use the same stream, log driving or boom companies are formed, the expense being prorated among the operators in proportion to their interest. These organizations were developed to settle disputes among operators, log owners, and boom owners. Where community driving is the custom, each log is marked or stamped on the end with a symbol to indicate the ownership. Logs are marked at one or both ends so that they may be readily recognized at the sorting boom and diverted to their respective owners. Marking is usually done by a hammer with raised letters or design. Some operators use several colors of paint and apply such symbols as a circle, bar, cross, or a letter. Some companies dab the ends of the logs or pulpwood with paint to indicate that they have been inspected for grade, size, or condition.

The operation of stream driving may be separated into the following parts, each of which requires special technique:

1. Breaking down landings. This may constitute 8% to 30% of the total driving charge and may require 1 to 2 weeks or more. Logs and cross ties are decked along high banks preparatory to being rolled, generally by hand, into the stream. Cross ties and pulpwood are often landed directly into the stream bed. The cheapest form of landing is on a frozen lake or stream. Some operators specify that logs, cross ties, and long pulpwood logs shall not be barked farther than 25 ft. from the edge of the driving stream. Four-foot pulpwood logs are generally piled in one or two rows along the bank. They are thrown into the stream by hand. Occasionally dynamite must be resorted to in order to break down frozen logs. In parts of eastern Canada high landings are used, the pulpwood logs being rolled down steep slopes from 30% to 70% or more and containing up to 1,250,000 b.f. in ex-

FIGURE 162. Logs of high value justify the considerable expense of follow-up crews salvaging them in stream driving.

treme cases. Slopes are cleared and the logs rolled into the stream before the spring thaw.

2. Operation of splash dams. Generally one man opens and closes the gates. During the time the dam is filling or emptying he may be engaged on other work unless there is likelihood of jams occurring at the gate. By means of a peavy he takes out each vertical log or plank composing the gate.

3. Sweeping. This consists of rolling in and dislodging logs that have strayed along the banks, on bars, and in shoals. The rear end of the drive, sometimes referred to as the follow-up crew, spends most of its time in sweeping. On some rivers the major cost of the drive is devoted to sweeping.

Stream driving is generally practiced for distances of 10 to 110 miles in northern Idaho and frequently for similar distances in Wyoming, Colorado, and northern Utah. The longest drives are in the Northeast and in eastern Canada, where drives of 75 to 150 miles are common.

For one large company which normally produces about 1,300,000 cords, the minimum driving distance is about 75 miles and the maximum 250 miles. Probably at least 80% of all pulpwood in eastern Canada is driven, and a large share of the pulpwood operations in northern New England and New York are stream-driven.

Within recent years more and more pulpwood is being driven in 4′ lengths. Previously it was customary to use 12′ to 20′ lengths. The tendency in New England and New York is to use 4′ lengths almost

FIGURE 163. The usual type of bateau employed in following logs on a drive.

entirely because smaller amounts are lost during the drive. The controlling factors in cutting lengths for stream driving are generally the width and driving capacities of the streams.

Some of the drives in eastern Canada may have up to 500,000 cords in one section on one river. In 1934 there were 300,000 cords in one drive to the mill on the Gatineau River.

In some operations 4′ pulpwood has been driven for 40% to 60% of the cost of driving 16′ lengths. On the upper Hudson River in New York, 4′ pulpwood can be driven with 25% of the head of water required for 16′ lengths, and approximately 100% can be delivered in one season. Sixteen-foot pulpwood driving needs a considerably greater head of water, and it required 2 to 3 years to deliver all the wood to the mill. Formerly, with 16′ lengths, only 20% of the total drive arrived at the mill during the first year, 60% during the second year, and 20% during the third year. Of 1,250,000 cords driven by one large company in Canada, 300,000 cords, or slightly more than 25%, were

4′ lengths, the remainder being 10′, 12′, 14′, 16′, and occasionally 20′ lengths.

The number of men needed on the drive varies greatly with conditions such as width of stream, volume of water flow, character of stream improvements, depth of stream, length of logs, and length of drive. A 30,000-cord job requires 50 men for 60 days.

MECHANIZED PULPWOOD CONVEYOR AND SORTING SYSTEM

There are five pulp mills located along the St. Maurice River above its confluence with the St. Lawrence River in Quebec, Canada, with five separate owners using the same stream for different-length pulpwood. A new mechanized sorting unit handles an average of 7000 cords or 217,000 logs per day of 20 hours over a drive period of 62 days. In a previous year, the driving period was 74 days during which more than 15 million logs were sorted. With the two mechanical sorters, 63 men do the work formerly performed by 150 to 175 men.

These sorting units supported on pontoons have inclined log conveyors measuring 60′ long by 20′ wide. A gap between the conveyor head shift and the live rolls allows the 4′ logs to fall through, while the 8′ to 16′ lengths are carried across the gap and bounced off to their proper destination.

CHAPTER XX

River and Lake Rafts and Booms

HISTORY AND DEVELOPMENT

As sled logging and river driving marked the first development of softwood logging from the earliest colonial days in the United States, so did rafting represent the second step. Along the entire Atlantic Seaboard from Maine to Florida and along the Gulf to Texas, notably on the larger rivers, rafting has been used to bring in softwood logs to the mills. Where mixed forests of both hardwoods and softwoods were involved, the latter, during more recent years, provided the necessary buoyancy for the heavier hardwoods. In fact, the advance settlements up the rivers of the North Atlantic Seaboard were very intimately connected with the development of rafting. The several coastal streams in Maine, particularly the Penobscot and the Kennebec, as well as the Connecticut, Hudson, Delaware, Susquehanna, and the more sluggish streams affected by tidewater along the South Atlantic and Gulf Coasts were used. Frequently rafts were floated from 180 to 250 miles to mills located at tidewater or along the lower reaches of these rivers. In the North, these rafts first took only the virgin white pine and then only the butt logs of the best trees. There was steady demand for ship masts and spars up to 100′ or more in length. Then, with the gradual disappearance of white pine, spruce in northern New England and New York and hemlock and hardwoods in the Middle Atlantic States were used in these rafts. From Virginia south, rafts were composed of longleaf, loblolly, and shortleaf pine. Until about 1880, these rafts were almost exclusively made up of logs. Then with the development and increase in the use of circular sawmills came lumber, timbers, cross ties, piling, and other products from both softwoods and hardwoods. Log rafting reached its peak along the Atlantic Seaboard about 1860 to 1870. Villages sprang up at the confluence of nearly every major tributary to these principal rafting streams. The coming of the railroads and the disappearance of the virgin-timber stands caused a definite recession in the importance of rafting.

With the decline of rafting on the Atlantic Seaboard and the disappearance of virgin forests, rafting on the Allegheny, Monongahela, and other tributaries of the Ohio, as well as on the upper reaches of the Mississippi, assumed very definite importance. Sawmills sprang up along the Mississippi River, notably at Keokuk and Clinton, as well as at La Crosse and St. Louis, supplied by white pine logs rafted from the upper Wisconsin tributaries of the Mississippi River. Log rafts were also employed in Michigan and Minnesota. Rafting has almost disappeared both in the Lake States and in the Northeast but is still widely used on the slow sluggish streams as well as the bays of the Coastal Plain along the Atlantic Seaboard and Gulf of Mexico.

Rafting was one of the principal occupations of the people along these several streams in the spring following the winter logging work in the woods. The same men turned to farming in the summertime. These men led hazardous and vigorous lives, working from dawn to dark, in piloting the rafts over treacherous rifts, in avoiding sand bars, rocky ledges or rock outcrops—also in steering the rafts around sharp curves and into the deepest channels. Rafts were frequently lost during the flood periods in the many treacherous places. Literally billions of feet of virgin timber were rafted to sawmills in the Lake States and along the east and southern coasts. Probably one of the earliest rafts was built by the men of the Sullivan–Clinton expendition during the Revolutionary War to float men down the Susquehanna River from Cooperstown, N. Y., to near Elmira during the year 1779.

Rafts were generally floated singly, in pairs or several together one before the other. On the Mississippi, they were often towed, or, more frequently than otherwise, the current furnished the motive power.

These rafts contained 50,000 to 300,000 b.f., and sometimes rafts of 500,000 b.f. were constructed where the streams were broad and waters deep and there were few hazards en route. The rafts were often from 50′ to 125′ wide and 100′ to 200′ or more long. Several tied together, one behind the other, gave the appearance of a series of two to ten rafts in one tow. Those using the current were generally piloted singly or doubly. During periods of high water the longest and widest rafts were floated with the least hazards and effected about twice the distance made during low or slack periods of water. Furthermore, rafts were much more easily steered during flood periods. These rafts traveled generally 30 to 40 miles per day. During flood periods 50 to 75 miles were often made per day. Rafting was always started just before or during freshets. It began as soon as the ice broke up in the spring. River men on the North Atlantic Seaboard counted on three distinct freshets as follows: (1) The ice freshet immediately following the spring thaw and the passage of the ice floes down the river, (2) the "strawberry freshet" generally in late June or July, and (3) the "pumpkin freshet" in October. The rivers were literally full of rafts during these freshets and especially in the springtime.

FIGURE 164. Rafts are extensively used in the coastal streams and tidal estuaries of the South and Southeast, especially on the Albemarle and Pamlico Sounds and tributaries. This illustrates a typical flat raft with each log fastened to a ⅝″ chain with a ring dog so that floaters will hold up the sinkers. Rafts generally contain about 2000 logs or about 100 M.b.f. The logs are hauled to the lower end of the jack ladder by means of cable and drum, operated by power as shown in the illustration.

Rafts were largely made of 16′ logs laid parallel to each other and longitudinally with the stream and held together by semicircular wooden pegs made of ash and fitted into auger holes. Later spikes or "dogs" were used to hold the rafts together, connected by chains, ropes, or cables. Sometimes small logs or skids were laid crosswise, augmented by spikes, dogs, ropes, chains, or cables. Many of the later forms of rafts had lumber, cross ties, sawn timbers, masts, piling, posts, and so on laid on top of the logs. Some had cabins or shacks for sleeping and eating, but generally the raft was moored to a tree or rock in a quiet bend of the river near some village where the men were housed and fed.

A steersman or captain had charge of each raft. The raftsmen often worked for their passage down the river at no cost to the lumber company.

Some single rafts had one sweep at the rear end; the larger ones had two or three or even more at each end, depending on the current, size of the raft, number of bends in the river, and other hazards. There was generally one man to each sweep if he was experienced and able to operate it effectively. Otherwise two and sometimes three men served on each sweep. These sweeps were generally made of small straight hemlock, pine, or other softwood saplings 30′ or longer and about 7″ in diameter at the large end, tapering down to about 3″ or 4″ at the handle end. The sweep operated on a swivel and had a wide board or two fitted into the large end to facilitate steering.

Most of these rafts depended on the current for floating and steering purposes, but tugs or regular river steamers were used to increase the speed on many of the rivers, notably on the Ohio and the Mississippi. Frequently rafts floated down with the current and were picked up by tugs and towed to their destination at tidewater or wherever the current was insufficient to convey them to the mills. In the South Atlantic and Gulf Coasts the streams are generally sluggish and affected by tidewater; therefore, tugboats and launches are frequently required to tow rafts from the woods operations to the sawmills.

Interspersed along the rivers of New York, Maine, and eastern Canada are many lakes, ponds, and flowage areas impounded by regulating dams where logs are assembled in rafts or booms and towed to the natural outlets, special sluiceways, or jack works where the logs may be loaded on railway cars. In bringing in pulpwood in eastern Canada, booms commonly collect the logs rather than rafts.

PIERS

Very important and frequently necessary features of water transportation in connection with booms, rafts, and holding grounds are piers placed singly or in series. They serve to direct, hold, or control booms of loosely floated logs or rafts on the larger streams or in estuaries or bays of the ocean. They are found principally in the Northwest, eastern Canada, and the Southeast.

In the Northwest and the Southeast they are largely used to hold and direct rafts towed to sawmills. On the St. Lawrence River and its tributaries in eastern Canada, they are employed primarily to direct, hold, or control booms and loosely floated pulpwood logs in connection with stream driving and floating.

Piers are constructed chiefly of piling in the tidal waters of the Southeast and Northwest. Piling is placed in series or in groups to hold and direct flat rafts that have been towed from the logging operations to the tidewater sawmills. In the Northeast, eastern Canada, and northern Idaho, and some other sections of the country, they are generally constructed of timber cribbing and occasionally of concrete or stone. Log cribs filled with stone are the type for shallow water and for temporary usage.

Sometimes piers are sunk in deep water, and chains or timber boom sticks connect them with the floating booms which control the floating pulpwood, saw logs, and so on. Sunken booms are sometimes spaced 500 ft. apart and placed in water that may be from 50 to more than 100 ft. deep. Drift or glance booms used to direct logs into sorting gaps must be reinforced by secure piers, because the strain of the wood, current, and wind pressure may be considerable. In the installation of piers and holding booms, consideration must be given to the following factors:

1. Volume of logs to be retained within the holding booms anchored to or held in place by a series of piers.

2. Velocity of water in stream, measured at high-water level. Lamothe has estimated that, where the maximum velocity is not over 2½ sec.-ft., the average depth of water at least 30′, and there is a good flow in the stream. Piers are not required in eastern Canada, and the holding boom with ample guys from shores are sufficient.

3. Velocity of wind. This is a very important factor on waters where there are wide open areas—it may be of even more importance on lakes than on rivers. A square mile of pulpwood containing about 60,000 cords exposed to a strong wind once broke a mile-long holding boom guyed with a 1″ cable every 200 ft.

TYPES OF FLAT RAFTS

Flat rafts are used more in the Pacific Northwest than in any other part of the country, as there is a constant traffic of logs from logging operations to sawmills located in the harbors, bays, and larger rivers. Rafts furnish a very inexpensive medium of transportation to these mills. The log rafts common in the Pacific Northwest generally vary in length from 72′ to 84′ for one raft section and up to 1000′ for a raft composed of 12 sections. The width of these rafts generally varies from 64′ to 76′. The average length of rafts on Puget Sound consisting of 8 sections is approximately 650′. These rafts are towed for a minimum of short tows from log dumps where they are unloaded from railroad cars to mills in local harbors up to a maximum of about 116 miles. Several billion feet have normally been towed every year in the Pacific Northwest, including Alaska and British Columbia. The average towing distance is approximately 65 miles. Tugboats of appropriate size are commonly employed in towing these logs to the mills, the towing being done from the front end by the use of a bridle and long cable which permits the tug to take a position as far in front of the raft as conditions require. Usually the tugboat is as far in front of the raft as the length of the raft, depending on the width of the stream, local traffic, and other towing conditions.

FIGURE 165. Typical raft of Douglas fir logs towed to the sawmill. This raft is made up of six sections, each of which is about 65′ long. The average raft contains about 300 M.b.f. of logs.

The board content of these rafts varies with the size of the logs and with the number of sections. An eight-section raft is typical. Each section would contain the following board footage:

(*a*) Small Douglas fir logs average 15,000 to 20,000 b.f. to the average section.

(*b*) The camp run of logs of the Hood Canal type average 35,000 to 45,000 b.f. per section.

(*c*) Large old-growth logs frequently run from 60,000 to 80,000 b.f. per section, camp run. Some of these sections of old-growth logs may exceed 100,000 b.f.

(*d*) Western hemlock logs average 30,000 to 40,000 b.f. per section.

(*e*) Average camp-run western red cedar logs run 40,000 to 50,000 b.f. per section.

(*f*) White fir logs are comparable in size and footage per section to western hemlock, and Sitka spruce logs are comparable to Douglas fir logs previously described.

Boom sticks are commonly cut 76′ to 84′ in length for the sides of the booms. Cross sticks in front and in the rear of the sections are generally cut 64′ to 76′ in length. Because of the limitations of certain canals and waterways, cross sticks are sometimes shortened to 64′.*

Log booming grounds of the Portland and near-by Columbia River areas at one time held 445 million b.f. of logs. In the Seattle and near-by Puget Sound area, there were 445 million b.f. of logs, mostly Douglas fir, stored in rafts awaiting sawing at the mills. At times, storage areas for log rafts were so full that some logging operations in Oregon had to be suspended until the mills could saw some of the surplus accumulation.

Flat rafts along the tidal streams of the South and Southeast are composed of southern pine, cypress and occasionally hardwood logs placed parallel to each other in groups of 30 to 50 logs or more. Sometimes a galvanized-iron chain is laid across the logs at right angles to their length and fastened to them with U-shaped "dogs" in order to hold them together. Boom sticks are always used along the sides and in front of the sections which are generally composed of 12- to 20-foot log lengths. Where hardwoods are towed with southern pine and cypress they are hooked to wooden floats or dogged on the cross chain or cable.

Along the Albemarle and Pamlico Sounds and other estuaries and inlets of the South Atlantic Coast there is considerable rafting of logs for distances up to 50 miles or more. Flat rafts of logs held together with a galvanized ⅝″ chain tow logs to many of the sawmills located on tidewaters as well as along the coastal rivers. Every log is fastened to this chain, which lies across the middle point of the logs and at right angles to the length. Sometimes ring dogs or U-shaped dogs are used so that the floaters will hold up the sinkers and keep the raft together. These chains are generally 25′ long; thus, each raft is made up of several sections 25′ long. Some of the largest rafts contain 2500 to 3000 logs or more. The average 16′ log contains 50 b.f. so that a 3000-log raft would contain approximately 150 M.b.f. Sometimes four chain

* In the Pacific Northwest, flat rafts are often broken up and individual logs selected for special purposes, for example: large No. 1 Douglas fir logs for veneer plants; large No. 1 western hemlock logs for flooring; low-grade (small) No. 3 logs of western hemlock for pulp and rayon plants; large high-grade western red cedar logs go to the cedar mills for manufacture into siding; low-grade cedar logs go to the shingle mills; small fir logs to Swedish gang or small mills for manufacture into dimension stock, small timbers, and the like. The cost of sorting and transporting the logs by water is exceedingly low, and so the logging companies can readily afford to do this selective separation of rafts at or near destination.

rafts are placed parallel to each other and towed by one tug. There must be favorable winds and tidal flows or natural currents for towing large rafts of this kind. The average raft contains only about 1800 to 2000 logs, or about 90,000 to 100,000 b.f.

Pulpwood cut in the Lake Superior district, and especially in the western part of the province of Ontario, is rafted in considerable volume to American mills along the south shore of Lake Superior and to other Canadian mills. The pulpwood is cut in either fall or winter, generally in 8′ 4″ lengths; it is hauled to the banks of the rivers which empty into Lake Superior or directly to the shores of Lake Superior itself. During the spring stream-driving season the pulpwood banked along these streams is driven to the mouths of the rivers where it is held in storage booms. There it is either towed in small rafts to the domestic Canadian mills located along the shore of Lake Superior or made into larger rafts for export to the American mills. Probably 700,000 to 800,000 cords are taken out annually in these Lake Superior rafts or loaded on vessels for consumption in the Canadian and American pulpwood mills. Rafting is exceedingly hazardous, except in favorable summer seasons.

One company rafted about 80,000 cords of pulpwood across Lake Superior in one year, mostly in flat rafts of 4000 to 8000 cords each. At the destination, an automatic conveyor elevated the 8′ sticks and dropped them into gondola cars for railroad haul to the pulp mill.

Considerable pulpwood is shipped by steamer and barge to the American mills. Some of the principal centers for rafting are at the mouths of the Pigeon River, the Pine River, the Jarvis River, the Current River at Port Arthur, the Kaministikwia River at Fort William, the Pearl River, the Wolf River, the Nipigon River, and the Black Sturgeon River.

After the pulpwood is caught and held in storage booms at the mouths of these rivers, rafts are made up by releasing the wood from the holding booms, permitting the current to carry it to the towing booms which are held in position in readiness for it. When an amount suitable for a raft has been collected, the towing booms are closed and tightened by winches on a small tugboat which literally "squeezes" the wood into an area approximately one fourth of that which it would normally cover if permitted to float freely on the surface of the water. After the boom is tightened about the raft, two more strings of booms are wrapped around it as an additional precaution. Then, when favorable weather occurs, the raft is started toward its destination. An offshore wind is generally necessary for the first day of towing in order that the raft may move a good distance from land. After the first day, the wind is not so important, although a head wind is considered difficult and sometimes dangerous. The raft proceeds at the rate of about 1 m.p.h. to its destination, where it is emptied into the storage booms at the pulp mills.

The rafting season continues only about 12 or 14 weeks, or from about the middle of June to the middle of September. If continuous rafting operations are permissible as a result of favorable weather conditions during the rafting season, about six trips can be made with one set of equipment from points along the Canadian shore to points on the American shore of Lake Superior.

The average raft consists of about 7500 cords. In unusual instances, rafts have contained as much as 15,000 cords, but that is not considered good practice. When the wood is floating freely, the area occupied is about 200 sq. ft. per cord, and, when tightened for towing, about 50 sq. ft. per cord. A good rafting equipment consists of a large supply of towing booms, usually about 1000 in number, two small 100-hp. tugboats, one large tugboat of about 500 hp. and a large quantity of cables. Each of the sticks composing the towing boom is about 24′ long and generally 3½′ to 4′ in top diameter. These boom sticks are connected by 1½″ chains and are suitably bored and capped.

PULPWOOD BOOMS

Booms * are widely used in pulpwood logging in eastern Canada and, to a somewhat limited extent, in Maine and New York, principally for guiding and holding logs. They serve the following purposes:

1. Guiding logs and preventing them from becoming stranded or piling up in jams on river drives.

2. Towing wood in bag or other rafts across lakes, rivers, tidal harbors, or estuaries.

3. Separating logs of various ownerships at sorting gaps.

4. Holding pulpwood in water storage before separation into smaller booms for delivery to the sawmill, for loading on barges or for later distribution to various destinations.

According to Koroleff, they may be classified according to their uses, as:

1. Glance or guiding booms.

* Taken largely from "Booms," by A. Koroleff, Canadian Pulp and Paper Association, Woodlands Section, Montreal; reprint from *Canadian Pulp and Paper Magazine,* undated.

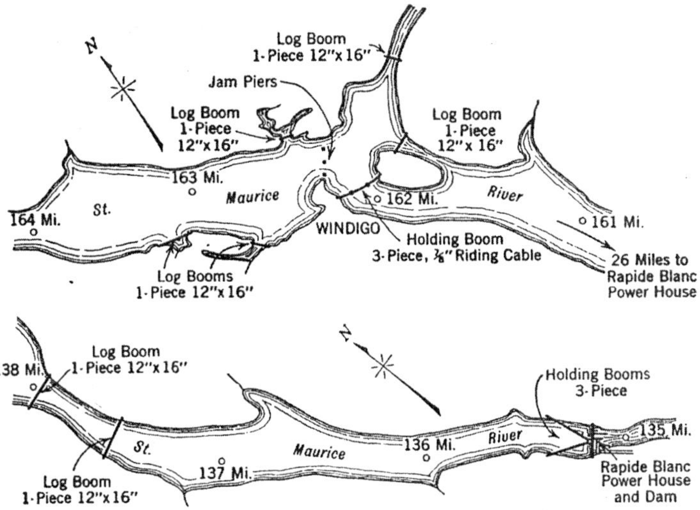

FIGURE 166. Diagram illustrates placement of seven different log booms, two holding booms, and jam piers along the St. Maurice River in eastern Canada which is used to float many thousand cords of pulpwood each spring. This illustrates the use of booms to deflect logs from low places and keep them in the main stream as well as to retain them until needed at the pulp mills.

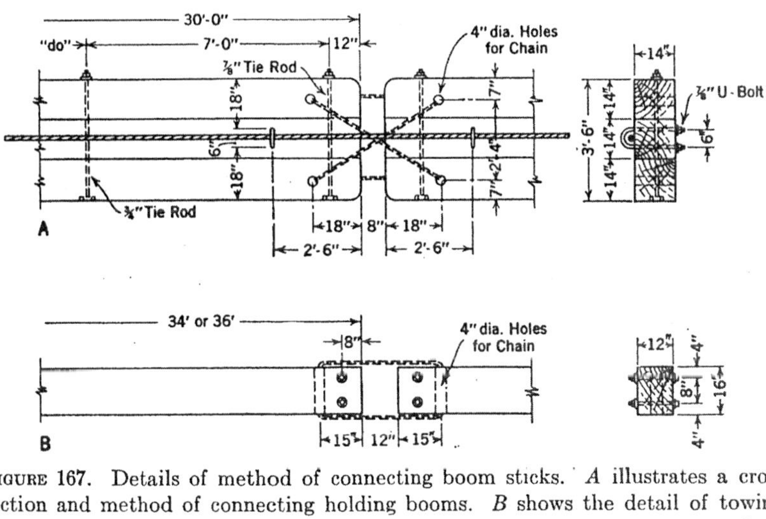

FIGURE 167. Details of method of connecting boom sticks. *A* illustrates a cross section and method of connecting holding booms. *B* shows the detail of towing and rafting booms used in eastern Canada and the Northeast. There are three pieces of 14" × 14" × 30' Douglas fir timbers bolted together and used for holding booms and one piece of 12" × 16" × 34' or 36' long used for the towing and rafting booms. The pressure exerted against holding booms used in impounding large quantities of saw logs or pulpwood is much greater than with towing and rafting booms.

2. Towing booms (bag booms, etc.).
3. Holding booms.

They may be stiff or limber. The limber ones, much more frequently used, consist of single logs of long sections commonly 20′ to 30′

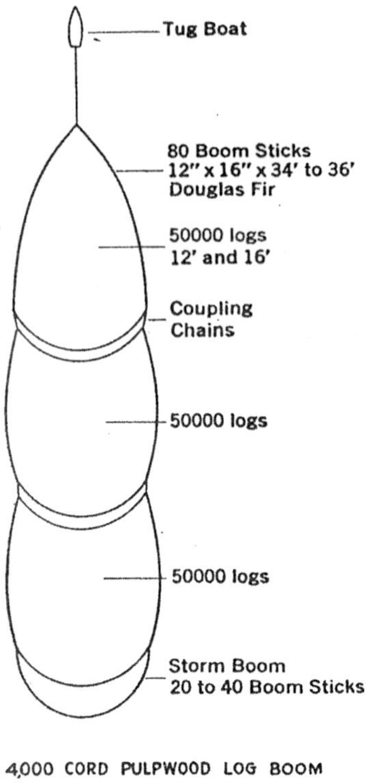

FIGURE 168. Sketch of bag boom holding approximately 4000 cords of spruce pulpwood used in towing approximately 150,000 logs on some of the larger rivers in Canada.

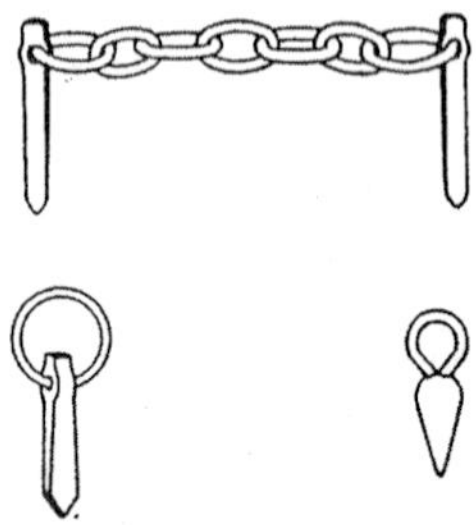

FIGURE 169. Three types of dogs used in rafting logs. The upper one is a chain dog, lower left a ring dog, and lower right an eye dog. Chains, cables, ropes, or saplings hold the logs together by means of these various types of dogs. In some regions, the "floaters" serve to hold up the "sinkers." In this way, hardwoods are sometimes rafted with softwoods. In the southern yellow pine region, rafting is practiced on some of the larger streams, and often some logs of the same species float much more readily than others.

long made of several logs or squared timbers and connected by short chains. Several hundred miles of booms are employed constantly in connection with pulpwood transportation in eastern Canada. On the St. Maurice River over a distance of about 70 miles there are 10 miles of booms and a bag raft containing 6000 to 7000 cords of pulpwood requiring about 1 mile of towing boom.

Koroleff segregates different types of booms as follows:

1. **Single-log glance booms.** These are for the larger drives, to keep pulpwood in the main stream. Generally 30′ spruce logs with 6″ to 10″ top diameters are used.

2. **Fin or rudder booms.** These serve as glance or diversion booms during the course of driving operations to keep pulpwood from being stranded or collecting in jams.

3. **Sheepshank auxiliary booms.** These are for handling wood on a drive or in a storage pond. Pulpwood logs serve as temporary boom sticks connected by ropes. They are inexpensive and temporary forms of booms.

4. **Sitka spruce log towing booms.** For long-distance towing of large rafts in rough weather, large heavy booms made of Sitka spruce and Douglas fir from British Columbia are widely used, especially in rafts on Lakes Superior and Huron for distances up to 150 miles. Boom sticks are generally average 20′ long and 30″ in diameter. New booms are only 45% to 50% submerged, but after about 8 years of service 55% to 80% of the volume is submerged.

5. **Strip booms.** These, employed for towing wood on small lakes, consist of large logs with three or four small ones spiked to each and parallel to them. They are used in Newfoundland for towing 4′ pulpwood in rafts up to 700 cords across lakes 5 to 10 miles in width.

6. **Three-log towing booms.** Generally three 30′ spruce or jack pine logs averaging 12″ to 14″ in diameter are bolted together longitudinally. They serve the same purpose as one large sawed timber and are used on Atibiti Lake.

7. **Keel towing booms.** These are for towing wood on large lakes in Newfoundland; they are similar to the Bathhurst boom.

8. **Bathhurst booms.** These are for towing logs on large lakes, tidal harbors, and estuaries. This boom is made of four logs or squared timbers placed around short core blocks and bolted together. It provides large bulk. It is fairly inexpensive and successfully holds logs together. According to Nix, the length of a Bathhurst boom is about 2000′ and the width about 300′ to 400′. The average water surface covered is about 600,000 sq. ft., and the number of cords about 2000. Used for 25 years; it was invented at Bathhurst, New Brunswick; hence the name.

There are several other types, including the Roberval swivel towing boom, the Oxford fenced towing boom, and others which are described in Koroleff's extensive treatment of the subject.

CHAPTER XXI

Ocean Rafts

HISTORY AND DEVELOPMENT

As early as the eighteenth century the idea existed of rafting timbers across the ocean from America to England. Two lumber rafts were actually towed to England, only to founder on the coast. In 1884, Captain H. R. Robertson designed and built the forerunner of the present ocean-going log rafts in a cradle on the ground at Victoria, Nova Scotia. It contained 110,000 lin. ft. of piling. It was destined for Boston, but a storm broke it up off the Maine coast. Another enlarged raft, moored in water, which required 15′ draft, was lost in a storm and located many months later, broken up and ridden with teredos off the coast of Norway. Robertson moved to the West Coast and constructed a raft in Coos Bay in 1891 which contained piling, and was lost while bound for San Francisco. Another raft, constructed in a cradle in 1894 at Astoria, also was destroyed by a storm. Finally, a raft was built and towed to San Francisco, 60% of the cargo being delivered. In 1896 Robertson built a large cradle in West Seattle for the Port Blakely Mill Company. This was copper-painted to protect it from teredos. Two rafts were built and successfully delivered in San Francisco, but in one year the teredos had ruined the cradle. He then moved his operations to the fresh waters of the Columbia River in 1897. A number of rafts were constructed and towed southward along the Pacific Coast, and only two were lost over a period of years.

In 1906 Benson and Evenson started an ocean rafting business at Wallace slough on the Columbia River. They used a simple cradle with a better center-locking device and also improved on the towing gear and systems devised by Robertson. The first raft contained piling and saw logs and several thousand board feet of sawn timbers and lumber. This raft was towed to San Diego, and a sawmill was constructed with the cargo. The raft arrived in perfect condition, after a journey of 1100 miles. Since 1906 four to five rafts have been towed annually to San Diego each carrying 4 million to 8 million b.f. of logs.

In 1911 Davis at Vancouver constructed the so-called Davis raft. This raft had a shallow draft and yet was constructed to carry large logs. It was used primarily for shorter hauls, particularly in interstate rafting between British Columbia and Washington and Oregon. Rafts were towed from points along the coast to sawmill centers 50 to 300 miles distant. Great momentum was given to the construction and use of the Davis raft when salvaged wind-thrown timber in the Olympic Peninsula was towed into Grays Harbor. One logging firm operating on the Siletz River dispatched 214 rafts containing a total of 46 million b.f. and 250 rafts containing 60 million b.f. over a period of 2 years.

Concurrently with the development of the Benson and Davis rafts, various other designs were introduced. A pontoon raft was patented, the principal feature of which is parallel steel cylinders which support the log raft on both sides.

Logs have never been rafted along the ocean on the Atlantic Seaboard to any extent owing to the small size of the logs and the location of mills in the timber at considerable distances from the coast. Lumber and timber rafts have been successfully built and operated in the North and Baltic Seas by the English, Swedes, and Norwegians. They have also been used in the Gulf of Finland.

At the present time the center of ocean rafting is the Pacific Northwest. The Benson and Davis are the principal types of rafts.

The Benson raft is cigar-shaped, built in a floating cradle which is constructed in two sections. On completion of the raft, one section of the cradle is released and towed to its mooring place along piling. The raft is then launched. The two sections of the cradle are then towed into place, the center lockers set, and the cradle is ready for the construction of another raft. When the raft is constructed, one side of the cradle is moored to a series of piling while a derrick moves back and forth on a running line parallel to the raft.

Logs towed alongside are elevated by the derrick and piled lengthwise, course on course. All sizes and lengths of logs are used, but the strength of the raft depends principally on a large proportion of tree lengths. Long logs give the necessary overlap and backbone which resist the wave action and ground swells during the ocean passage.

The towing season is generally June 15 to September 15, and 15 days are usually required to tow the raft from the mouth of the Columbia

River to the bar at San Diego. Generally five rafts are constructed yearly, and the average raft holds 4 million to 8 million b.f. The rafts are equipped fore and aft, for safety, with two range lights which burn 21 days without servicing. In each raft there are 175 tons of chains. The rate of towing speed is generally 4 to 5 m.p.h. For towing, 1½″ steel hawsers are used. It generally requires 11 men from 6 to 7 weeks to construct a raft, five rafts constituting a season's work.

FIGURE 170. Approximately 6 million b.f. of Douglas fir logs in a cigar-shaped ocean-going raft constructed on the Columbia River in the Pacific Northwest. On the right are the frames or cradles used in constructing the raft which is securely bound by large quantities of heavy chains and brought to destination by sea-going tugs.

The Davis raft is constructed by means of a cable passing over and under a tier of logs at each end of the tier and fastened to boom sticks at each side. The logs are laced firmly to form a floor on the two sides of boom sticks, which are chained end to end throughout the length of the raft and are held a fixed distance apart by cross logs known as swifters. A load of loose logs is piled on the log floor and retained in position by means of cables or chains called top binders which are fastened to the boom sticks. Floor and top binders are made fast to the boom sticks by trip hooks. These rafts are up to 800′ long and 80′ wide and contain up to 2 million b.f. This type has the advantage of simplicity. It may be constructed in 15′ to 20′ of water where wind currents and tide do not interfere. Almost any convenient size or length may be used, provided the correct proportions are followed. The Davis raft is generally made in still water.

A crane mounted on a scow or preferably a raft places the logs as required. The top binders are drawn tight by a donkey engine or a crane. These rafts are widely used in British Columbia.

CONDITIONS FOR OCEAN-GOING RAFTS

Two principal types of ocean-going rafts are used under two different shipping conditions. The Benson raft is used for long-distance coastwise rafting of large numbers of logs; the Davis raft, for shorter distances where the rafts are subject to fewer dangers and difficulties, as in shallow inlets and bays of the ocean which are typical of the Pacific Coast. The Davis raft is used where the boom raft would be unsuitable under adverse weather conditions and generally where smaller quantities of logs are to be towed.

If a raft must be quickly built, the Benson type is out of the question. The Davis and other types of rafts are more readily and quickly constructed. When small quantities of logs—50,000 to 500,000 b.f.—are to be rafted, the Benson raft is not feasible. The Benson rafts are especially required when distances up to 1100 miles are involved, and when sawmills located at considerable distances must be furnished with a steady continuous supply of raw material. Sawmills at such points as San Diego can utilize slabs, edgings, trimmings, cull lumber, sawdust, and other sawmill refuse to excellent advantage because they are located in regions without local timber resources. Furthermore, logs are moved cheaply by rafts from points near logging operations to distant manufacturing centers. Under these circumstances the use of log rafts is justified and encouraged.

POWER USED IN TOWING

Companies that specialize in towing log rafts have generally built especially designed ocean-going tugs. These are usually steam tugs which have long been in favor for towing heavy loads at relatively slow speeds. The gasoline motor must generally operate at high speed in order to tow efficiently and deliver the required horsepower; therefore, it is not so popular in ocean rafting. The steam tug has the advantage of engines which turn very slowly and deliver the needed power to tow the heavy rafts at speeds that seldom exceed 5 m.p.h. As an auxiliary to the power, heavy hawsers used as tow lines act as power equalizers and make possible a more steady, continuous, and powerful pull at slow speeds. The low-powered tugs are increasing in popularity for towing the medium- and small-size ocean rafts.

More than 4250 M.b.f. of cedar and spruce logs entered Vancouver Harbor in one tow of two rafts, after traveling a distance of more than 500 miles from the Queen Charlotte Islands in one month's time. These Davis-type rafts were 800′ long and 80′ wide and had a draught of 30′.

CHAPTER XXII

Flumes

Flumes are board conduits or troughs, generally V-shaped and partly filled with running water. In logging practice, they are used to transport saw logs, lumber, pulpwood, cross ties, fuel wood, acid wood, poles,

FIGURE 171. Inexpensive transportation and very complete utilization afforded along a flume which transported cross ties, mine timbers, saw logs, and lagging in Montana. Note the brush carefully piled for burning to insure fire protection. The lodgepole pine forests have been cut under various silvicultural systems to insure continuous productivity of the land. Some of the most complete forms of utilization are practiced under the conditions of the good markets obtained at the smelters and copper mines in Montana.

and posts by floating (or sliding on steep pitches) from loading points usually at skidways or log decks in the woods to landings at destination. The terminus, often called an "elephant," may be a log pond at a sawmill, a logging railroad, a common carrier, a stream, or other means of further transportation. Though flumes are employed, to a limited extent, for logging purposes, they are still more extensively used for irrigation and for hydraulic-power projects.

They are generally restricted to the medium-size and shorter forms of forest products. Unusually large, heavy, crooked, or long logs are not adapted to flume transportation.

FIGURE 172. End of the flume bringing round and split hardwood bolts employed in the hardwood distillation process near Kingsport in eastern Tennessee. Flumes are occasionally used for bringing pulpwood, logs, cord wood, and fuel wood down the steep ravines and along the watercourses of the Southern Appalachian Mountains. Note the devices for shunting off the 5′ cord wood at the terminus of the flume.

Flumes are generally of the right-angle V-shaped type varying from 12″ to 60″ on each side with a "backbone" at the base or vertex of the flume made of 6″ × 6″ or 4″ × 6″ timbers sawn diagonally.

Flumes are found only in mountainous regions where other forms of log transportation, generally a haul road or railroad, would be prohibitively expensive. They are generally a "last resort" method. They are usually 2 to 10 miles long, although one flume terminating at Madera, Calif., was 55 miles long.

Relatively few flumes are in use, as conditions advantageous to them are comparatively limited. They were found chiefly in the Northern Rocky Mountains, especially in Idaho. They have also been used in eastern Canada, the Adirondacks, the Southern Appalachians, California, and the Northwest. In no section of the country are they so common as other methods of transportation. They often serve as an auxiliary to other forms of logging such as railroads, chutes, tractors, and stream driving.

The capacity of flumes is generally limited by the rate of water flowage available and the possibility of handling the material at the terminus of the flume. The latter is the chief governing factor where logs, lumber, ties, and other products are to be loaded on railroad cars.

Trucks have occasionally replaced flumes. The advisability of using flumes depends on costs compared with other methods.

TYPES AND CONDITIONS OF USE

Flumes are generally made with the sides placed at a 90° angle. In the United States rectangular box flumes are seldom used for logging, although they are the prevailing type for irrigation and waterpower projects. Conditions necessary for their use may be summarized as follows:

1. A steady flow of water should be available at the head or intake of the flume. On long flumes additional water should be available at several points along the route from feeder * flumes, streams, or splash dams, or all three. The loss of water from leakage, splashing, and overflowing from jams is often considerable.

2. Steep rough topography and rocky ravines or canyons favor flumes where the construction cost of railroads or other possible means of transport would be excessive. Flumes are often built along the sides of steep canyons or gulches or along the base of steep ravines where extensive rock blasting for railroad construction or truck roads would be expensive, and steep grades would make motor or rail transportation impracticable.

3. Heavy stands of medium or small-size softwood timber, the forests of northern Idaho with their heavy acreage yield, for instance, are ideal for flume logging.

4. The flume location must provide for only slight horizontal curves in order to furnish passage for even short logs and poles up to 20′ to 30′ long.

5. Flumes are especially adapted to short distances where a considerable and rapid drop in elevation is involved.

6. Available streams are too small, precipitous, or rocky to be driven.

* A feeder or subsidiary flume may be used for water supply for the main flume or for log transport or both. Feeder flumes are generally of smaller size and lighter construction than main flumes.

ADVANTAGES

For investigating the best method of logging a given area the advantages of flumes may be summarized as follows:

1. When conditions are favorable to their construction and operation, flumes are a very economical means of transportation, the least expensive under the circumstances. They require little space for cuts or for trestlework compared to alternative means of transport.

2. Inexpensive local materials resulting in low-cost construction are usually available. Portable mills are generally installed to furnish the lumber. Sometimes a mule or horse-pack train takes in the portable mill parts, and the flume is built from the top down, the trees close by being used for the lumber and the smaller trees for struts, legs, braces, and such. The lumber is floated down the flume under construction as required.

3. The cost of maintenance, repairs, and operation is low. Flumes are generally renewed every 7 to 12 years, the average being about 10 years. Unskilled and low-paid labor may be used for repairs and upkeep. Generally a single patrolman is required for each 1 to 4 miles of flume.

4. Flumes can fit the topography, a wide grade variation being permissible from almost level to steep short pitches up to 20% to 45% or more. In precipitous places the logs or lumber may travel more rapidly than the water itself, and steel strips may be required to prevent excessive wear on the faces of the flume.

5. A large volume of product can be delivered with a definite degree of regularity in a relatively short time.

6. Flumes are independent of weather conditions. They are not generally used during the wintertime in cold climates, but they can be operated continuously both day and night and do not need to be shut down during hot, dry, or rainy seasons.

7. No empty return haul is involved as with motor trucks, tractors, railroads, and similar means of transportation.

DISADVANTAGES

The following factors must be carefully considered in the installation of flumes; although all of them may be of vital importance, they are not generally of frequent occurrence in the regions where flumes may be used, and seldom are more than one or two found to be applicable.

1. The water flowage available in streams at the intake or from feeders may not be sufficient for constant operation. Reliability of available water is of great importance; a constant supply is a prime necessity in flume operation.

2. On portions of the flume where the water flow is slow or sluggish, jams may be frequent or serious, requiring a stoppage of the flume operation and resulting in loss of materials (logs, mine timbers, and the like). Jams are relieved usually by throwing the logs and other timber to one side. These may be recovered later, but often at considerable expense. The disadvantage applies only generally to lumber, lagging, stulls, and other small forms of products such as cordwood and rived fuel wood.

3. No means is provided for hauling in men and supplies, as with railroads, motor trucks, and other forms of transportation. Tote roads must be built to the regular camps at or near the head of the flume. This factor, however, does not usually apply to branch or feeder flumes.

4. Piers or legs may settle, especially after frost leaves the ground in the spring, and the grade of the flume may be seriously disturbed. Flumes are sometimes subject to disturbance or destruction by windfalls, forest fires, floods, rock slides and other natural causes.

5. Flumes are limited to relatively short-length logs (seldom over 20′) and small sizes (seldom over 40″ in diameter); thus, for the large timber of the West, flumes, generally speaking, are not suitable. Crooked logs are also floated with difficulty and are inclined to cause jams.

6. Heavy hardwood logs are not suited to fluming. Rived hardwood bolts, however, have been flumed in the Southern Appalachian Mountains.

REGIONS WHERE USED

The most famous flumes were (1) the McShane flume transporting lodepole pine and Douglas fir cross ties 25 miles from the Big Horn Mountains to the Burlington Railroad at Sheridan, Wyo.; (2) the Allen flume which conveyed poles, lagging, cordwood, and mine props from the Atlantic slopes of the Continental Divide on the Deerlodge National Forest in western Montana over the Divide to the Pacific slope about 16 miles to the loading point on the Northern Pacific Railroad near Anaconda, Mont.; (3) the Madera flume, 55 miles in length —the longest in the United States—conveying up to 400 M.b.f. of lumber daily from the Madera Sugar Pine Company's mill in the

Sierra Mountains to the drying and reworking plant on the Southern Pacific Railroad at Madera, Calif.; (4) the Bridal Veil flume at Bridal Veil, Ore., which sloped precipitously about 2200 ft. in 4 miles' distance and was in use for over 25 years.

Northern Idaho * presents ideal conditions for flumes because of (1) size of softwood timber generally cut to 16′ lengths or shorter, (2) steep gulchlike ravines with heavy stands of timber, (3) abundance of available streams and ample water flows. In flume operation in this region, "hot logging" is generally practiced from about May 1 to December 1; that is, the logs are not decked or banked between skidding and fluming operations. Logging is started as soon as the deep snows melt in the spring.

Flumes have been operated in the Southern Appalachians for spruce pulpwood, chestnut tannic acid wood, various species of hardwoods in 5′ bolts for destructive distillation purposes and for lumber. Fluming is not practiced nearly so much, however, in this region as in the past.

Flumes have been used in eastern Canada for spruce pulpwood, both in 4′ and 16′ lengths. They are often applicable under unusual conditions. In this region the Valin Shipshaw flume was installed to eliminate bad river driving and costly transportation. Logs formerly were driven down the narrow and winding Valin River with many rapids and then hauled upstream to the mill. An 11-mile flume across country to the Shipshaw River and then a drive down 20 miles to the existing flume that led to the mill proved to be successful.

CONSTRUCTION

Before a flume is constructed, the forested area to be served by it must be surveyed carefully with the plan of locating the flume near or below the main body of timber and securing an adequate grade for flowage. Competent engineers are employed for the larger and more expensive flumes. For the shorter and more temporary ones, however, rough approximations are used both for the location and construction of the flume.

* Practically every large operation in northern Idaho has had a flume; for example, the Humbird Lumber Company on Priest River, the Diamond Match Company on the Pend Oreille River, the Panhandle Lumber Company on Spirit Creek, the Winton Lumber Company near Coeur d'Alene, the Edward Rutledge Timber Company on Marble Creek, the St. Maries Lumber Company on Mica Creek, the McGoldrick Lumber Company on the Coeur d'Alene River, and the Potlatch Forests, Inc., on Beaver Creek.

The five cardinal principles to be observed in flume construction are:

1. Unnecessary expense in supports and trestle construction must be avoided by keeping the flume as close to the ground as possible. On some operations the flume is supported on horizontally laid logs di-

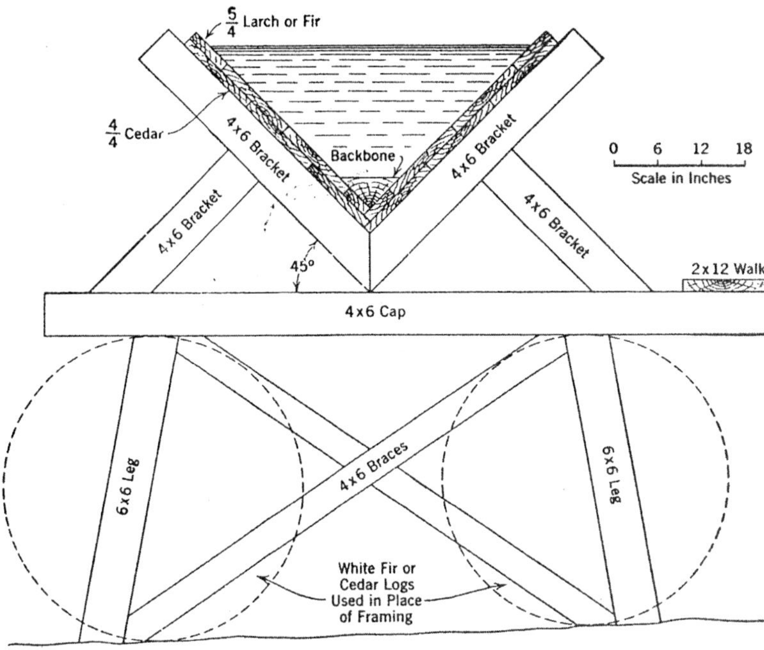

FIGURE 173. Details of a typical large-size log flume used in northern Idaho, the main flume having 5′ sides and branches 3½′ sides. The dotted circular lines indicate that, in crossing streams, ravines and some of the low places, logs may be used in place of legs. The water level shown in this sketch is found only on comparatively level stretches of the flume. On the steeper grades the water level recedes, and on the steepest pitches the logs do not float. On some of the steepest places logs travel faster than the water so that the flume must be reinforced by steel strips to prevent undue abrasion.

rectly on the ground or in stream bottoms. The flume should follow the topography as long as the constant downgrade is maintained. Frequently a stream bed is a more favorable location than a sidehill.

2. Except on steep grades the water should be maintained within 8″ of the top of the flume to get best results.

3. The flume should be located as near as possible to the center of the timber supply in the region to be served. For side gulches, feeder flumes can be constructed. Long and expensive log hauls to the flume chiefly by horse skidding or chutes or tractors should be avoided.

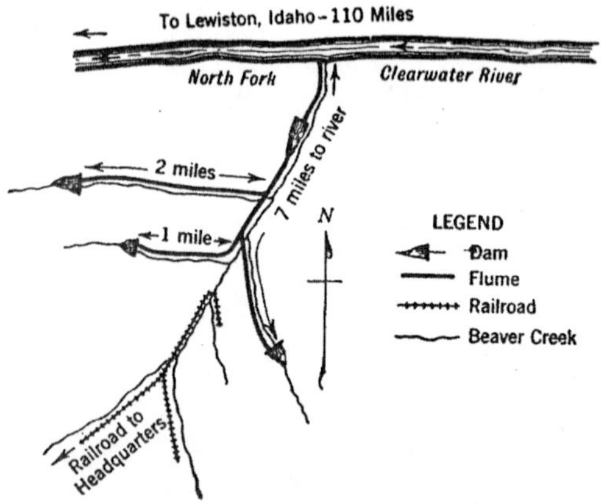

FIGURE 174. Diagram illustrating stream driving on the Clearwater River, with flumes to take the logs down the lower reaches of Beaver Creek while the logs from the upper portion of the same creek were hauled uphill on a logging railroad. The lower portion of Beaver Creek was so steep and rocky that railroad construction and operation would be prohibitively expensive. Small dams served to supply water at the headwaters of the flume and its feeders as well as to facilitate the progress of the logs down the main flume.

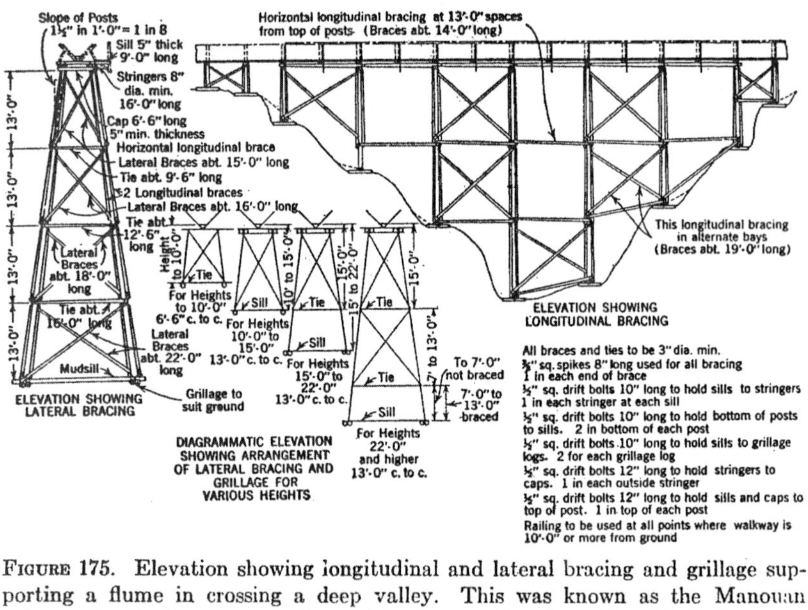

FIGURE 175. Elevation showing longitudinal and lateral bracing and grillage supporting a flume in crossing a deep valley. This was known as the Manouan flume for floating spruce pulpwood in eastern Canada. *After W. J. Carson.*

4. In order to prevent damage by windfalls, forest fires, rock slides, and such, a 70′ to 100′ right of way should be cleared of all standing timber, dead snags, and other obstructions. The width depends largely on the height of the trees along the right of way.

5. Easy and quick loading and emptying are essential for low-cost operation.

Flumes are generally constructed of rough lumber obtained from inferior local species. For example, in fluming western white pine timber, they are constructed of low-priced local species such as white fir, western larch, cedar, and Douglas fir. Logs are often used instead of legs whenever available. Logs placed horizontally are frequently cheaper than trestles or bent construction. Cribs are also used to support the flume. Small round timbers are employed for braces, legs, and trestlework. The sides of the flume consist of two layers of 1″ or 1½″ boards with the joints staggered to prevent undue leakage. Boards are nailed from the outside, the nails being clinched in the direction of the water flow.

Trestles have been built to a height of 110′ above the ground. Flumes involving the construction of trestles must have a stable and secure foundation. Piers are often made of concrete to support trestles and to prevent settling. Round timbers cut from the local forest are utilized in trestle construction rather than sawed timbers.

Table XLIII gives the sizes of sawn lumber and timber customarily used for right-angle V-shaped flumes with 3′ to 5′ sides.

TABLE XLIII

Parts	*Size*	*Length*
Inside facing	¾″ × 8″ to 16″	10′ to 16′
Outside facing	¾″ × 8″ to 16″	10′ to 16′
Brackets	4″ × 6″	3′ to 6′
Braces	2″ × 4″ and 2″ × 6″	Various
Caps	4″ × 6″	6′ 6″ to 8′
Logs	4″ × 6″ and 6″ × 6″	2′ to 16′

The flume legs are generally placed 10′ to 16′ apart, depending on the size of the flume. When the legs are more than 16′ in length, the trestle form of construction is adopted.

Grades must be maintained on a constant and uniform base if possible. Construction costs will be lessened, however, by following the topography. Main flumes are generally constructed on a grade of 1% to 4% with an average of 2% on many flumes. Grades up to 20% to 45% are sometimes used. In the latter case logs travel at speeds of 40 m.p.h. or more. Two ¼″ steel plates of varying length and width are applied on each face to prevent undue wear and tear on the steepest slopes. Logs slide on the face of the flume on steep grades rather than float, the flume thus becoming, in effect, a chute. It re-

FIGURE 176. Two views of the terminus of a flume in northern Idaho. The upper one shows the flume in operation; the lower one, the type of construction and the sloping apron at the foot of the flume to divert the logs to the landing from which they are loaded onto railroad cars. A portable sawmill was hauled in by pack train and erected at the intake to cut the lumber for this flume.

quires about 15 minutes for the average board to go down the Bridal Veil flume in Oregon, a drop of 2200′ over a distance of 4 miles. Some feeder flumes average 17% grade. Feeder flumes are generally very much steeper in grade than main flumes.

Curves are determined by the maximum length of logs to be transported in the flumes. Generally flumes must have very gradual curves.

Logs float higher on gentle than on steep grades; therefore curves can be sharper on the former. The maximum should be a 30° curve for 20′ lengths under average conditions. Extra boards or "runners" are sometimes placed on the outside of curves to guide the logs and assist in preventing jams.

OPERATION

Flumes are fed by men rolling in logs from a landing or deck along the flume or directly from skidways or chutes. In "hot logging," logs are loaded directly as received from the tractors or horse skidding or from chutes. Cross ties and pulpwood are often banked along the flume and thrown in by one or two men. In lumber flumes, lumber is fed directly into the flume from the rear end of the sawmill. Sometimes logs, cross ties, and such are fed from a reservoir of splash dam at the intake.

The amount of water required varies considerably with the size of the flume, grades involved, leakage, distance, and similar factors.

There are three main parts to flume operation:

1. Feeding at the intake or from landings along the route.
2. Running the flume, preventing jams, and reporting and repairing serious leaks or breakage.
3. Emptying at the outlet or "elephant."

Patrolmen or guards are kept in touch with these three phases of operation by telephone. A boardwalk is generally placed alongside the flume for the patrol.

Patrolmen are kept constantly on the flume. One company has a man for every 2 miles; another employs two men, one for feeding and one for patrolling the flume for a total distance of 4 miles, whereas still another company has seven men to patrol and operate a flume 7 miles in length. The number required to patrol a flume depends on the frequency and likelihood of jams, the steepness of grades, and the stability and permanence of construction.

The capacity of a flume is determined by (1) the amount of water available, (2) grades, (3) facilities at the terminus.

Splash dams are used to provide water both at the head of the flume and at various points along the route. Dams are also used to hold and carry logs over level places along the flume and to impound enough water to carry the logs over the lower reaches of the flume.

Depreciation may be figured on the basis that flumes last about 12 years with more or less constant repairs. The depreciation cost per 1000 b.f. is low on an operation where 190 million b.f. are transported over 15 miles of flume including feeder flumes within 5 years. Where a small volume of footage is conducted over a flume, depreciation charges may be high.

Under average conditions, the costs of flume operation, including depreciation and upkeep, are very low. Provided conditions are favorable for their installation and operation, they are regarded as a very economical method of log transportation. Most mistakes, however, have been made in attempting to construct and operate flumes where conditions were not favorable.

Leakage is an important consideration in flume operation. It depends largely on the type of construction and age of the flume. The average leakage in a flume carrying 5 to 10 sec.-ft. of water is about 0.45 sec.-ft. per mile.

EXAMPLES

An example of a small, inexpensive, and temporary type of flume is one near Sylva, N. C., used for bringing round and split 5′ bolts of spruce to a large near-by pulp mill. This is a V-shaped board flume 4½ miles long built in 12′ sections down a steep mountain gorge where it would be impracticable and too expensive to construct a haul road (estimated to cost $2000 per mile). There is a drop in elevation of 1600 ft. Boards 1½″ × 12″ form the trough; the legs are made of 2″ × 6″ and 4″ × 6″ pieces sawn locally, as well as round timbers cut from the adjacent forest. Braces are of 2″ × 2″, 2″ × 4″, 1″ × 6″, and round sticks.

A flume built in northern Idaho primarily for fluming western white pine was 6 miles long. This flume was built in 16′ sections with joints every 8 ft. and 4″ × 4″ brackets every 4 ft. Two thicknesses of 1″ lumber were used in the sides of a 54″ V flume, the grades varying from 4/10% to 33% with an average grade of 12%, and a maximum curvature of 8°, so that the longest timber that could be flumed without extra assistance was 45′. The highest trestle above the ground was 45′. The average height of the trestle above the ground was 9′. The deepest cut was 18′, the longest cut 300′.

Seven men can build 1 mile of flume in 20 days, providing the material is at hand and no trestle construction must be undertaken. These seven men are: one foreman, two men setting up pieces, two men spiking sides, one man on brackets, and one handyman. There are 2755 lb. of nails used per mile. As the side walls became worn, they were relined with 2″ × 10″ × 16′ boards.

CHAPTER XXIII

Barges and Steamers

Logs loaded on barges are transported to the mills in the lower Mississippi hardwood region, in eastern Canada, on the Great Lakes, and in the Southeast along the estuaries where stream currents, tidal influence, and winds prevent successful rafting. Barging is practiced to a larger extent with hardwoods in the lower Mississippi Valley and southern hardwood bottomlands than in other sections of the United States because of the failure of hardwoods to float satisfactorily and the high cost of rail transport. River mills which conduct logging operations along streams occasionally use barges to transport hardwood logs both upstream and downstream to the mill.

Barges used in hardwood transport are generally 130′ long, 32′ wide, and 6′ deep. A barge of this size generally carries about 400 logs of an average footage of 200 b.f. each or a total of 80 M.b.f.* (With 16′ logs this would mean a log run of 5.) Barges in eastern Canada have carried 300 to 350 cords or more of spruce pulpwood per load. One pulp mill in South Carolina receives about 250 cords daily by barge from distances up to 230 miles.

Barges are towed by tugs or powerboats upstream with two loaded barges at a rate of 2½ m.p.h., or downstream a distance of 100 miles with two loaded barges at 6 m.p.h.

One company operating a sawmill in eastern North Carolina, cutting 90 M.b.f. per day, brings in practically all logs by barge. Ten barges are used in this operation, each having a capacity of about 50 M.b.f. of logs, or approximately 1000 logs averaging 50 b.f. each. These barges are towed by one tug 30 to 60 miles or more. It requires about 12 hours for towing a loaded barge 60 miles in tidal water. Another company in this region operates 7 barges from 160′ to 200′ long. Each carries 35,000 to 100,000 b.f. logs for distances up to 125 miles to the

* One company operating on the Mississippi and tributary rivers in Arkansas and Louisiana used barges carrying 100,000 to 200,000 b.f. of hardwood logs. On a round trip of 160 miles, towing a loaded barge with 125,000 b.f. of hardwood downstream required 2 days' time, not including unloading at the sawmill.

sawmill. Loads are governed by the depth of the channels rather than by the capacity of the barges. There is much barging of logs and pulpwood along the intercoastal channel between Virginia and Florida.

Above Mobile, barges 34′ wide and 85′ long and 8½′ deep are used to carry 50,000 to 75,000 b.f. of hardwood logs each to sawmills at Mt. Vernon, Ala. Barges are used because no highways or railroads are available, and hardwoods cannot be rafted.

Figure 177. Barges are used extensively in the Southeast to bring logs from the woods to the sawmills. They have supplanted rafts to some extent because of the hazards of storms and sinkage. Pulpwood is also barged extensively in the Southeast.

Pulpwood loaded on barges in Canada is shipped along the St. Lawrence River and across Lake Ontario to points in northern New York where much Canadian pulpwood is consumed. About 300,000 cords of pulpwood are barged up the St. Lawrence River annually in loads of about 1000 to 1500 cords per barge.

There is considerable intracoastal traffic in logs, pulpwood, poles, piling, and woods products along the Atlantic Seaboard. Much pulpwood is transported on tramp steamers from the St. Lawrence River and points in New Brunswick to Portland, Me., Baltimore, and Boston. Steamers loading about 2500 to 3000 cords of 4′ spruce pulpwood are preferred. Boats seldom carry less than 2000 cords or more than 3500 cords. One company ships about 100,000 cords to the United States by this method. The pulpwood is loaded on the vessel from pontoons by the ship's tackle. It is loose floated to these pontoons and stacked by hand ready for a chain sling which lifts about one cord in an average load into the ship's hold.

Within recent years, ocean steamers, each loaded with about 3000 cords of 4′ spruce pulpwood, have delivered much wood at North Atlantic ports from Finland, Russia, and Sweden.

Throughout eastern Canada, spruce pulpwood is transported from the outlets of the main rivers tributary to the St. Lawrence to destinations up and down the river, as well as throughout the Great Lakes region, by means of a lake-type steamer. These boats generally carry 1400 to 1450 cords with a 16′ draft and 1000 to 1100 cords with a limited canal draft of 12′ 6″. The latter loads are used in order for these boats to pass along canals from the St. Lawrence River to the Great Lakes region, and vice versa.

A boat with a capacity of 1400 cords is loaded in about 35 to 40 hours with two crews of 25 men each, working day and night shifts. Unloading the same boat load at the pulp mills is generally accomplished in 16 to 18 hours. These boats carry their own booms and winches for unloading, and their operation is thus very economical so that the freight charge per cord is low in comparison with rail rates from similar points of origin and destination.

In addition to the afore-mentioned lake-type boat, barges carrying about 300 cords each are frequently used to move wood, especially in harbors and bays where the depth of water is insufficient to permit the lake-type boat to operate. These barges are generally towed, two or three at a trip, by a tug. The barges draw 8 ft. of water when loaded.

Hardwood-veneer logs and saw logs are frequently barged from South Atlantic points, especially Virginia and North Carolina, to points along the North Atlantic, especially to Pennsylvania, New York, and New Jersey ports.

Fuel wood is often carried along the rivers on flat barges holding from 25 to 100 cords apiece. Ocean-going barges with holds, each carrying 400 to 600 cords of oak and pine 4′ cordwood, have for years brought fuel wood from Virginia to New York City. There is also some marine traffic in pulpwood, cross ties, piling, poles, and other direct products of logging jobs along the Atlantic and Gulf Coasts.

There is considerable movement of Canadian pulpwood to the northeastern states, particularly to the New York pulpwood mills. One mill receives more than 11,000 cords annually, shipped on large lake boats and ocean-going steamers holding an average of 1000 cords each (varying from 750 to 1250 cords per boat). This wood is shipped approximately 1200 miles from Richibucto, Buctouche, and Shiptigan, after being cut in eastern New Brunswick, to Oswego, N. Y., where it is transferred on barges for shipment to inland mills. Boats make the trip to Oswego in approximately 8 days, and all boats carry a large deck load. The wood is largely sap-peeled 4′ spruce pulpwood. The deck load is thrown by hand from the steamers onto the barges and piled on the deck. The wood in the hold of the steamers is piled onto racks or slings holding three quarters of a ccrd and swung out into the barges by deck winches and chain slings. The time consumed in unloading the steamers to barges is four 16-hour days per steamer, working 50 men. It requires 7 hours for the barge trip from Oswego to the

FIGURE 178. The usual lake type of boat used in the transportation of pulpwood on St. Lawrence River in eastern Canada. These boats carry 1400 to 1450 cords with a 16′ draft and 1000 to 1100 cords with a canal draft of 12′ 6″ to provide for passage through the canals from the St. Lawrence River to the Great Lakes and in reverse direction.

mill, where 15 men, including the hoist operator and checker, unload approximately 20 cords per hour. The wood is piled onto racks holding three quarters of a cord each on the barge and then hoisted by the chain slings and a 10-ton Browning hoist into the pulpwood hopper with a conveyor bottom from which it is dumped onto the chain conveyor which takes the wood to the top of the storage pile.

Pulpwood cut in the Lake Superior district for water transportation originates principally in the western part of the province of Ontario.* A total of 700,000 to 800,000 cords is taken out annually in this region for consumption in either the Canadian mills or for export to Michigan and other American mills. The cutting operations are carried on in the fall and early winter; in middle and late winter, pulpwood is hauled to the banks of rivers which empty into Lake Superior or to the shore

* The principal loading harbors for steamers and barges on the north shore of Lake Superior are Fort William, Little Trout Bay, Port Arthur, Borden Island, Black Bay, Nipigon Bay, Squaw Harbor, Heron Bay, and Michibicoten Harbor.

of Lake Superior itself. When the spring floods come, the pulpwood on the banks of the rivers is watered and driven to the mouth of the river, where it is caught and held in the storage booms. From this point it either goes in small rafts to the domestic Canadian mills, which are invariably located close to the mouth of the river, or else is made ready for export.

The only exception to this general routine of winter cutting and spring driving is on operations which may be conducted in timber tributary to a river which can be driven in either the spring or the late summer. Operations of this nature are not very frequent, but, where they do occur, pulpwood may be cut, peeled, driven, and delivered in summer months only. This makes a very special and advantageous type of operation.

After the wood is caught and held in the storage booms at the mouth of the rivers, it may be transported to its destination either by vessel or by raft. If by vessel, all the wood in a certain district is usually assembled in a harbor, where it is held until it can be loaded out. The harbor must be well protected from storms in all directions, so that the loading operation may be carried on in all kinds or weather.

The vessels used for transporting pulpwood may be the ordinary lake steamboats of varying size or tow barges. Tugboats must be used in conjunction with tow barges.

The capacity of a steamboat used for carrying pulpwood may vary from 300 to about 1500 cords (averaging 1200). The capacity of barges varies widely from 300 to 2500 cords.

To take the wood out of the water and put it aboard the vessel, a pulpwood loader is used as shown in Figure 179. A pulpwood loader requires a crew of eight men, including the engineer to operate it. The wood is loaded onto the boat at the rate of about 40 cords per hour, and once it is on the boat it has to be straightened out in even tiers by the stowing crew of about 20 men, so that the maximum amount may be loaded on the boat.

After the boat is loaded, it proceeds to its destination where it is unloaded with the aid of the derricks on the boat or by some special unloading machine of the grab-bucket type. Either method may be used to unload the wood into the water or on a dock. Unloading into the water proceeds much faster than unloading on the dock.

The season for transporting the wood by vessel on Lake Superior lasts from the opening of navigation, usually about June 1, until the close of navigation, about November 15. Thus there is a much longer season for transport by vessel than by rafting.

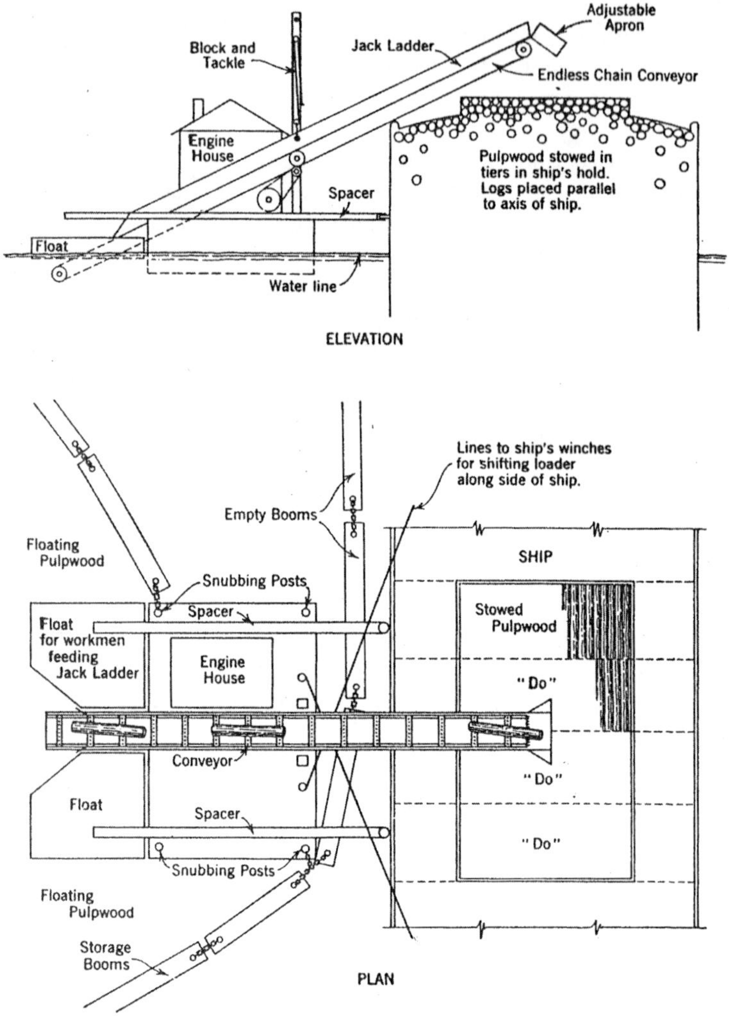

FIGURE 179. A ground plan and elevation sketch of a pulpwood conveyor or jack ladder used on Lake Superior to elevate floating pulpwood from a boom to the hold of a vessel for marine transportation. Many hundreds of thousands of cords of pulpwood are transported by barges as illustrated, or in other forms of vessels, from holding or storage booms to pulp mills in eastern Canada, the Great Lakes, and the northeastern part of the United States.

The cost of delivering wood by boat from points on the Canadian shore of Lake Superior to points on the American shore is higher than the cost of delivering wood by raft between the same points.

In order to deliver wood by vessels, it must be first assembled in a sheltered harbor, and this cost usually runs around 25 cents per cord; then the wood must be loaded on the vessels with a pulpwood loader and stowed carefully by hand.

PART VII

SUMMARY

CHAPTER XXIV

Summary of Regional Logging Methods*

This chapter is intended to give a general summary of the forest conditions and logging methods in each of the leading regions where the lumber industry reaches some prominence.

The location and size of timber stands, topography, climatic conditions, and, therefore, the methods of skidding and main haul in each logging region vary greatly. There are some exceptions to the general prevailing practice as described in the following statements. Furthermore, methods have changed rapidly, and it is likely that constant improvements will be introduced to change the procedures from time to time. However, certain fundamental conditions in each region will doubtless continue for some years. The relative importance of various forest products is changing in some regions. For example, the South and Northwest have advanced rapidly in pulpwood production during recent years.

The regions are presented in the approximate order of importance from the standpoint of volume of standing saw timber. Western British Columbia conditions are somewhat similar to those described in the Northwest Douglas fir region. From the viewpoint of current and past lumber production, southern pine has been the leading species since before 1900. Also, from the standpoint of future growth possibilities, southern pine with its large areas and rapid growth is the most important.

Fifteen regions are described in this way, including eastern Canada, which is important as both a present and a future source of pulpwood supplies. These regions are as follows:

The **Pacific Coast** region includes the three states of Oregon, Washington, and California. The same general type of forest found in

* In checking and correcting this chapter, the author has received assistance from Professors J. Kenneth Pearce, W. H. Schaeffer, Myron Krueger, W. G. Edwards, J. H. Rich, and Messrs. Earl Schindler, L. A. Nix, A. Koroleff, Arthur R. Spillers, W. C. Hammerle, J. E. MacCaffrey, and several men in the U. S. Forest Service.

Washington extends into British Columbia and Alaska, but the percentage of Douglas fir gradually decreases, and that of West Coast hemlock increases to the North. The five subregions of the Pacific Coast region are:

1. *Northwest Douglas fir* in Oregon and Washington and extending into British Columbia.

2. *Ponderosa pine* in Oregon and Washington east of the Cascade range and extending into west central Idaho.

3. *California pine* (sugar and ponderosa) principally in the Sierra Nevada range from the Oregon line to Sequoia National Park and the southern end of the Sierras.

4. *California redwood,* principally in northwestern counties of Humboldt and Mendocino.

5. *Southeastern Alaska,* along the coast.

The general **Southern** region includes the nine South Atlantic and Gulf Coast states from Virginia to Texas, inclusive, and extends into Arkansas, Tennessee, and Oklahoma, as well as the southern bottomland hardwoods, Appalachian hardwoods, and cypress.

6. *Southern pine* from Virginia to Florida, then west to Texas and north to Arkansas, Tennessee, and Oklahoma, inclusive.

7. *Appalachian hardwoods* along the Southern Appalachian Mountains from southern Pennsylvania and West Virginia to northern Alabama and extending into both the Southeast and the Central States regions.

8. *Southern bottomland hardwoods* found throughout region 6 along the river bottoms and coastal swamps.

9. *Cypress* largely in swamps along the Atlantic and Gulf Coasts and streams leading to them from Virginia to Texas, inclusive, extending up the Mississippi Valley to Arkansas. This region often overlaps or blends with region 8.

The general **Rocky Mountain** region includes the following three subregions:

10. *Northern Rocky Mountains,* including northern Idaho, western Montana, and northeastern Washington.

11. *Central Rocky Mountains,* including Wyoming, Colorado, Utah, southern Idaho, and western South Dakota.

12. The *Southwest,* including New Mexico and Arizona.

The general **Northern region** includes the following three subregions:

13. The *Northeast,* including New England, New York, and northern (and including most of) Pennsylvania.

14. The *Lake States,* including Michigan, Wisconsin, and Minnesota.

15. *Eastern Canada.*

Logging operations in these 15 forest regions are largely conducted to supply small sawmills producing 10 M.b.f. or less per day, except in the first five subregions mentioned, namely, the (1) Northwest Douglas fir; (2) ponderosa pine, east of the Cascade range; (3) California pine; (4) redwood; and (5) southeastern Alaska. In these five subregions are found most of the huge sawmills operating largely in virgin timber, and, consequently, except in Alaska, the location of most of the logging railroads, and the cable-hauling skidding machines (high-lead and skyline systems). In two of the above regions, products other than saw logs are of first importance, namely, cross ties in the central Rocky Mountains and pulpwood in eastern Canada. Pulpwood has advanced importantly in volume in the South and in the Pacific Northwest.

SUMMARY OF MINOR AND MAJOR TRANSPORTATION METHODS

Table XLIV shows the location, species, principal methods of minor transportation (skidding) and major transportation, and principal products in each of the forest regions.

TABLE XLIV

Region and Leading Species	*Minor Transportation (Skidding Methods)*	*Major Transportation (Main Log Haul)*	*Principal Products*
1. *Northwest Douglas fir* (including British Columbia). Douglas fir, West Coast hemlock, white firs (Abies), western red cedar, Sitka spruce	1. High lead (direct) 2. Cold decking and swinging to railroads or truck roads 3. Skyline or aerial yarders 4. Tractors and arches 5. Combinations of 4 and others	1. Motor trucks 2. Railroads 3. Rafting 4. Combinations of two or more of above	Saw logs, pulpwood, veneer logs, shingles, fuel wood, cross ties, and piling
2. *Ponderosa pine* in eastern Oregon and Washington. Also some western larch, lodgepole pine, white fir, and Douglas fir	1. Tractors and arches 2. Tractors and pans 3. Horses with skidding tongs 4. Mobile cranes 5. Combinations of above	1. Railroads 2. Motor trucks 3. Combinations of above	95% saw logs for lumber and box shooks. Also poles, piling, cross ties, and fuel wood
3. *California pine*. Ponderosa and sugar pines. Also true firs (white and red), Douglas fir, and incense cedar	1. Tractors with and without arches 2. Donkeys—steam and gasoline 3. Combinations of 1 and 2	1. Motor trucks 2. Railroads 3. Combined 1 and 2	Saw logs for lumber and box shooks Veneer logs, fuel wood, posts and shakes
4. *California redwood*. Redwood with 6% to 10% Douglas fir and miscellaneous	1. Skyline cableway or high lead 2. Tractors increasing in use 3. Combinations of 1 and 2	1. Railroads 2. Motor trucks	95% saw logs for lumber, cross ties, shingles, posts, and mine timbers
5. *Southeastern Alaska*. West Coast hemlock, Sitka spruce, small per cent of western red cedar and Alaska cedar	1. High lead 2. Skyline 3. Tractors 4. Combinations of above	1. Rafting 2. Motor trucks to tidewater	Saw logs chiefly for lumber for canneries and mines. Piling and cross ties. Good pulpwood future

TABLE XLIV (*Continued*)

Region and Leading Species	*Minor Transportation (Skidding Methods)*	*Major Transportation (Main Log Haul)*	*Principal Products*
6. *Southern pine.* Four species of southern pine—sold in markets as shortleaf and longleaf pine. Also southern hardwoods in various proportions	1. Tractors, direct and with allied equipment 2. Ground-line skidders 3. Animal bunching 4. Combinations of above. Also combinations of skidding and loading	1. Motor trucks 2. Railroads 3. Wagons 4. Barges and rafting 5. Combinations of 1 and 2	Saw logs for lumber. Naval stores, pulpwood, cross ties, poles and piling, posts. Fuel wood
7. *Southern Appalachian hardwoods.* Red and white oaks, yellow poplar, hard maple, ash, cherry, basswood, beech, hemlock, shortleaf pine, white pine, black gum, red gum, and spruce	1. Tractor skidding 2. Animal skidding 3. Combinations of above	1. Motor trucks 2. Railroads 3. Tractors 4. Flumes 5. Chutes 6. Stream driving and floating 7. Combinations of above	Saw logs for lumber, cross ties, pulpwood, veneer logs, cooperage stock, mine timbers, cordwood for acid (distillation) wood, tannin extract, and fuel wood
8. *Southern bottomland hardwoods.* Red gum, white and red oaks, tupelo gum, cypress, cottonwood, hickory, yellow poplar, and ash	1. Tractor skidding 2. Mud boats 3. Overhead-cable skidding 4. Ground-line skidding 5. Animal skidding 6. Combinations of above	1. Railroads 2. Motor trucks 3. Tractors 4. Rafting 5. Barging 6. Combinations of two or more methods	Saw logs, veneer logs, stave bolts, cross ties, pulpwood, fuel wood
9. *Cypress.* Cypress and swamp hardwoods, chiefly red and tupelo gum and some water oak, ash, and cottonwood	1. Overhead cableway 2. Ground-line skidding 3. Tractors 4. Animal skidding	1. Railroads 2. Motor trucks 3. Rafting	Saw logs (90%), cross ties, piling, rived shingles, and posts

10. *Northern Rocky Mountains.* Western white and ponderosa pines. Some western red cedar, western larch, Douglas fir, spruce, white fir, and lodgepole pine	1. Tractors 2. Horse skidding 3. Combinations of 1 and 2	1. Motor trucks 2. Railroads 3. River driving and towing 4. Combinations of two or more of the above	Saw logs, cross ties, cedar poles, mine timbers, pulpwood, fuel wood, ranch timbers (corral timbers, fence posts, and railing, miscellaneous)
11. *Central Rocky Mountains.* Engelmann spruce and lodgepole pine with some ponderosa pine, Douglas fir, and white fir	1. Tractors 2. Horse skidding	1. Motor trucks 2. Stream driving 3. Sled haul 4. Combinations of 1 and 2	Cross ties, saw logs, poles and piling, mine timbers, fuel wood, and ranch timbers
12. *Southwest* (including New Mexico and Arizona). Ponderosa pine (84%), Douglas fir, and true firs	1. Tractors 2. Horse skidding	1. Railroads 2. Motor trucks 3. Combinations of above	Saw logs (80%), cross ties, mine timbers, piling, ranch timbers, and fuel wood
13. *Northeast.* White pine, hard maple, yellow birch, hemlock, beech, spruce, white ash, elm, cherry, and basswood	1. Tractor skidding 2. Horse skidding	1. Motor trucks 2. Sled haul 3. Stream driving 4. Chutes 5. Combinations of above	Saw logs, pulpwood, veneer logs, cross ties, piling, and fuel wood
14. *Lake States.* Hemlock, yellow birch, aspen, hard maple, white pine, spruce, jack pine, red oak, beech, ash, elm, and basswood	1. Tractors 2. Horse skidding	1. Motor trucks 2. Sleds (tractor- and horse-drawn) 3. Railroads	Saw logs, pulpwood, cross ties, veneer logs, mine timbers, piling, distillation wood, fuel wood
15. *Eastern Canada.* Spruce and balsam fir, jack, white, and red pines; yellow and paper birch; hard maple; and larch	1. Horse skidding (about 90%) including bobbing (draying) and chaining 2. Tractors	1. Sled haul 2. Motor trucks 3. Stream driving 4. Chutes 5. Generally combinations of 1. 2. and 3	Pulpwood, saw logs, fuel wood, cross ties, veneer logs, poles, mine props, and piling

INDEX